Quantum Mechanics

The Manchester Physics Series

General Editors

F. MANDL: R. J. ELLISON: D. J. SANDIFORD

Physics Department, Faculty of Science,
University of Manchester

Properties of Matter:	B. H. Flowers and E. Mendoza
Optics: *Second Edition*	F. G. Smith and J. H. Thomson
Statistical Physics: *Second Edition*	F. Mandl
Electromagnetism: *Second Edition*	I. S. Grant and W. R. Phillips
Statistics:	R. J. Barlow
Solid State Physics: *Second Edition*	J. R. Hook and H. E. Hall
Quantum Mechanics:	F. Mandl
Particle Physics:	B. R. Martin and G. Shaw

QUANTUM MECHANICS

F. Mandl

*Department of Theoretical Physics,
University of Manchester*

John Wiley & Sons

CHICHESTER   NEW YORK   BRISBANE   TORONTO   SINGAPORE

Other Wiley Editorial Offices

John Wiley & Sons, Inc., 605 Third Avenue,
New York, NY 10158-0012, USA

Jacaranda Wiley Ltd, G.P.O. Box 859, Brisbane,
Queensland 4001, Australia

John Wiley & Sons (Canada) Ltd, 22 Worcester Road,
Rexdale, Ontario M9W 1L1, Canada

John Wiley & Sons (SEA) Pte Ltd, 37 Jalan Pemimpin 05-04,
Block B, Union Industrial Building, Singapore 2057

Library of Congress Cataloging-in-Publication Data:

Mandl, F. (Franz), (date)
    Quantum mechanics / F. Mandl.
        p.   cm.—(The Manchester physics series)
    Includes bibliographical references (p.   ) and index.
    ISBN 0-471-92971-9 (cloth)—ISBN 0-471-93155-1
    (paper)
    1. Quantum theory. I. Title. II. Series.
QC174.12.M35   1992
530.1′2—dc20   91-24255
                                                  CIP

British Library Cataloguing in Publication Data:

A catalogue record for this book is
available from the British Library.

ISBN 0 471 92971 9 (cloth)
ISBN 0 471 93155 1 (paper)

Typeset in Times 10/12 pt by Techset Composition Ltd, Salisbury, Wiltshire
Printed and bound in Great Britain by
Biddles, Guildford, Surrey

Contents

Editors' Preface to the Manchester Physics Series

The first book in the Manchester Physics Series was published in 1970, and other titles have been added since, with total sales world-wide of more than a quarter of a million copies in English language editions and in translation. We have been extremely encouraged by the response of readers, both colleagues and students. The books have been reprinted many times, and some of our titles have been rewritten as new editions in order to take into account feedback received from readers and to reflect the changing style and needs of under-graduate courses.

The Manchester Physics Series is a series of textbooks at undergraduate level. It grew out of our experience at Manchester University Physics Department, widely shared elsewhere, that many textbooks contain much more material than can be accommodated in a typical undergraduate course and that this material is only rarely so arranged as to allow the definition of a shorter self-contained course. In planning these books, we have had two objectives. One was to produce short books: so that lecturers should find them attractive for under-graduate courses; so that students should not be frightened off by their encyclopaedic size or their price. To achieve this, we have been very selective in the choice of topics, with the emphasis on the basic physics together with some instructive, stimulating and useful applications. Our second aim was to produce books which allow courses of different length and difficulty to be selected, with emphasis on different applications. To achieve such flexibility we have encouraged authors to use flow diagrams showing the logical connections between different chapters and to put some topics in starred sections. These cover more advanced and alternative material which is not required for the understanding of later parts of each volume. Although these books were conceived as a series, each of them is self-contained and can be used independently of the others. Several of them are suitable for wider use in other sciences. Each author's preface gives details about the level, prerequisites, etc., of his volume.

We are extremely grateful to the many students and colleagues, at Manchester and elsewhere, whose helpful criticisms and stimulating comments have led to many improvements. Our particular thanks go to the authors for all the work they have done, for the many new ideas they have contributed, and for discussing patiently, and often accepting, our many suggestions and requests. We would also like to thank the publishers, John Wiley & Sons, who have been most helpful.

F. MANDL
R. J. ELLISON
January, 1987   D. J. SANDIFORD

Author's Preface

In writing an undergraduate textbook on quantum mechanics, it is necessary to select severely from this huge subject. Over the years, I have lectured on quantum mechanics at all levels, from introductory undergraduate to advanced postgraduate. Perhaps surprisingly, many of these courses shared a common approach. It is this approach I am presenting here. I am concentrating on general principles and methods. My aim is to display the structure of quantum mechanics clearly. Students who have worked through this book should be able to follow quantum-mechanical arguments in books and (not too advanced) papers, and to cope with simple cases themselves. For this purpose, I employ ideas and methods of wide applicability without taking them to their full generality. For example, symmetry arguments are more prominent and angular momentum is treated from a more general viewpoint than is usual at this level.

The title of this book, Quantum Mechanics rather than Quantum Physics, reflects its emphasis on principles and methods rather than applications. Of course, applications are essential for readers to test their understanding of the formalism and to bring it to life. I have chosen applications from atomic physics. This is an important branch of physics since it underlies molecular and solid state physics. Many of its ideas can, with some modifications, be taken over to other fields such as nuclear or particle physics. It also has the virtue of simplicity: we know the forces which act inside atoms. Although not primarily a book on atomic physics, the applications cover many of the important topics of atomic physics.

No knowledge of quantum mechanics is assumed in this book but the depth of treatment presupposes a certain maturity. It would help most readers to have met some elementary wave mechanics before, making it suitable for third-year UK undergraduate physics courses though good second-year students should find much of it useful and stimulating. In the US, it may be used at senior undergraduate or first-year graduate level. It should also be of interest to experimental research workers who require a good grasp of quantum mechanics without the full formalism needed by the professional theorist. Many good books cater for the latter but very few fill the gap between elementary and advanced accounts. I take as my starting point wave functions instead of the abstract and more general Dirac state vectors, as more appropriate to the introductory level of this book. The elegance and formal simplicity of the Dirac formalism easily create an illusion of understanding. Dirac state vectors are not introduced until Chapter 5 and never used in their full generality, although an account of the Dirac formalism is given in Chapter 12 which can be studied with benefit any time after Chapter 3.

Two features facilitate the use of this book for courses of different length and difficulty. Firstly, a flow diagram (on the inside of the front cover) shows the logical connections of the chapters which make it possible to study chapters in different order and to omit some topics altogether. Secondly, about a quarter of the text forms 'starred' sections, and some material, insufficient to justify a separate section, is printed on tinted background. Material distinguished in either of these ways is not required for the understanding of material later in the book. Typically, I have covered the material in Chapters 1 to 11, including about half the starred sections, in 40 lectures of 50 minutes each.

The problems at the end of each chapter form an important part of the book. Their purpose is to enable readers to test their understanding and to introduce some interesting physics which is enlarged on in the solutions at the end of the book.

I would like to thank the many students and colleagues who, over the years, have influenced my understanding and teaching of quantum mechanics, and in particular Tony Phillips and David Sandiford who read the whole manuscript and suggested many improvements, some of a major kind. I am grateful to Sandy Donnachie for encouraging me to write this book. My special thanks go to my wife Betty for producing the computer-generated graphs and for much other help with the preparation of this book. Without her support, it would never have seen the light of day.

June 1991   FRANZ MANDL

CHAPTER

Basic concepts

As the reader is no doubt aware, classical physics fails to describe the behaviour of atoms and sub-atomic systems. Treating an atom like a solar system scaled down to atomic dimensions and applying classical physics to it in no way predicts the observed phenomena. This should not be surprising. The laws of classical physics were obtained from the study of macroscopic systems, and we have no right to expect these laws to hold for systems many orders of magnitude smaller. Indeed, many of the observed properties of atoms are startlingly at variance with the predictions of classical physics and are intuitively wholly incomprehensible.

Examples of such perplexing behaviour are:

(i) *Energy quantization* Atoms possess *discrete* energy levels, i.e. the observed values of the energy of an atom do not form a *continuum*, as expected from classical physics. These discrete energy levels show up when atoms make transitions between discrete atomic states with well-separated energies, for example in the Franck–Hertz experiment or in the optical line spectra of atoms.

(ii) *Angular momentum quantization* In the Stern–Gerlach experiment a beam of atoms, each possessing a permanent magnetic moment, traverses an inhomogeneous magnetic field. Classically, one would expect the atoms to be deflected into a fan-like *continuum* of directions. In fact, they are deflected into a *discrete* number of different directions only. The explanation of this phenomenon, as we shall see, is that the component of the atom's angular momentum in the direction of the magnetic field is quantized, i.e. it can only assume certain discrete values.

(iii) *Barrier penetration*   In the alpha particle decay of heavy nuclei, the alpha particles tunnel through a potential barrier. According to classical mechanics, the alpha particle could not escape from the nucleus as it does not have enough kinetic energy to surmount the barrier. In penetrating the barrier, it passes through a region in which its classical kinetic energy is negative!

From the point of view of classical physics and from our experience with everyday objects, these phenomena appear crazy, i.e. quite unintelligible. In contrast, quantum mechanics is outstandingly successful in providing detailed descriptions of them, in quantitative agreement with experiments. Moreover, quantum mechanics gives correct descriptions (i.e. in agreement with observations) of the properties of atoms, molecules and nuclei, provided relativistic effects are allowed for.

The difficulties of learning and understanding quantum mechanics are largely conceptual. We have no direct experience of atoms and molecules, and we must not visualize them as tiny scaled-down versions of classical macroscopic objects. To argue by analogy in this way is usually totally misleading. These conceptual difficulties lead one to start a systematic account of quantum mechanics with a more abstract mathematical formulation.

In this chapter, I shall develop some of the basic mathematical formalism of quantum mechanics. In Chapter 2 this formalism will at once be applied to some very simple situations to get a feeling for how it all works: to see how the formalism is handled and to illustrate its physical interpretation. In particular, I shall be able, already at this early stage, to give the quantum-mechanical explanations of the three startling unclassical phenomena discussed above. These examples will take us a long way towards a quantum-mechanical way of thinking. One builds up a sort of intuitive picture of what the mathematics means.

The first two chapters are seen to be complementary. The first develops the mathematical formalism, the second shows how it is used and what it means. A reader who has difficulties mastering Chapter 1 on first studying it should proceed to Chapter 2 and then have another go at the first chapter. In order to facilitate the study of Chapter 1, I shall start by considering the very simplest system: a particle of mass m and no internal structure, i.e. a so-called point-particle. Later I shall extend these ideas to many-particle systems and particles with spin.

1.1 THE STATE OF A SYSTEM

In quantum mechanics, the state of a system is specified by a wave function ψ. For a system consisting of a single point particle, ψ is a complex function of the position coordinate $\mathbf{r}$ of the particle and the time t: $\psi = \psi(\mathbf{r}, t)$ specifies the state of the system at time t. The temporal development of the wave function is given by the equation of motion for the system. For non-relativistic quantum mechanics this is the Schrödinger equation (discussed in section 1.3) which is the quantum-mechanical analogue of Newton's equations in classical mechanics.

It is a fundamental assumption of quantum mechanics that *all* information about a system at a given instant of time t can be derived from the wave function ψ. In the next section we shall see how the physical properties of a system in the state ψ are determined. Here I only wish to remind the reader that for a particle in the state $\psi(\mathbf{r}, t)$

$$P(\mathbf{r}, t)\, d^3\mathbf{r} = |\psi(\mathbf{r}, t)|^2\, d^3\mathbf{r} \tag{1.1}$$

gives the probability that at time t the particle is in a volume element $d^3\mathbf{r}$ at the point $\mathbf{r}$. The definition of probability at once implies the following normalization condition for ψ:

$$\int |\psi(\mathbf{r}, t)|^2\, d^3\mathbf{r} = 1. \tag{1.2}$$

The integration in Eq. (1.2) will be over all space in general. If the particle is confined to a certain region of space (for example, within a box of volume V) then the range of integration in Eq. (1.2) is restricted to this region. Unless otherwise indicated, integrals will always be over all regions of space accessible to the particle, i.e. over the whole configuration space of the system. A wave function satisfying the condition (1.2) is said to be normalized or normed. Later we shall relax the condition (1.2): we shall wish to consider wave functions which cannot be normed in this way. For example, for the plane wave

$$\psi(\mathbf{r}, t) = \exp[i(\mathbf{k} \cdot \mathbf{r} - \omega t)]$$

the normalization integral (1.2), taken over all space, diverges. For the present we shall only consider states ψ which can be normed according to Eq. (1.2).

Eq. (1.1) illustrates two basic features of quantum mechanics.

First, according to Eq. (1.1) the position of a particle (at a given time t) is not uniquely determined but only given by a probability distribution even when the particle is in a definite state, i.e. in a state fully specified according to quantum mechanics. This is in contrast to classical mechanics where the properties of a system in a definite state are uniquely determined and probabilities only occur for incompletely specified systems; for example, in statistical mechanics the state of a system is specified by macroscopic average quantities like temperature and density.

Secondly, Eq. (1.1) relates the probability density $P(\mathbf{r}, t)$ which is an observable quantity to the wave function of the system, $\psi(\mathbf{r}, t)$, which is not observable. Nevertheless, wave functions play a central role in quantum mechanics. In particular, they satisfy the *linear superposition principle*: if $\psi_1 \equiv \psi_1(\mathbf{r}, t)$ and $\psi_2 \equiv \psi_2(\mathbf{r}, t)$ are two possible states of a system, so is

$$\psi = c_1\psi_1 + c_2\psi_2 \tag{1.3}$$

where c_1 and c_2 are complex numbers. For the state (1.3), Eq. (1.1) leads to the probability density

$$P(\mathbf{r}, t) = |\psi|^2 = |c_1|^2|\psi_1|^2 + |c_2|^2|\psi_2|^2 + 2\,\mathscr{Re}\{c_1^*c_2\psi_1^*\psi_2\}, \tag{1.4}$$

i.e. probabilities are not additive: Eq. (1.4) contains an interference term which stems from the fact that the relation (1.1) between probabilities and wave functions is a quadratic one. This feature is typical of quantum mechanics: observable quantities depend quadratically on the unobservable wave functions, leading to interference effects. Such effects do not occur in classical mechanics. They are, of course, characteristic of all wave motion. For example, light intensities are not additive whereas light amplitudes are. Analogously, one refers to wave functions as probability amplitudes.

From the superposition of amplitudes it follows that the relative phases of different parts of an amplitude [e.g. of ψ_1 and ψ_2 in Eq. (1.3)] are all-important: they affect critically the interference pattern. On the other hand, multiplying all wave functions by the same phase factor $e^{i\alpha}$ (where α is a real number) produces no observable effects.

For electromagnetic waves there are no interference effects if waves without well-defined phase relationships between them are superposed. For example, in a Young double-slit experiment an interference pattern occurs if both slits are illuminated with light from the same source. If this source is replaced by two incoherent light sources, each illuminating one slit only, the interference pattern disappears and the light intensities are additive. In quantum mechanics too, we distinguish these two cases: a system in a *pure state* specified by a wave function, and a system in a *mixed state*, i.e. a system in a mixture of pure states without well-defined phase relationships between them. Such a mixture cannot be specified by a single wave function. The properties of a mixed state follow from those of pure states, as will be illustrated later. We shall mainly be considering pure states and, as is usual, refer to them simply as states.

1.2 OBSERVABLES

In the last section I stated that all physical properties of a system can be derived from its wave function. We must now see how this is done.

Quantities such as the position, momentum or energy of a particle, which can be measured experimentally, are called observables. In classical physics, observables are represented by ordinary variables. In quantum mechanics, observables are represented by operators, i.e. by quantities which operate on a wave function giving a new wave function. If $\hat{A}$ denotes an operator, $\hat{A}$ will transform a state $\psi(\mathbf{r}, t)$ into another state, called $\hat{A}\psi(\mathbf{r}, t)$.* We shall have explicit examples of operators in a moment. It is of course not possible to derive the operators which represent observables in any fundamental way from classical mechanics. However, a simple prescription exists which establishes the bridge between the classical quantities and the operators. The operators corresponding to the position coordinate $\mathbf{r}$ and the momentum $\mathbf{p}$ of a particle are, respectively, given by

$$\hat{\mathbf{r}} = \mathbf{r}, \qquad \hat{\mathbf{p}} = -i\hbar\nabla. \tag{1.5}$$

Thus the operator $\hat{\mathbf{r}}$ is the operation of multiplying a wave function by $\mathbf{r}$:

$$\hat{\mathbf{r}}\psi(\mathbf{r}, t) = \mathbf{r}\psi(\mathbf{r}, t). \tag{1.6}$$

Similarly the operator $\hat{\mathbf{p}}$ is the operation of taking the gradient of a wave function and multiplying it by $-i\hbar$:

$$\hat{\mathbf{p}}\psi(\mathbf{r}, t) = -i\hbar\nabla\psi(\mathbf{r}, t). \tag{1.7}$$

More generally, the operator corresponding to a function of $\mathbf{r}$ and $\mathbf{p}$, $F(\mathbf{r}, \mathbf{p})$, is given by

$$\hat{F} = F(\hat{\mathbf{r}}, \hat{\mathbf{p}}) = F(\mathbf{r}, -i\hbar\nabla). \tag{1.8}$$

For example, in classical mechanics the energy of a point particle of mass m, moving in a potential $V(\mathbf{r})$, is given by the Hamiltonian function

$$H(\mathbf{r}, \mathbf{p}) = \frac{1}{2m}\mathbf{p}^2 + V(\mathbf{r}),$$

leading to the energy operator

$$\hat{H} = H(\hat{\mathbf{r}}, \hat{\mathbf{p}}) = -\frac{\hbar^2}{2m}\nabla^2 + V(\mathbf{r}). \tag{1.9}$$

I shall take as the fundamental connection between the observable properties of a system and its state the following postulate:

For a system in the normed state $\psi(\mathbf{r}, t)$, the expectation value at time t of the observable A, represented by the operator $\hat{A}$, is given by

$$\langle\hat{A}\rangle = \int \psi^*(\mathbf{r}, t)\hat{A}\psi(\mathbf{r}, t)\, d^3\mathbf{r}. \tag{1.10}$$

This expression is the basis of all comparisons between the predictions of quantum mechanics and experiment. Another reason for taking it as the cornerstone of our discussion is that it leads to or suggests many important consequences. Before coming on to these, I want to make some comments on Eq. (1.10).

---

* Initially and if I want to stress that a quantity is an operator, I shall mark it by a circumflex accent: $\hat{A}$ will denote the operator corresponding to the variable A. Later on, the accent will be omitted, unless doing so could lead to confusion.

Firstly, if for the observable A we take the position coordinate $\mathbf{r}$, Eq. (1.10) becomes

$$\langle \hat{\mathbf{r}} \rangle = \int \mathbf{r} |\psi(\mathbf{r}, t)|^2 \, \mathrm{d}^3\mathbf{r}. \tag{1.11}$$

This result is in agreement with the probability distribution (1.1), which is very comforting.

Secondly, the order of the factors in the integrand of Eq. (1.10) must, in general, not be changed: they must be in the order shown. For example, for the momentum operator $\hat{\mathbf{p}} = -i\hbar\nabla$ we have

$$\psi^*(-i\hbar\nabla)\psi \neq -i\hbar\nabla(\psi^*\psi) = -i\hbar[(\nabla\psi^*)\psi + \psi^*\nabla\psi].$$

This is a general feature of operators: they are usually non-commuting quantities: the order of factors matters.

Thirdly and most importantly, Eq. (1.10) is a statistical statement. It says nothing about the result of a single observation but only about the expectation value, i.e. the mean value, obtained from many repeated measurements. Before each of these measurements we must prepare the system to be in the state $\psi(\mathbf{r}, t)$. Averaging over the results of these measurements, we obtain a mean value. It is this mean value which is given by Eq. (1.10). In Chapter 3 we shall return to the problem of how to prepare a system to be in a definite state.

I now come on to consider some of the consequences of Eq. (1.10). In this section, we shall be concerned with the properties of the system at one instant of time only. For the present we can therefore forget about the time dependence of Eq. (1.10), treating ψ and $\hat{A}$ as the state and the operator at the instant of time considered.

Observables such as momentum or energy are real quantities. Hence the expectation value (1.10) must be real for any state ψ:

$$\int \psi^* \hat{A}\psi \, \mathrm{d}^3\mathbf{r} = \int \psi(\hat{A}\psi)^* \, \mathrm{d}^3\mathbf{r} = \int (\hat{A}\psi)^*\psi \, \mathrm{d}^3\mathbf{r}. \tag{1.12}$$

An operator which satisfies condition (1.12) for all states ψ is called *Hermitian* or *self-adjoint*. We conclude that observables must be represented by Hermitian operators.

The condition (1.12) for $\hat{A}$ to be Hermitian is equivalent to the condition that for any two states ψ_1 and ψ_2

$$\int \psi_1^* \hat{A}\psi_2 \, \mathrm{d}^3\mathbf{r} = \int (\hat{A}\psi_1)^*\psi_2 \, \mathrm{d}^3\mathbf{r}. \tag{1.13}$$

To derive this condition, we consider the state

$$\psi = c_1\psi_1 + c_2\psi_2 \tag{1.14}$$

where c_1 and c_2 are arbitrary complex numbers. Substitution of Eq. (1.14) into Eq. (1.12) gives

$$\sum_{m,n=1}^{2} c_m^* c_n \left[\int \psi_m^* \hat{A} \psi_n \, d^3\mathbf{r} - \int (\hat{A}\psi_m)^* \psi_n \, d^3\mathbf{r} \right] = 0$$

and since this holds for arbitrary c_1 and c_2 Eq. (1.13) follows. In deriving this result, we assumed that

$$\hat{A}(c_1 \psi_1 + c_2 \psi_2) = c_1 \hat{A} \psi_1 + c_2 \hat{A} \psi_2. \tag{1.15}$$

An operator with this property is called a linear operator. The operators which were introduced in Eq. (1.8) for observables clearly have this property. With one exception, which we shall not be considering, all operators in quantum mechanics are linear.

We next ask: can there be a state ψ for which the result of measuring the observable A is unique, i.e. measuring A always leads to the same value? In general, we know, repeated measurements will lead to a spread of results with the mean $\langle \hat{A} \rangle$. As a measure of this spread we shall take the standard deviation ΔA, defined by

$$(\Delta A)^2 = \int \psi^* (\hat{A} - \langle \hat{A} \rangle)^2 \psi \, d^3\mathbf{r}. \tag{1.16}$$

If in the state ψ A has a unique value, ΔA must vanish. Since for the Hermitian operator $\hat{A}$ Eq. (1.12) holds and $\langle \hat{A} \rangle$ is real, we can rewrite expression (1.16) in the form

$$(\Delta A)^2 = \int \psi^* (\hat{A} - \langle \hat{A} \rangle)(\hat{A} - \langle \hat{A} \rangle)\psi \, d^3\mathbf{r}$$

$$= \int [(\hat{A} - \langle \hat{A} \rangle)\psi]^* [(\hat{A} - \langle \hat{A} \rangle)\psi] \, d^3\mathbf{r}$$

$$= \int |(\hat{A} - \langle \hat{A} \rangle)\psi|^2 \, d^3\mathbf{r}. \tag{1.17}$$

It follows that $\Delta A = 0$ if and only if

$$\hat{A}\psi = a\psi \tag{1.18}$$

for some number a, and

$$\langle \hat{A} \rangle = a.$$

Eq. (1.18) is an important equation. It tells us that the operator $\hat{A}$ acting on the state ψ simply multiplies ψ by the number a. A state ψ for which Eq. (1.18) holds is called an *eigenstate* (or *eigenfunction*) of $\hat{A}$, and a is the corresponding *eigenvalue*. Eq. (1.18) is known as an eigenvalue equation.

From Eqs. (1.17) and (1.18) we reach the following very important conclusions: For a system in an eigenstate ψ of the Hermitian operator $\hat{A}$, measuring the observable A necessarily gives the corresponding eigenvalue for the result of the measurement, i.e. $\Delta A = 0$. Conversely, if for a system in the state ψ measuring A gives the unique result a (i.e. $\Delta A = 0$ for this state), then ψ is an eigenstate of $\hat{A}$ and a is the corresponding eigenvalue.

It is apparent from this result, and the reader is no doubt aware, that eigenfunctions and eigenvalues play a central role in quantum mechanics. Solutions of specific eigenvalue problems will occur in Chapter 2 and later. Here I shall discuss some general properties.

First, we see that the eigenvalues of a Hermitian operator are real. If $\hat{A}$ is Hermitian, $\langle \hat{A} \rangle$ is real for any state; in particular for an eigenstate with the eigenvalue a, for which $\langle \hat{A} \rangle = a$. Hence a is real.

Next, we shall show that if ψ_1 and ψ_2 are two eigenfunctions of the Hermitian operator $\hat{A}$, belonging to different eigenvalues a_1 and a_2, i.e.

$$\hat{A}\psi_n = a_n\psi_n, \qquad n = 1, 2, \quad a_1 \neq a_2, \tag{1.19}$$

then

$$\int \psi_1^* \psi_2 \, \mathrm{d}^3\mathbf{r} = 0.$$

To prove this result we premultiply Eq. (1.19) by ψ_m^* and integrate, obtaining

$$\int \psi_m^* \hat{A}\psi_n \, \mathrm{d}^3\mathbf{r} = a_n \int \psi_m^* \psi_n \, \mathrm{d}^3\mathbf{r}. \tag{1.20}$$

Taking the complex conjugate of this equation, using the fact that a_n is real, and interchanging the labels m and n gives

$$\int \psi_n (\hat{A}\psi_m)^* \, \mathrm{d}^3\mathbf{r} = a_m \int \psi_m^* \psi_n \, \mathrm{d}^3\mathbf{r}.$$

Applying the Hermiticity condition (1.13) on the left-hand side of this equation we can rewrite it as

$$\int \psi_m^* \hat{A}\psi_n \, \mathrm{d}^3\mathbf{r} = a_m \int \psi_m^* \psi_n \, \mathrm{d}^3\mathbf{r}. \tag{1.21}$$

Finally, subtraction of Eq. (1.20) from Eq. (1.21) leads to

$$(a_m - a_n) \int \psi_m^* \psi_n \, \mathrm{d}^3\mathbf{r} = 0. \tag{1.22}$$

It follows from this equation that

$$\int \psi_m^* \psi_n \, \mathrm{d}^3\mathbf{r} = 0 \qquad \text{if } a_m \neq a_n. \tag{1.23}$$

Eq. (1.23) is called an *orthogonality relation* and two states satisfying such a relation are said to be *orthogonal*. Correspondingly, the integral in Eq. (1.23) is called the *scalar product* of the state ψ_m with ψ_n. This is in analogy with ordinary vector analysis: two vectors $\mathbf{u}_1$ and $\mathbf{u}_2$ are orthogonal if their scalar product vanishes:

$$\mathbf{u}_1 \cdot \mathbf{u}_2 = 0. \tag{1.24}$$

It is very helpful in visualizing the more complicated mathematics which we are now using to import such geometrical concepts from vector analysis. Just as $\mathbf{u}_1$ and $\mathbf{u}_2$ are vectors in ordinary space, so one can think of ψ_1 and ψ_2 as vectors in a function space. This analogy is possible and helpful because the linear superposition principle also applies to ordinary vectors. The analogue of Eq. (1.3) states that from two vectors $\mathbf{u}_1$ and $\mathbf{u}_2$ we can form a new vector

$$\mathbf{u} = c_1\mathbf{u}_1 + c_2\mathbf{u}_2. \tag{1.25}$$

Here the coefficients c_1 and c_2 are real numbers since ordinary vector analysis deals with real quantities, whereas we are now dealing with essentially complex quantities. In spite of this difference and others, some of which will be mentioned later, there are many parallels between the two situations and we shall frequently employ the geometrical language of vector analysis to make clearer the mathematics of quantum mechanics.

It may happen that the eigenvalue a of the Hermitian operator $\hat{A}$ possesses more than one eigenfunction. If ϕ_1 and ϕ_2 are two eigenfunctions belonging to the eigenvalue a, then so does any linear combination

$$\psi = c_1\phi_1 + c_2\phi_2. \tag{1.26}$$

The eigenvalue a is called s-fold *degenerate* if there exist exactly s linearly independent eigenfunctions $\phi_1, \phi_2, \ldots, \phi_s$ with this eigenvalue.* Any other eigenfunction ψ belonging to this eigenvalue can then be expressed as a linear superposition

$$\psi = \sum_{n=1}^{s} c_n\phi_n. \tag{1.28}$$

Taking $a_1 = a_2$ in Eq. (1.22), one sees that two degenerate eigenfunctions are not necessarily orthogonal. However, from the linearly independent eigenfunctions $\phi_1, \ldots, \phi_s$, one can always form s linear combinations $\psi_1, \ldots, \psi_s$ which are orthogonal (see problem 1.4.) Furthermore, we can always normalize the eigenfunctions: if ψ satisfies $\hat{A}\psi = a\psi$, so does const. $\times \psi$. Combining these results, we conclude that we can always choose *all* the eigenfunctions of a

---

* The functions $\phi_1(\mathbf{r}), \phi_2(\mathbf{r}), \ldots, \phi_s(\mathbf{r})$ are linearly independent if the condition

$$\sum_{n=1}^{s} c_n\phi_n(\mathbf{r}) = 0 \qquad \text{for all values of } \mathbf{r} \tag{1.27}$$

implies $c_1 = c_2 = \cdots = c_s = 0$.

Hermitian operator to be normed and mutually orthogonal, i.e. to satisfy the conditions

$$\int \psi_m^* \psi_n \, d^3\mathbf{r} = \delta_{mn} \tag{1.29}$$

where the Kronecker delta is defined by

$$\delta_{mn} = \begin{cases} 1, & m = n, \\ 0, & m \neq n. \end{cases} \tag{1.30}$$

Eq. (1.29) is known as an *orthonormality relation*, and states satisfying it are called *orthonormal*.

In discussing the eigenfunctions of a Hermitian operator, I have said nothing about how many such eigenfunctions there are. A most important property of a Hermitian operator $\hat{A}$ is that the totality of its linearly independent eigenfunctions $\psi_1, \psi_2, \ldots, \psi_n, \ldots$ (which will be taken as orthonormal) form a *complete set*. By this one means that if ψ is *any* state of a system and A is an observable of the system, then ψ can be expanded in terms of the eigenfunctions of the corresponding Hermitian operator $\hat{A}$:

$$\psi = \sum_n c_n \psi_n. \tag{1.31}$$

Multiplying this equation by ψ_m^*, integrating and using the orthonormality relation (1.29), one obtains the expansion coefficients

$$c_m = \int \psi_m^* \psi \, d^3\mathbf{r}. \tag{1.32}$$

We have here assumed that the totality of eigenvalues

$$a_1, a_2, \ldots, a_n, \ldots \tag{1.33}$$

(a_n belongs to ψ_n) is countable, i.e. the eigenvalues form a discrete set. One refers to this set as the spectrum of eigenvalues, in this case as a discrete spectrum. Later we shall also meet operators with continuous spectra, where the eigenvalues take on all values in a continuous interval. An example of the expansion theorem (1.31)–(1.32), which the reader will have met, is the expansion in a Fourier series of a periodic function. The normed trigonometric functions of the same periodicity constitute the corresponding complete set of orthonormal functions. The expansion theorem (1.31)–(1.32) can be proved in considerable generality but the proof is very difficult and we shall assume it.*

---

* I am altogether adopting an extremely cavalier attitude to all aspects of mathematical rigour, for example as regards the convergence of infinite series like the eigenfunction expansion (1.31). For some enlightenment and references the reader is referred to Gottfried, pp. 52–60, and Merzbacher, pp. 150–157.

The expansion theorem (1.31)–(1.32) for the eigenfunctions of the Hermitian operator $\hat{A}$, corresponding to the observable A, allows us to deduce the probability distribution for the results of measuring A. The expectation value of A in the normed state ψ is given by

$$\langle \hat{A} \rangle = \int \psi^* \hat{A} \psi \, d^3 \mathbf{r}. \tag{1.34}$$

Expanding ψ and ψ^* by means of Eq. (1.31), we obtain

$$\langle \hat{A} \rangle = \sum_m \sum_n c_m^* c_n \int \psi_m^* \hat{A} \psi_n \, d^3 \mathbf{r}. \tag{1.35}$$

Using the eigenvalue equation

$$\hat{A} \psi_n = a_n \psi_n \tag{1.36}$$

and the orthonormality relation (1.29), we can write

$$\int \psi_m^* \hat{A} \psi_n \, d^3 \mathbf{r} = a_n \int \psi_m^* \psi_n \, d^3 \mathbf{r} = a_n \delta_{mn}.$$

Hence Eq. (1.35) reduces to

$$\langle \hat{A} \rangle = \sum_n |c_n|^2 a_n. \tag{1.37}$$

First, let us assume that all the eigenvalues a_n are non-degenerate. Remember that for a system in the state ψ_n a measurement of A necessarily gives the value a_n. Eq. (1.37) for the mean value $\langle \hat{A} \rangle$ then leads us to interpret

$$P(a_n) = |c_n|^2 = \left| \int \psi_n^* \psi \, d^3 \mathbf{r} \right|^2 \tag{1.38}$$

as th probability that a measurement of A on a system in the state ψ produces the r sult a_n. Furthermore, we can rewrite the normalization condition

$$\int \psi^* \psi \, d^3 \mathbf{r} = 1, \tag{1.39}$$

using the expansion (1.31) and the orthonormality relations (1.29), in the form

$$\sum_n |c_n|^2 = 1. \tag{1.40}$$

On account of Eq. (1.38), the last equation can be written

$$\sum_n P(a_n) = 1, \tag{1.41}$$

from which we must conclude that the *only* values which can be obtained when measuring A are the eigenvalues $a_1, a_2, \ldots$. Other values, e.g. somewhere in between a_1 and a_2, can never be observed. These conclusions hold whatever the state of the system and are a consequence of the assumption that the eigenfunctions of an observable form a complete set. They are quite unexpected from the point of view of classical physics where observables always take on continuous ranges of values. They are, of course, overwhelmingly confirmed by experimental evidence. We mention two examples from atomic physics: the discrete energy levels of atoms which are responsible for the line spectra, and the quantization of angular momentum which manifests itself in the Stern–Gerlach experiments and the Zeeman effect.

In the case of degenerate eigenvalues, only some details of the above arguments need modifying. If the eigenvalue a is s-fold degenerate, say

$$a_{r+1} = a_{r+2} = \cdots = a_{r+s} = a, \tag{1.42}$$

then the probability of observing the value a in the state ψ is

$$P(a) = \sum_{n=r+1}^{r+s} |c_n|^2. \tag{1.43}$$

With this change, our earlier conclusions continue to hold.

We shall next consider the implications of these results for measurements. Suppose that in a measurement of an observable A of a system one obtains for it the value a. The statement that A has the value a must mean that an immediate remeasurement of A necessarily gives the same value a. The proviso that the second measurement is performed immediately after the first ensures that between the two measurements the value of A has not altered, either due to the system changing with time according to its equations of motion or due to the system having been disturbed in some way, for example by some other measurement. If we know that the second measurement of A is certain to produce the value a, then after the first measurement the system must be in an eigenstate of $\hat{A}$ with the eigenvalue a. If $a = a_n$ is a non-degenerate eigenvalue, the state ψ' of the system after the first measurement must be the eigenstate ψ_n:

$$\psi' = \psi_n. \tag{1.44}$$

If a is the s-fold degenerate eigenvalue (1.42), the state ψ' of the system after the first measurement must be a superposition of the corresponding eigenstates $\psi_{r+1}, \psi_{r+2}, \ldots, \psi_{r+s}$. We shall postulate that if the system was in the state

$$\psi = \sum_n c_n \psi_n \tag{1.45}$$

before the first measurement, then after this measurement its wave function is given by

$$\psi' \propto \sum_{n=r+1}^{r+s} c_n \psi_n \tag{1.46}$$

i.e. apart from the overall normalization of ψ', the amplitudes $c_{r+1}, \ldots, c_{r+s}$ of the states ψ and ψ' are the same. The reasons for this assumption will become clear when dealing with the simultaneous measurements of several observables.

We have arrived at a perplexing result. Before the first measurement the wave function of the system is given by Eq. (1.45); after the measurement by Eqs. (1.44) or (1.46). As a result of the measurement the wave function has changed discontinuously from ψ to ψ'. One talks of the *collapse of the wave function*. According to quantum mechanics, the time development of a system is governed by the Schrödinger equation (to be discussed in section 1.3). This is a first-order differential equation in the time and hence the wave function changes continuously with time. It is difficult to reconcile this continuous time evolution of the wave function with its abrupt change when an observation is made. A measurement involves the interaction of the system with the measuring apparatus. Presumably the latter obeys the laws of quantum mechanics so that 'system plus measuring apparatus' make up a composite system which is itself described by a sort of 'grand Schrödinger equation'. This equation predicts a continuous time development for 'system plus measuring apparatus'. So is it the wave function of the composite system which collapses when the observer looks at the measuring apparatus? To answer in the affirmative does not necessarily help. We could next include part of the observer (how far? up to the retina? right into the brain?) in the composite system to be described by a wave function. When a measurement has been made and a definite value has been found for the observable, the wave function has collapsed. Several attempts have been made to resolve these difficulties. None of them are sufficiently convincing to be generally accepted. It is a curious dilemma of quantum mechanics that it provides well-defined prescriptions for predicting the results of experiments, with excellent agreement between theory and experiment; yet we have difficulties in understanding some of the basic implications of the theory.

The above formalism shows how to determine the state of a system, i.e. how to prepare it to be in a definite state. If measuring A produces a non-degenerate eigenvalue a_n as result, the state ψ' of the system immediately after the measurement is fully determined as $\psi' = \psi_n$. If the result is a degenerate eigenvalue a, one only knows that ψ' is a superposition of the degenerate eigenfunctions of $\hat{A}$ with this eigenvalue. The determination of ψ' then requires further measurements of other observables. These must be carefully chosen so as not to obliterate our knowledge of the value of A. We shall return to this in Chapter 3.

1.3   THE SCHRÖDINGER EQUATION

In the previous section we considered the properties of a system at one instant of time. Next, we discuss the time development of a system. This can be expressed in various ways. In the formulation we shall use, known as the

Schrödinger picture, the wave function of a system changes with time and for an isolated system this is the only time dependence, i.e. the observables have no explicit time dependence. [For a system which is not isolated, the observables can depend on the time. For example, for a charged particle in an oscillating electric field, the potential energy term in Eq. (1.9) will have such a dependence $V = V(\mathbf{r}, t)$.] In the Schrödinger picture, the time development of a system, described by the Hamiltonian operator H, is given by the Schrödinger equation

$$i\hbar \frac{\partial \psi}{\partial t} = H\psi. \tag{1.47}$$

(From now on, I shall usually omit the circumflex accent for operators.) For a system of a single particle, to which we are again restricting ourselves, $\psi = \psi(\mathbf{r}, t)$. The Schrödinger equation (1.47) is the quantum-mechanical equation of motion of the system. It is a first-order differential equation in time. Given the state $\psi(t_0)$ of the system at some initial time t_0, the state $\psi(t)$ at later times is determined and follows from Eq. (1.47) by integration.

The normalization condition (1.2), which states that the particle is somewhere in space, must hold at all times, i.e. we require that

$$\frac{\partial}{\partial t} \int \psi^*(\mathbf{r}, t)\psi(\mathbf{r}, t) \, \mathrm{d}^3\mathbf{r} = 0. \tag{1.48}$$

We can easily show that Eq. (1.48) follows from the Schrödinger equation. Differentiating the left-hand side of Eq. (1.48) and substituting for $\partial\psi/\partial t$ and $\partial\psi^*/\partial t$ from the Schrödinger equation, one obtains

$$\frac{1}{-i\hbar} \int [(H\psi)^*\psi - \psi^*(H\psi)] \, \mathrm{d}^3\mathbf{r}. \tag{1.49}$$

This expression vanishes since the Hamiltonian operator H must satisfy the Hermiticity condition (1.12). Hence Eq. (1.48) holds.

We shall now consider systems for which H has no explicit time dependence. We shall look for solutions of the Schrödinger equation which are of the form

$$\psi(\mathbf{r}, t) = \phi(\mathbf{r}) \, e^{-iEt/\hbar}. \tag{1.50}$$

Substituting this expression in the Schrödinger equation (1.47), we obtain

$$H\psi(\mathbf{r}, t) = E\psi(\mathbf{r}, t) \tag{1.51}$$

or equivalently

$$H\phi(\mathbf{r}) = E\phi(\mathbf{r}). \tag{1.52}$$

Eq. (1.52) tells us that a solution of the form (1.50) of the Schrödinger equation is an eigenfunction of the Hamiltonian H with the eigenvalue E: in the state (1.50) the system has the energy E. Eq. (1.52) is called the *time-independent*

Schrödinger equation, in contrast to the *time-dependent* Schrödinger equation (1.47). These two equations have quite different characters. The time-dependent equation (1.47) is the equation of motion of the system; given the initial wave function, it can always be integrated. On the other hand, the time-independent equation (1.52) is an eigenvalue problem for the Hamiltonian operator H of the system, and the results of the last section about eigenvalues and eigenfunctions apply. In particular, the energy eigenvalues $E_1, E_2, \ldots$, are real and the corresponding eigenfunctions $\phi_1(\mathbf{r})$, $\phi_2(\mathbf{r}), \ldots$, form a complete set of orthonormal functions.*

The energy eigenstates are called *stationary states* because for them the physical properties of a system are constant in time. For example, for the energy eigenstate

$$\psi_n(\mathbf{r}, t) = \phi_n(\mathbf{r})\, e^{-iE_n t/\hbar} \qquad (1.53)$$

the probability distribution (1.1) reduces to the time-independent probability distribution

$$P(\mathbf{r})\, d^3\mathbf{r} = |\phi_n(\mathbf{r})|^2\, d^3\mathbf{r}. \qquad (1.54)$$

More generally, for a system in the state ψ_n, the expectation value of an observable A at the time t is given by

$$\langle A \rangle_t = \int \psi_n^*(\mathbf{r}, t) A \psi_n(\mathbf{r}, t)\, d^3\mathbf{r}$$

$$= \int \phi_n^*(\mathbf{r})\, e^{iE_n t/\hbar} A \phi_n(\mathbf{r})\, e^{-iE_n t/\hbar}\, d^3\mathbf{r}$$

$$= \int \phi_n^*(\mathbf{r}) A \phi_n(\mathbf{r})\, d^3\mathbf{r}. \qquad (1.55)$$

If the operator A has no explicit time dependence, $\langle A \rangle_t$ is constant in time.

Since the energy eigenfunctions form a complete orthonormal set, an arbitrary state $\phi(\mathbf{r})$ of the system can be expanded in terms of them:

$$\phi(\mathbf{r}) = \sum_n c_n \phi_n(\mathbf{r}) \qquad (1.56a)$$

---

* We are assuming a completely discrete spectrum of energy eigenvalues. In practice, one often meets mixed spectra, partly discrete and partly continuous. For example, atoms possess discrete energy levels. In addition, at sufficiently high excitation energies, atoms become ionized and the ionized electron can have any energy. These ionized states of the atom correspond to the continuous energy spectrum. Formally, the theory for continuous spectra is very similar to that for the discrete case; roughly put, the summations over discrete sets of states are replaced by integrations over continuous sets. I shall not write down the rather cumbersome expressions for this case.

with

$$c_n = \int \phi_n^*(\mathbf{r})\phi(\mathbf{r}) \, d^3\mathbf{r}. \tag{1.56b}$$

We can use this result to integrate the time-dependent Schrödinger equation (1.47) for an arbitrary initial state

$$\psi(\mathbf{r}, t_0) = \phi(\mathbf{r}). \tag{1.57}$$

We know the time dependence of each state (1.53). Since the Schrödinger equation (1.47) is linear, a linear superposition of its solutions is also a solution. It follows that

$$\psi(\mathbf{r}, t) = \sum_n c_n \phi_n(\mathbf{r}) \, e^{-iE_n(t-t_0)/\hbar} \tag{1.58}$$

is the solution of the time-dependent Schrödinger equation (1.47) which evolves from the initial state $\psi(\mathbf{r}, t_0) = \phi(\mathbf{r})$, defined by Eqs. (1.56).

In a stationary state, a system has a definite energy and the expectation value of an observable A which has no explicit time dependence is constant in time. In contrast, for a superposition of stationary states, the energy of the system will in general not have a definite value and $\langle A \rangle_t$ will usually change with time. As an example, consider the superposition of two states given by

$$\psi(\mathbf{r}, t) = c_1 \phi_1(\mathbf{r}) \, e^{-iE_1 t/\hbar} + c_2 \phi_2(\mathbf{r}) \, e^{-iE_2 t/\hbar}. \tag{1.59}$$

(With ϕ_1 and ϕ_2 orthonormal and $|c_1|^2 + |c_2|^2 = 1$, ψ is normed.) Measuring the energy of the system in this state gives either the value E_1 or E_2: we can interpret $|E_1 - E_2|$ as the energy uncertainty of the system in the state (1.59). For this state, $\langle A \rangle_t$ is given by

$$\langle A \rangle_t = |c_1|^2 A_{11} + |c_2|^2 A_{22} + c_1^* c_2 A_{12} \, e^{i\omega t} + c_2^* c_1 A_{21} \, e^{-i\omega t} \tag{1.60}$$

where

$$\omega = (E_1 - E_2)/\hbar \tag{1.61}$$

and

$$A_{mn} = \int \phi_m^*(\mathbf{r}) A \phi_n(\mathbf{r}) \, d^3\mathbf{r}. \tag{1.62}$$

We use the Hermiticity condition $A_{21} = A_{12}^*$, which follows from Eq. (1.13). This allows us to simplify Eq. (1.60) to

$$\langle A \rangle_t = |c_1|^2 A_{11} + |c_2|^2 A_{22} + 2 \, \mathcal{R}e[c_1^* c_2 A_{12} \, e^{i\omega t}]. \tag{1.63}$$

This expectation value oscillates between the extreme values

$$|c_1|^2 A_{11} + |c_2|^2 A_{22} \pm 2|c_1^* c_2 A_{12}|$$

with the period

$$T = \frac{2\pi}{|\omega|} = \frac{2\pi\hbar}{|E_1 - E_2|}. \tag{1.64}$$

This oscillatory behaviour is typical of non-stationary states. The period T characterizes how quickly the system changes with time. The larger the energy uncertainty $|E_1 - E_2|$, the more rapidly do the properties of the system in the state (1.59) change.

1.4   MANY-PARTICLE SYSTEMS

The extension of the formalism to many particles is straightforward. A system of N particles is described by a wave function $\psi(\mathbf{r}_1, \ldots, \mathbf{r}_N, t)$, where $\mathbf{r}_k$ is the position coordinate of the particle labelled k ($= 1, 2, \ldots, N$).

We shall assume that all the particles are different. For identical particles (for example, two electrons) the ordering of the particle labels must be immaterial (e.g. whether we label the two electrons in a helium atom 1 and 2 or, in the reverse order, 2 and 1). This imposes symmetry restrictions on the wave functions of identical particles, as we shall see in section 4.4.

The formalism and its interpretation for the N-particle system are very similar to those for the one-particle case. Hence we can be quite brief. For a system in the state $\psi(\mathbf{r}_1, \ldots, \mathbf{r}_N, t)$, we interpret

$$|\psi(\mathbf{r}_1, \ldots, \mathbf{r}_N, t)|^2 \, d^3\mathbf{r}_1 \ldots d^3\mathbf{r}_N \tag{1.65}$$

as the probability that, at the time t, the particle labelled 1 is in the volume element $d^3\mathbf{r}_1$ at the position $\mathbf{r}_1$, the particle labelled 2 is in $d^3\mathbf{r}_2$ at $\mathbf{r}_2$, and so on for all N particles. This imposes the normalization condition

$$\int |\psi(\mathbf{r}_1, \ldots, \mathbf{r}_N, t)|^2 \, d^3\mathbf{r}_1 \ldots d^3\mathbf{r}_N = 1, \tag{1.66}$$

integration being over the whole configuration space of the system. For a system in the normed state $\psi = \psi(\mathbf{r}_1, \ldots, \mathbf{r}_N, t)$, the expectation value at time t of an operator A is given by

$$\langle A \rangle = \int \psi^* A \psi \, d^3\mathbf{r}_1 \ldots d^3\mathbf{r}_N. \tag{1.67}$$

The operators representing observables are defined by the natural generalizations of Eqs. (1.5) and (1.8); for the position and momentum coordinates of the kth particle

$$\hat{\mathbf{r}}_k = \mathbf{r}_k, \quad \hat{\mathbf{p}}_k = -i\hbar\nabla_k, \tag{1.68}$$

and for an observable $F(\mathbf{r}_1, \ldots, \mathbf{r}_N, \mathbf{p}_1, \ldots, \mathbf{p}_N)$

$$\hat{F} = F(\mathbf{r}_1, \ldots, \mathbf{r}_N, -i\hbar\nabla_1, \ldots, -i\hbar\nabla_N). \tag{1.69}$$

Thus for a system with the classical Hamiltonian $H(\mathbf{r}_1, \ldots, \mathbf{r}_N, \mathbf{p}_1, \ldots, \mathbf{p}_N)$, the Schrödinger equation (1.47) becomes

$$i\hbar \frac{\partial}{\partial t} \psi(\mathbf{r}_1, \ldots, \mathbf{r}_N, t) = H(\mathbf{r}_1, \ldots, \mathbf{r}_N, -i\hbar\nabla_1, \ldots, -i\hbar\nabla_N)\psi(\mathbf{r}_1, \ldots, \mathbf{r}_N, t). \tag{1.70}$$

Although the formalism for N particles is straightforward, the resulting equations are extremely difficult to solve in general. An exception occurs for a system of non-interacting particles. For two non-interacting particles the Hamiltonian is of the form

$$H = H_1 + H_2 \tag{1.71}$$

where H_1 and H_2 are the Hamiltonian operators for particles 1 and 2 respectively. For example, if the particle labelled k ($= 1, 2$) has the mass m_k and moves in the potential $V_k(\mathbf{r}_k)$, then

$$H_k \equiv H_k(\mathbf{r}_k, -i\hbar\nabla_k) = -\frac{\hbar^2}{2m_k}\nabla_k^2 + V_k(\mathbf{r}_k). \tag{1.72}$$

The eigenfunctions and eigenvalues of the Hamiltonian (1.71) can at once be written down in terms of those of the single-particle Hamiltonians H_1 and H_2. Let

$$H_1 \phi_m^{(1)}(\mathbf{r}_1) = E_m^{(1)} \phi_m^{(1)}(\mathbf{r}_1), \qquad m = 1, 2, \ldots \tag{1.73a}$$

$$H_2 \phi_m^{(2)}(\mathbf{r}_2) = E_m^{(2)} \phi_m^{(2)}(\mathbf{r}_2), \qquad m = 1, 2, \ldots \tag{1.73b}$$

i.e. $\phi_m^{(k)}(\mathbf{r}_k)$ and $E_m^{(k)}$, $m = 1, 2, \ldots$, are the eigenfunctions and eigenvalues of the Hamiltonian H_k for the particle labelled k ($= 1, 2$). It follows from Eqs. (1.73) for any product of eigenfunctions $\phi_m^{(1)}(\mathbf{r}_1)\phi_n^{(2)}(\mathbf{r}_2)$ that

$$\begin{aligned} H\phi_m^{(1)}(\mathbf{r}_1)\phi_n^{(2)}(\mathbf{r}_2) &= (H_1 + H_2)\phi_m^{(1)}(\mathbf{r}_1)\phi_n^{(2)}(\mathbf{r}_2) \\ &= [H_1\phi_m^{(1)}(\mathbf{r}_1)]\phi_n^{(2)}(\mathbf{r}_2) + \phi_m^{(1)}(\mathbf{r}_1)[H_2\phi_n^{(2)}(\mathbf{r}_2)] \\ &= (E_m^{(1)} + E_n^{(2)})\phi_m^{(1)}(\mathbf{r}_1)\phi_n^{(2)}(\mathbf{r}_2), \end{aligned}$$

i.e. the product function

$$\phi_{mn}(\mathbf{r}_1, \mathbf{r}_2) = \phi_m^{(1)}(\mathbf{r}_1)\phi_n^{(2)}(\mathbf{r}_2) \tag{1.74}$$

is an eigenfunction of the Hamiltonian $H = H_1 + H_2$ belonging to the eigenvalue

$$E_{m,n} = E_m^{(1)} + E_n^{(2)}. \tag{1.75}$$

Eq. (1.75) is what one would expect for a system of two non-interacting particles: its energy is the sum of the energies of the individual particles.

The generalization of these results to a system of N non-interacting particles, with the Hamiltonian

$$H = H_1 + H_2 + \cdots + H_N, \tag{1.76}$$

is obvious.

Realistically, the Hamiltonian (1.76) is usually augmented by terms representing the interactions between particles. It is these terms which make it so difficult to solve the Schrödinger equation, so that one must employ approximate and computational methods. In these situations it is very helpful to understand the qualitative features of a system, such as its angular momentum and other symmetry properties. Later in this book we shall often consider both these aspects: symmetries and approximation methods.

PROBLEMS 1

1.1 The normalized energy eigenfunction of the ground state of the hydrogen atom is given by

$$\psi(r) = C \exp(-r/a_0)$$

where a_0 is the Bohr radius and C is a constant. For this state, calculate:
(a) the constant C;
(b) the probability that the electron is within a sphere of radius a_0;
(c) the most probable value of the radial coordinate r;
(d) the expectation value of r;
(e) the expectation value of the potential energy of the electron;
(f) the standard deviation of r;
(g) the expectation value of the x-component of momentum of the electron.
The first excited state of the hydrogen atom has the wave function

$$\phi(r) = A(1 + \lambda r) \exp(-r/2a_0)$$

where A and λ are constants. Find λ.

$$\left(\text{A helpful integral:} \int_0^\infty e^{-\alpha r} r^n \, dr = \frac{n!}{\alpha^{n+1}}, \quad n > -1. \right)$$

1.2 The one-dimensional motion of a particle, restricted to the region $0 \leqslant x \leqslant L$, is described by the wave function

$$\psi(x) = C \sin \frac{\pi x}{L}$$

where C is a normalization constant. Obtain the probability that the particle is (i) in the interval $\frac{1}{2}L \leqslant x \leqslant L$, (ii) in the interval $\frac{1}{4}L \leqslant x \leqslant \frac{3}{4}L$.

1.3 (i) If $\psi_0(\mathbf{r}) = zf(r)$ is a normalized wave function, find the constant C such that $\psi_1(\mathbf{r}) = C(x + iy)f(r)$ and $\psi_{-1}(\mathbf{r}) = C(x - iy)f(r)$ are also normed to unity.
(ii) Show that these wave functions $\psi_m(\mathbf{r})$, $m = 1, 0, -1$, are eigenstates of the z-component $\hat{l}_z$ of the orbital angular momentum operator $\hat{\mathbf{l}} = \hat{\mathbf{r}} \wedge \hat{\mathbf{p}}$, and find the corresponding eigenvalues λ_m, $m = 1, 0, -1$.

(iii) If a system is in the state $\phi(\mathbf{r}) = (z - ix)f(r)/\sqrt{2}$, calculate the probabilities $P(\lambda_m)$, $m = 1, 0, -1$, that a measurement of l_z yields the results λ_1, λ_0 and λ_{-1} respectively. What is the probability that a measurement of l_z on the system in the state $\phi(\mathbf{r})$ yields some other value?

1.4 From two linearly independent functions $\phi_1(\mathbf{r})$ and $\phi_2(\mathbf{r})$, construct linear combinations

$$\psi_1(\mathbf{r}) = a\phi_1(\mathbf{r}), \qquad \psi_2(\mathbf{r}) = b\phi_1(\mathbf{r}) + c\phi_2(\mathbf{r})$$

which are orthonormal, i.e.

$$\int \psi_m^*(\mathbf{r})\psi_n(\mathbf{r})\, d^3\mathbf{r} = \delta_{mn}, \qquad m, n = 1, 2.$$

1.5 The Schrödinger equation of two interacting particles of masses m_1 and m_2 is

$$i\hbar \frac{\partial}{\partial t} \psi(\mathbf{r}_1, \mathbf{r}_2, t) = \left[-\frac{\hbar^2}{2m_1}\nabla_1^2 - \frac{\hbar^2}{2m_2}\nabla_2^2 + V(|\mathbf{r}_1 - \mathbf{r}_2|) \right]\psi(\mathbf{r}_1, \mathbf{r}_2, t). \quad (1.77)$$

Show that this equation is separable if one introduces the centre-of-mass and relative coordinates

$$\mathbf{R} = (m_1\mathbf{r}_1 + m_2\mathbf{r}_2)/(m_1 + m_2), \qquad \mathbf{r} = \mathbf{r}_1 - \mathbf{r}_2,$$

and writes $\psi(\mathbf{r}_1, \mathbf{r}_2, t) = \Phi(\mathbf{R}, t)\phi(\mathbf{r}, t)$. Interpret the resulting equations.

1.6 Use the Schrödinger equation

$$i\hbar \frac{\partial}{\partial t} \psi(\mathbf{r}, t) = \left[-\frac{\hbar^2}{2m}\nabla^2 + V(\mathbf{r}) \right]\psi(\mathbf{r}, t)$$

to show that

$$-\frac{d}{dt}\int_V \rho(\mathbf{r}, t)\, d^3\mathbf{r} = \int_V \mathbf{\nabla}\cdot\mathbf{j}(\mathbf{r}, t)\, d^3\mathbf{r} = \int_S \mathbf{j}(\mathbf{r}, t)\cdot d\mathbf{S} \quad (1.78)$$

where

$$\rho(\mathbf{r}, t) = |\psi(\mathbf{r}, t)|^2, \quad \mathbf{j}(\mathbf{r}, t) = \frac{i\hbar}{2m}(\psi\mathbf{\nabla}\psi^* - \psi^*\mathbf{\nabla}\psi); \quad (1.79)$$

V is any finite volume, bounded by the surface S, and $d\mathbf{S}$ is a vector element of surface, of magnitude equal to the area of the element and directed along the outward normal to the element.

Interpret $\rho(\mathbf{r}, t)$, $\mathbf{j}(\mathbf{r}, t)$ and Eq. (1.78).

Simple examples

In this chapter I shall illustrate the ideas and the formalism of Chapter 1. I shall show how quantum mechanics accounts for the existence of quantized energy levels, for barrier penetration, and for the results of the Stern–Gerlach experiment. The examples we shall discuss will serve to introduce in a simple context many ideas which are important in realistic situations.

2.1 ONE-DIMENSIONAL SQUARE-WELL POTENTIAL

As a first example we shall consider the one-dimensional motion of a particle of mass m in the one-dimensional square-well potential defined by

$$V(x) = \begin{cases} -V_0(<0), & |x| < a \\ 0, & |x| > a \end{cases} \qquad (2.1)$$

and shown in Fig. 2.1(a). In spite of the restriction to one dimension, this problem displays many of the typical quantum-mechanical features of more realistic three-dimensional problems. The discontinuities of the square-well potential at $x = \pm a$ represent another unnatural aspect of this potential. As a result, the particle experiences an infinite force $F = -\mathrm{d}V/\mathrm{d}x$ at the positions $x = \pm a$. Such infinite forces do not occur in nature. We can think of the square-well potential as the limit of a more realistic smoothly varying short-range potential, sketched in Fig. 2.1(b). This potential has steep but not infinitely steep walls with rounded-off corners, and the force $F = -\mathrm{d}V/\mathrm{d}x$ acting on the

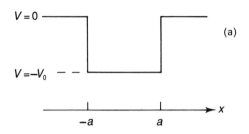

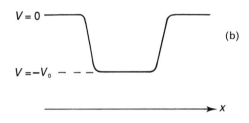

Fig. 2.1   (a) The one-dimensional square-well potential, Eq. (2.1). (b) A more realistic short-range potential $V(x)$.

particle is always finite and is directed towards the centre of the well, i.e. we are dealing with an attractive potential.

According to classical mechanics, we distinguish two situations. For $E < 0$, the particle is confined to the inner region, $|x| < a$, since in the outer regions, $|x| > a$, its kinetic energy $(-|E|)$ would be negative. Hence for $E < 0$, the outer regions are not accessible to the particle: it is localized within the well: we are dealing with a bound state of the particle. For $E > 0$, no such restriction applies: the particle can be both outside or inside the well. We shall see that according to quantum mechanics the situation is very similar, with a typical quantum-mechanical refinement!

In this section we shall consider the case of negative energies: $E < 0$.

In one dimension, the time-independent Schrödinger equation is given by [see Eqs. (1.9) and (1.52)]

$$\left[-\frac{\hbar^2}{2m} \frac{d^2}{dx^2} + V(x) \right] \phi(x) = E\phi(x). \tag{2.2}$$

For the potential (2.1), this equation takes on different forms in the three regions:

$$\frac{d^2\phi}{dx^2} - \kappa^2 \phi = 0, \qquad \text{for } x < -a \text{ or } x > a, \tag{2.3a}$$

and

$$\frac{d^2\phi}{dx^2} + q^2 \phi = 0, \qquad \text{for } |x| < a, \tag{2.3b}$$

where the real parameters κ and q are defined by

$$\kappa = + \left[\frac{2m}{\hbar^2} |E| \right]^{1/2}, \quad q = + \left[\frac{2m}{\hbar^2} (V_0 - |E|) \right]^{1/2}. \tag{2.4}$$

The general solution of Eq. (2.3a) is a linear superposition of the *real exponential* functions

$$e^{\kappa x}, \quad e^{-\kappa x} \quad (|x| > a) \tag{2.5a}$$

whereas the general solution of (2.3b) is a superposition of the *oscillatory* terms

$$\cos qx, \quad \sin qx \quad (|x| < a). \tag{2.5b}$$

The difference in the nature of the solutions (2.5a) and (2.5b) is very important and is quite general. If for a potential $V(x)$ we define a local wave number $k(x)$ by

$$k(x) = \left\{ \frac{2m}{\hbar^2} [E - V(x)] \right\}^{1/2}, \tag{2.6}$$

we can write the Schrödinger equation (2.2)

$$\frac{d^2\phi}{dx^2} + [k(x)]^2 \phi = 0. \tag{2.7}$$

In regions where $k(x)$ is real [i.e. $E > V(x)$] the solutions of this equation are oscillatory; in regions where $k(x)$ is imaginary [i.e. $E < V(x)$] they are non-oscillatory.

We consider the solutions of the Schrödinger equation (2.3a) in the outer regions. For $x > a$, the general solution is given by

$$\phi = C e^{-\kappa x} + C' e^{\kappa x}, \quad x > a,$$

when C and C' are constants of integration. For this wave function the probability density $|\phi(x)|^2$, of finding the particle at x, tends to infinity as $x \to \infty$, if $C' \neq 0$. Clearly this does not make sense. We must take $C' = 0$ so that the wave function reduces to

$$\phi = C e^{-\kappa x}, \quad x > a, \tag{2.8a}$$

which allows a sensible probability interpretation. The same argument, applied to the outer region $x < -a$, gives for the wave function in this region

$$\phi = A e^{\kappa x}, \quad x < -a \tag{2.8b}$$

where A is another constant of integration.

For the wave functions (2.8), the probability density $|\phi(x)|^2$ tends to zero as $x \to \pm \infty$. In other words, these wave functions represent a localized state of the particle, we are dealing with a bound state. We started from the explicit solutions (2.5) of the Schrödinger equation for the square-well potential. We can generalize the above arguments and impose as the general boundary conditions for a wave function $\phi(x)$ to represent a bound state the conditions

$$\phi(x) \rightarrow 0, \qquad \text{as } x \rightarrow \pm \infty. \tag{2.9}$$

Solving the Schrödinger equation (2.2) for a potential $V(x)$, with the boundary conditions (2.9) imposed, then gives the bound states of this potential.

The wave functions (2.8a) and (2.8b) lead to the non-zero probability density

$$|\phi(x)|^2 = \text{const. } e^{-2\kappa|x|}, \qquad |x| > a, \tag{2.10}$$

in the outer regions. This is a most startling result: according to quantum mechanics, the particle can penetrate into the classically forbidden regions! The probability density (2.10) falls off more steeply as κ increases, that is, from Eqs. (2.4), as $|E|$ increases. The larger the binding energy $|E|$ of the particle, the more tightly it is bound in the well and the less far it penetrates into the forbidden outer regions, just as one would expect. As the reader may realize, we have here a mechanism which allows a particle to tunnel through a region in which its classical kinetic energy is negative and which is classically not accessible. We shall return to this explanation of barrier penetration in the next section.

In the inner region $|x| < a$ the Schrödinger equation (2.3b) has the general solution

$$\phi = B \sin qx + B' \cos qx, \qquad |x| < a, \tag{2.8c}$$

where B and B' are two further integration constants.

The four constants A, B, B' and C in the wave functions (2.8a) to (2.8c) are determined by matching these wave functions at the boundaries $x = \pm a$ of the three regions and by normalizing the wave function. This matching problem would not occur for a continuous potential. It arises only because of the discontinuities of the square-well potential at $x = \pm a$. Similar matching problems occur in other situations involving discontinuities. For example, in electrostatic problems involving materials of different relative permittivities or in the propagation of light through media with different refractive indices the electric and magnetic fields must be appropriately matched at the boundaries between the media. We shall always assume that the wave function $\phi(x)$ is finite, for its interpretation as a probability amplitude to make sense. We see from the Schrödinger equation (2.2) that at a point x where $V(x)$ has a finite discontinuity $d^2\phi/dx^2$ is finite (though discontinuous). By integrating the Schrödinger equation across the discontinuity, one shows that $d\phi/dx$ and ϕ are continuous; for example, consider the discontinuity of the square-well potential at $x = a$. Since $d^2\phi/dx^2$ is finite at this point,

$$\left(\frac{d\phi}{dx}\right)_{a+\varepsilon} - \left(\frac{d\phi}{dx}\right)_{a-\varepsilon} = \int_{a-\varepsilon}^{a+\varepsilon} \frac{d^2\phi}{dx^2}\, dx \rightarrow 0$$

as $\varepsilon \to 0$, i.e. $\mathrm{d}\phi/\mathrm{d}x$ is continuous at $x = a$. From the continuity of $\mathrm{d}\phi/\mathrm{d}x$ that of $\phi(x)$ follows in the same way. A similar argument applies to the square-well discontinuity at $x = -a$, and more generally. We can therefore state as the general matching condition: at a point x where the potential $V(x)$ has a finite discontinuity

$$\phi(x) \quad \text{and} \quad \mathrm{d}\phi/\mathrm{d}x \text{ are continuous.} \tag{2.11a}$$

The extension of this result to three dimensions is obvious. If $V(\mathbf{r})$ has a finite discontinuity across a surface S, then $\phi(\mathbf{r})$ and the derivative $\partial\phi(\mathbf{r})/\partial n$ along the normal n to the surface are continuous at every point P on S. Since the derivatives of $\phi(\mathbf{r})$ at P in the tangent plane to S are continuous (since $V(\mathbf{r})$ is continuous in these directions), we can state the general matching condition:

$$\phi(\mathbf{r}) \text{ and the gradient } \nabla\phi(\mathbf{r}) \text{ are continuous} \tag{2.11b}$$

at every point on the surface S.

Applying the matching conditions (2.11a) to the wave functions (2.8) at $x = \pm a$, we obtain four conditions

$$C\,\mathrm{e}^{-\kappa a} = B \sin qa + B' \cos qa \tag{2.12a}$$

$$-\kappa C\,\mathrm{e}^{-\kappa a} = qB \cos qa - qB' \sin qa \tag{2.12b}$$

$$A\,\mathrm{e}^{-\kappa a} = -B \sin qa + B' \cos qa \tag{2.12c}$$

$$\kappa A\,\mathrm{e}^{-\kappa a} = qB \cos qa + qB' \sin qa. \tag{2.12d}$$

These are four linear *homogeneous* equations for the four unknown quantities A, B, B' and C. In general they only possess the trivial solution $A = B = B' = C = 0$. For a non-trivial solution to exist, these equations must satisfy a consistency condition which is easily derived directly. Dividing Eq. (2.12a) by (2.12c) gives

$$\frac{C}{A} = \frac{tB + B'}{-tB + B'} \tag{2.13a}$$

where we wrote the abbreviation

$$t = \sin qa/\cos qa. \tag{2.14}$$

Similarly Eqs. (2.12b) and (2.12d) give

$$\frac{C}{A} = \frac{tB' - B}{tB' + B}. \tag{2.13b}$$

Equating the two expressions (2.13a) and (2.13b) gives the consistency condition

$$(tB + B')(tB' + B) = (tB' - B)(-tB + B'),$$

which reduces to

$$(t^2 + 1)BB' = 0.$$

This equation has two solutions: $B = 0$ or $B' = 0$, and only for these values do Eqs. (2.12) possess non-trivial solutions. Thus there exist two types of solution.

(i) For $B' = 0$, Eq. (2.13a) gives $C = -A$ and the wave functions (2.8) reduce to

$$\left.\begin{array}{ll} \phi(x) = B \sin qx, & |x| < a, \\ \phi(x) = -C\,e^{\kappa x},\quad x < -a; & \phi(x) = C\,e^{-\kappa x}, \qquad x > a. \end{array}\right\} \quad (2.15a)$$

These are the *odd* solutions, since the $\phi(x)$ are odd functions, i.e.

$$\phi(-x) = -\phi(x). \tag{2.16a}$$

(ii) For $B = 0$, Eq. (2.13a) gives $C = A$ and the wave functions (2.8) become

$$\left.\begin{array}{ll} \phi(x) = B' \cos qx, & |x| < a, \\ \phi(x) = C\,e^{\kappa x},\quad x < -a; & \phi(x) = C\,e^{-\kappa x}, \qquad x > a. \end{array}\right\} \quad (2.15b)$$

These are the *even* solutions, since the $\phi(x)$ are even functions, i.e.

$$\phi(-x) = +\phi(x). \quad * \tag{2.16b}$$

These solutions must satisfy the matching conditions (2.11a) at $x = \pm a$. Applied to Eqs. (2.12) or directly to Eqs. (2.15a), the matching conditions (2.11a) for the odd solutions become

$$C\,e^{-\kappa a} = B \sin qa \tag{2.17a}$$

$$-\kappa C\,e^{-\kappa a} = qB \cos qa. \tag{2.17b}$$

To satisfy these *two* conditions, we must consider the energy E as well as the ratio C/B as unknown variables. The ratio of the two equations then gives the consistency condition

$$\tan qa = -q/\kappa. \tag{2.18a}$$

If this equation holds, the two equations (2.17a) and (2.17b) give the same value for the ratio C/B. The individual amplitudes C and B are then determined from the normalization condition for the wave function

$$\int_{-\infty}^{\infty} |\phi(x)|^2 \, dx = 1.$$

---

* This division into odd and even solutions follows from the fact that the square-well potential itself is an even function: $V(-x) = V(x)$. These very important symmetry questions will be discussed in Chapter 4.

Eq. (2.18a) determines the values of the energy E for which the matching conditions hold. (Remember Eqs. (2.4) define q and κ in terms of E.) The solutions of Eq. (2.18a) are the energy eigenvalues of bound states of the particle in the square-well potential; Eqs. (2.15a) are the corresponding energy eigenfunctions.

To obtain the energy eigenvalues, we introduce the dimensionless parameters

$$\xi = qa, \qquad \eta = \kappa a. \tag{2.19}$$

From Eqs. (2.4) these are related by

$$\xi^2 + \eta^2 = \frac{2m}{\hbar^2} V_0 a^2, \tag{2.20}$$

and Eq. (2.18a) becomes

$$\tan \xi = -\frac{\xi}{\eta} = -\xi \left[\frac{2m}{\hbar^2} V_0 a^2 - \xi^2 \right]^{-1/2}.$$

Nowadays one will probably solve this equation on a computer but one gains more insight from a graphical solution. Fig. 2.2 shows the curves

$$y = \tan \xi, \quad y = -\xi \left[\frac{2m}{\hbar^2} V_0 a^2 - \xi^2 \right]^{-1/2} \tag{2.21}$$

for one value of the parameter $2mV_0a^2/\hbar^2$. The intersections of these two curves give the energy eigenvalues for this value of $2mV_0a^2/\hbar^2$. We see from the figure that for

$$\left[\frac{2m}{\hbar^2} V_0 a^2 \right]^{1/2} < \frac{\pi}{2} \tag{2.22a}$$

there exists no odd bound state; for

$$\frac{\pi}{2} < \left[\frac{2m}{\hbar^2} V_0 a^2 \right]^{1/2} < \frac{3\pi}{2} \tag{2.22b}$$

there exists one odd bound state, and so on.

One can understand these results by considering a square well potential of range a and increasing its depth from $V_0 = 0$. To begin with, the potential is not sufficiently attractive to give an odd bound state. At $V_0 a^2 = \hbar^2 \pi^2 / 8m$, the potential is *just* strong enough to support one odd bound state. As we increase V_0 further, the binding energy $|E|$ of this state increases: the particle becomes more firmly bound. At $V_0 a^2 = 9\hbar^2 \pi^2 / 8m$, the well is just capable of supporting two odd bound states, and so on.

For the square-well potential we have now demonstrated one of the startling results of quantum mechanics: bound states only occur for certain values of the energy: *the bound-state energy E is quantized.*

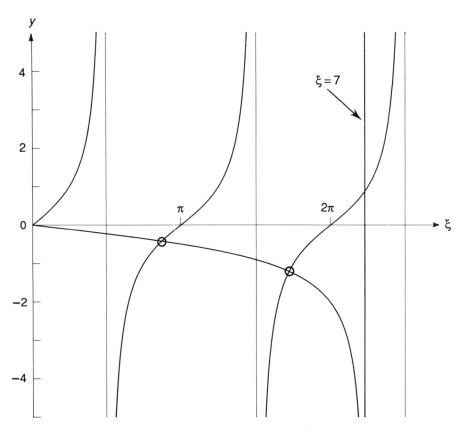

Fig. 2.2   The curves $y = \tan \xi$ and $y = -\xi(2mV_0a^2/\hbar^2 - \xi^2)^{-1/2}$ for $\sqrt{(2mV_0a^2/\hbar^2)} = 7$. Their intersections, at the points marked by circles, correspond to the two odd bound states which occur for this value of $\sqrt{(2mV_0a^2/\hbar^2)}$.

The restriction to a discrete set of energies has its origin in the boundary conditions (2.9). To show this for the odd solutions, we replace Eqs. (2.15a) by the more general solution of the Schrödinger equations (2.3):

$$\left. \begin{array}{ll} \phi(x) = B \sin qx, & |x| < a, \\ \phi(x) = -C\,e^{\kappa x} - C'\,e^{-\kappa x}, & x < -a, \\ \phi(x) = C\,e^{-\kappa x} + C'\,e^{\kappa x}, & x > a. \end{array} \right\}$$

For an arbitrary negative value of E and an arbitrary constant B, we can always satisfy the matching conditions (2.11a) at $x = \pm a$, since these conditions now give two linear *inhomogeneous* equations for C/B and C'/B. These can always be satisfied, but in general with $C' \neq 0$, i.e. the wave functions become infinite as $x \to \pm \infty$. Only for certain special values of E—the energy eigenvalues—will C' be zero and $\phi(x) \to 0$ as $x \to \pm \infty$. Similar arguments hold for more general potentials $V(x)$.

Analogous results hold for the even bound-state solutions (2.15b). The energy eigenvalues are now the solutions of

$$\tan qa = \kappa/q \tag{2.18b}$$

and one again obtains energy quantization with a finite number of bound states. The details are left as an exercise for the reader (see problem 2.3).

Combining the results for the even and odd solutions, we sketch in Fig. 2.3 the first three energy eigenfunctions, ordered according to increasing energy eigenvalues: $E_0 < E_1 < \cdots$. (We are assuming that the potential well is sufficiently attractive to support at least three bound states.) $\phi_n(x)$ is the eigenfunction belonging to the energy eigenvalue E_n. The states with even (odd) values of n are even (odd) solutions. We see from the figure that the ground state ($n = 0$) possesses no node, the first excited state ($n = 1$) possesses one node, and so on with similar results for all eigenfunctions: the nth excited state $\phi_n(x)$

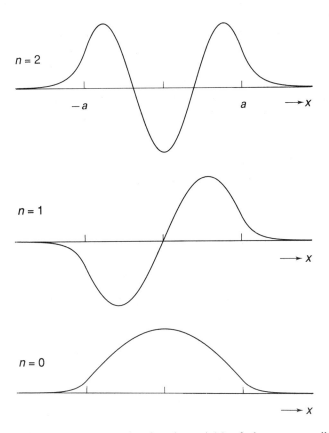

Fig. 2.3   The first three energy eigenfunctions $\phi_n(x)$ of the square-well potential. (The wave functions are not normalized.)

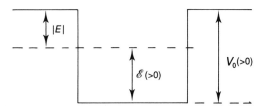

Fig. 2.4   The bound state energy: (i) $\mathscr{E}(>0)$ when measured from the bottom of the well; (ii) $E(<0)$ when measured from the top of the well.

possesses n nodes. (In counting nodes, we are excluding the 'end points' $x = \pm\infty$ where the eigenfunctions also vanish, i.e. here and in what follows we shall always mean *internal* nodes.) This result can be shown to be a general property of one-dimensional eigenvalue problems: with the energy eigenvalues ordered in a monotonic increasing sequence, as above, the nth eigenfunction $\phi_{n-1}(x)$ possesses $(n-1)$ nodes. For the convenience of later reference, we shall call this the node theorem.

Finally, we consider the limit $V_0 \to \infty$ as the square well becomes infinitely deep. This potential is known as the infinite square well. It is now convenient to measure energies from the bottom of the well instead of from the top as we have done so far. Let $\mathscr{E}(>0)$ denote the energy of a bound state measured from the bottom of the well, and let $E(<0)$ denote the energy of the same level measured from the top of the well, as so far; then

$$\mathscr{E} = V_0 - |E| = V_0 + E,$$

as illustrated in Fig. 2.4, and Eqs. (2.4) for κ and q become

$$\kappa = \left[\frac{2m}{\hbar^2}(V_0 - \mathscr{E})\right]^{1/2}, \quad q = \left[\frac{2m}{\hbar^2}\mathscr{E}\right]^{1/2}. \tag{2.23}$$

Hence $\kappa \to \infty$ as $V_0 \to \infty$, and we see from Eq. (2.10) that the energy eigenfunctions $\phi(x)$ vanish in the outer regions $|x| > a$. It follows from Eq. (2.17a) that, in the limit $\kappa \to \infty$,

$$\sin qa = 0 \quad \text{for odd solutions.}$$

In a similar way the continuity of the even solutions (2.15b) at $x = a$ gives, in the limit $\kappa \to \infty$,

$$\cos qa = 0 \quad \text{for even solutions.}$$

Thus the eigenfunctions always satisfy the boundary conditions

$$\phi(-a) = \phi(a) = 0. \quad * \tag{2.24}$$

---

* At an infinite discontinuity of $V(x)$ the matching conditions (2.11a) become modified. We see from the above that the wave function is still continuous. However, the argument we used earlier to derive the continuity of $d\phi/dx$ breaks down. We shall see that $d\phi/dx$ is discontinuous at $x = \pm a$.

Our eigenvalue problem is thus reduced to the Schrödinger equation (2.3b) together with the boundary conditions (2.24). This is just the problem of a vibrating string of length $2a$ with fixed ends. Its normed eigenfunctions are the fundamental and harmonic modes

$$
\left.
\begin{aligned}
\phi_n(x) &= \frac{1}{\sqrt{a}} \cos\left(\frac{n\pi x}{2a}\right), & n &= 1, 3, \ldots \\
\phi_n(x) &= \frac{1}{\sqrt{a}} \sin\left(\frac{n\pi x}{2a}\right), & n &= 2, 4, \ldots
\end{aligned}
\right\}
\tag{2.25}
$$

The corresponding energy eigenvalues are, from Eq. (2.23),

$$
\mathscr{E}_n = \frac{\hbar^2}{2m}\left(\frac{n\pi}{2a}\right)^2 \qquad n = 1, 2, \ldots . \tag{2.26}
$$

The eigenfunctions (2.25) are orthonormal, i.e.

$$
\int_{-a}^{a} \phi_m(x)\phi_n(x)\, \mathrm{d}x = \delta_{mn},
$$

and form a complete set: expansion in terms of them is just a Fourier series expansion. The nth eigenfunction $\phi_n(x)$ possesses $(n-1)$ internal nodes, in agreement with the node theorem.

A particle in an infinite square-well potential is necessarily confined to the region $|x| < a$: we are talking of a particle inside a one-dimensional box with rigid impenetrable walls. Such a particle possesses bound states only and there are an infinite discrete set of them, as we have just seen. In contrast, we found that a well of finite depth possesses only a finite number of bound states. In addition, it possesses a continuum of unbound states. We shall study these in the next section.

2.2   POTENTIAL BARRIER

We shall now study the one-dimensional square potential barrier, that is the potential

$$
V(x) = \begin{cases} V_0(>0), & |x| < a \\ 0, & |x| > a \end{cases} \tag{2.27}
$$

which is shown in Fig. 2.5. This purely repulsive potential cannot support any bound states, so that one deals here with quite different situations from the bound-state problem of the last section.

Quantum-mechanically, the non-existence of bound states for the barrier

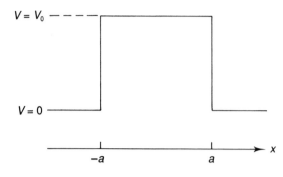

Fig. 2.5   The one-dimensional potential barrier (2.27).

potential (2.27) follows from the Schrödinger equation (2.2). Its solutions in the external regions $|x| > a$ are linear superpositions of

$$e^{ikx}, \quad e^{-ikx}$$

where the wave number k is related to the energy $E(>0)$ of the particle by

$$k = \sqrt{(2mE/\hbar^2)}. \tag{2.28}$$

These plane-wave solutions extend all the way to $x = \pm\infty$ with non-decreasing amplitudes, so they cannot represent a bound state. Furthermore, such solutions cannot be normalized, so that the probability interpretation of $|\phi(x)|^2$ for the position of the particle, etc., breaks down. Several procedures exist for dealing with this. An obvious one, by analogy with classical wave problems, is to think of truly plane waves as an idealization; more realistically one should consider wave packets. We shall come back to this approach in section 2.6.2. Here we shall employ a different approach. The plane waves

$$\phi(x) = A\, e^{ikx} \tag{2.29a}$$

$$\phi(x) = A\, e^{-ikx} \tag{2.29b}$$

(with A a constant amplitude) are eigenfunctions of the momentum operator $\hat{p}_x = -i\hbar\, d/dx$, with eigenvalues $\pm\hbar k$ respectively. Although these wave functions cannot be normalized to unity over all space, we can interpret Eqs. (2.29a) and (2.29b) as representing beams of particles of uniform density, $|A|^2$ particles per unit volume, and travelling with momentum $\hbar k$ in the positive and negative x-directions respectively. For example, we could have a source of particles, such as an accelerator, situated at a large negative value of x ($x = -\infty$), followed by an energy selector and beam collimator, producing a mono-energetic beam of particles travelling in the positive x-direction. We shall consider such a beam, represented by Eq. (2.29a), incident on the potential barrier. In other words, we are dealing with a one-dimensional scattering problem. Classically, *all* particles of the beam will be reflected at the left-hand edge ($x = -a$) of the barrier, if $E < V_0$; *all* will be transmitted through the barrier if $E > V_0$.

Quantum-mechanically, the situation is very different. From the discussion in the last section, we expect particles to be reflected by the barrier and transmitted through it at all energies, both $E < V_0$ and $E > V_0$. The solutions of the Schrödinger equation (2.2) in the outer regions are

$$\phi = A\,e^{ikx} + A'\,e^{-ikx}, \qquad x < -a \qquad (2.30)$$

$$\phi = C\,e^{ikx}, \qquad x > a \qquad (2.31)$$

with A' and C constants. The second term on the right-hand side of Eq. (2.30) corresponds to the particles reflected by the barrier. Eq. (2.31) contains no term in e^{-ikx} since there is no source at $x = \infty$ of particles travelling in the negative x-direction. In the inner region $|x| < a$ the wave function has oscillatory or real exponential character depending on whether $E > V_0$ or $E < V_0$. We shall write it

$$\phi = B\,e^{iqx} + B'\,e^{-iqx} \qquad |x| < a, \quad \text{if } E > V_0, \qquad (2.32a)$$

$$\phi = B\,e^{-\kappa x} + B'\,e^{\kappa x}, \qquad |x| < a, \quad \text{if } E < V_0, \qquad (2.32b)$$

where B and B' are constants and

$$q = + \left[\frac{2m}{\hbar^2}(E - V_0)\right]^{1/2}, \quad \kappa = + \left[\frac{2m}{\hbar^2}(V_0 - E)\right]^{1/2}. \qquad (2.33)$$

The amplitudes A, A' and C in Eqs. (2.30) and (2.31) are related to particle currents. The current density (particles per unit area normal to the direction of travel per unit time) of the incident beam is given by

$$j_I = \frac{\hbar k}{m}|A|^2. \qquad (2.34a)$$

The current densities of the reflected and transmitted beams are similarly defined by

$$j_R = \frac{\hbar k}{m}|A'|^2, \qquad j_T = \frac{\hbar k}{m}|C|^2. \qquad (2.34b)$$

The conservation of particles (e.g. no particles are absorbed within the barrier) is expressed in the condition

$$j_I = j_T + j_R. \qquad (2.35a)$$

Defining the reflection and transmission coefficients by

$$R = j_R/j_I, \quad T = j_T/j_I \qquad (2.36)$$

we can rewrite (2.35a) as

$$T + R = 1. \tag{2.35b}$$

To calculate the transmission and reflection coefficients, we must match the wave functions (2.30), (2.31) and either (2.32a) if $E > V_0$, or (2.32b) if $E < V_0$, at $x = -a$ and $x = a$. Continuity of ϕ and of $d\phi/dx$ at $x = -a$ and at $x = a$ gives four linear equations in A, A', B, B' and C. For a given amplitude A of the incident beam, these are four *inhomogeneous* equations for the four unknown amplitudes A', B, B' and C. Hence they possess a solution for *all* values of the energy $E(>0)$; i.e. there exists a continuum of positive-energy solutions.

The detailed derivation of these solutions is straightforward and is left as an exercise for the reader (see problem 2.6). For the case $E < V_0$, one finds

$$\frac{C}{A} = e^{-i2ak} \frac{2k\kappa}{2k\kappa \cosh 2a\kappa - i(k^2 - \kappa^2) \sinh 2a\kappa} \tag{2.37}$$

and hence for the transmission coefficient

$$T = \left|\frac{C}{A}\right|^2 = \frac{(2k\kappa)^2}{(k^2 + \kappa^2)^2 \sinh^2(2a\kappa) + (2k\kappa)^2}. \tag{2.38a}$$

Calculating the reflection coefficient R directly from the amplitudes A' and A gives

$$R = \left|\frac{A'}{A}\right|^2 = \frac{(k^2 + \kappa^2)^2 \sinh^2(2a\kappa)}{(k^2 + \kappa^2)^2 \sinh^2(2a\kappa) + (2k\kappa)^2}, \tag{2.38b}$$

in agreement with the conservation relation (2.35b).

Eq. (2.38a) provides the quantitative quantum-mechanical explanation of barrier penetration: particles of energy E can tunnel through a barrier of height $V_0 > E$. Although this result totally disagrees with classical particle mechanics, it is typical of all wave motion. For example, light can pass through a thin layer of air between two parallel slabs of glass, even when total internal reflection occurs at the first air–glass interface. A real exponential disturbance (called an evanescent wave) propagates into the air gap and this gives rise to an oscillating wave in the second slab of glass.*

The transmission coefficient (2.38a), i.e. the probability that a particle penetrates through the barrier, decreases as the height V_0 or the thickness $2a$ of the

---

* See, for example, Lipson and Lipson, section 4.3.2.

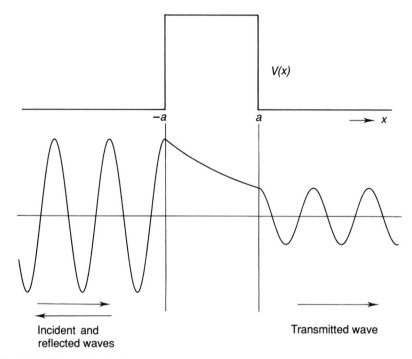

Fig. 2.6   The qualitative features of the real and the imaginary parts of the wave function in the case of barrier penetration.

barrier increases. For $2a\kappa \gg 1$, Eq. (2.38a) reduces to the very small probability for barrier penetration

$$T = \frac{16(k\kappa)^2}{(k^2 + \kappa^2)^2}\, \mathrm{e}^{-4a\kappa}, \qquad 2a\kappa \gg 1. \tag{2.39}$$

The behaviour of this expression is dominated by the exponential term which is very sensitive to the thickness $2a$ of the barrier and the height of the barrier above the particle energy. [Remember that $\kappa \propto \sqrt{(V_0 - E)}$.] Fig. 2.6 shows the qualitative features of the wave function for the barrier penetration problem.

   Examples of barrier penetration are the tunnelling of electrons through a thin insulating layer separating two pieces of metal or of pairs of super-conducting electrons (Josephson tunnelling), and the inversion of the ammonia molecule (the basic mechanism for the ammonia maser). Historically the first example of barrier penetration was, of course, the spontaneous emission of

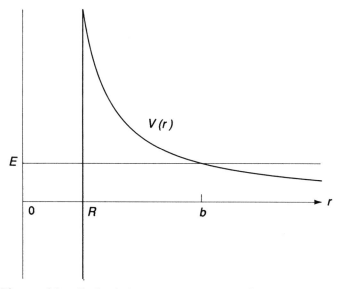

Fig. 2.7   The repulsive Coulomb barrier $V(r) = (Z - 2)2e^2/4\pi\varepsilon_0 r$ through which an alpha particle of energy E must tunnel in order to escape from the nucleus. R is the nuclear radius.

alpha particles by some heavy nuclei. In this case, the barrier is the repulsive Coulomb potential

$$V(r) = \frac{(Z - 2)2e^2}{4\pi\varepsilon_0 r}, \qquad r > R,$$

between the residual nucleus (atomic number $Z - 2$) and the alpha particle, once it has escaped from the nucleus, i.e. for r greater than the nuclear radius R. For $r < R$, the very strong attractive short-range nuclear forces dominate. The resulting potential is sketched in Fig. 2.7. An alpha particle of energy E must tunnel through the Coulomb barrier in order to escape from the nucleus. We are now dealing with a three-dimensional problem and a barrier of variable width and height. As one might guess, the dominating exponential factor in the transmission coefficient (2.39) now becomes replaced by

$$T = \exp\left\{ -2 \int_R^b dr \left(\frac{2m}{\hbar^2}\right)^{1/2} [V(r) - E]^{1/2} \right\}$$

where m is the mass of the alpha particle. The life time τ of an alpha-radioactive nucleus is inversely proportional to the transmission coefficient T. (Remember that $1/\tau$ is the decay probability per unit time and is therefore proportional to the barrier penetration probability T.) The huge range of observed lifetimes for different nuclei is explained by the exponential dependence of T on the energy E of the alpha particle relative to the height of the barrier and on the width of the barrier. One finds that a factor 2 in E can lead to a factor 10^{20} in the lifetime.

We briefly consider the case of particles with energy $E > V_0$ incident on the potential barrier (2.27). We see from Eqs. (2.32) that making the replacement $\kappa \to -iq$ in Eq. (2.37) gives the ratio C/A for this case, i.e.

$$\frac{C}{A} = e^{-i2ak} \frac{2kq}{2kq \cos 2aq - i(k^2 + q^2) \sin 2aq};$$

hence the transmission coefficient is given by

$$T = \left| \frac{C}{A} \right|^2 = \frac{(2kq)^2}{(k^2 - q^2)^2 \sin^2 2aq + (2kq)^2} \qquad (2.40)$$

where k and q are defined by Eqs. (2.28) and (2.33).

Eq. (2.40) also gives the transmission coefficient for particles with energy $E(>0)$ incident on the attractive square-well potential (2.1), if we replace V_0 by $(-V_0)$, i.e. define the wave number q in Eq. (2.40) as

$$q = \left[\frac{2m}{\hbar^2} (E + V_0) \right]^{1/2}.$$

Eq. (2.40) and the corresponding result for the reflection coefficient $R = 1 - T$ disagree with classical mechanics according to which all particles are transmitted ($T = 1$) and none reflected ($R = 0$) for the regimes we are now considering. Eq. (2.40) does predict complete transparency of the barrier ($T = 1, R = 0$) if $\sin 2aq = 0$, i.e. if

$$2a = n \frac{\pi}{q}, \qquad n = 1, 2, 3, \dots.$$

This is the condition for the width $2a$ of the barrier or well to be an exact multiple of half the deBroglie wavelength $\lambda = 2\pi/q$ of the particles within the barrier or well. An analogous effect, known as the Ramsauer–Townsend effect, is observed in the scattering of low-energy electrons by rare-gas atoms. At certain energies no scattering occurs, i.e. the incident electron beam emerges as though no target atoms were present. One can understand this as the wave function of the electrons inside the atom being of such a form that it matches on to the *undistorted* incident plane wave at the surface of the atom. The analysis of the Ramsauer–Townsend effect will be given in section 11.5.1.

2.3   ANGULAR MOMENTUM

Angular momentum, important in classical physics, is of special significance in quantum mechanics. This is due to the fact that the angular momentum properties of a system are largely independent of the details of the forces acting.

(The deeper explanation of this is that the angular momentum properties of a system are related to its rotational symmetry properties, as we shall see later.) For example, for a single particle moving in a central potential $V(r)$, a part of the wave function is fully determined by its angular momentum properties—and vice versa—independently of the exact form of the potential. More generally, the angular momentum properties and hence the qualitative features of the energy level schemes and the spectroscopy of many-electron atoms can be understood without solving the Schrödinger equation. In this section we shall consider the angular momentum of a single particle. The results which we shall derive for this simple case are typical of more complex situations and are basic to these generalizations.

We know from Eqs. (1.5) and (1.8) that the operator corresponding to the classical angular momentum $\mathbf{l} = \mathbf{r} \wedge \mathbf{p}$ of a particle is given by

$$\hat{\mathbf{l}} = \mathbf{r} \wedge (-i\hbar\nabla). \tag{2.41}$$

Angular momentum is best treated in terms of spherical polar coordinates r, θ, ϕ. These are shown in Fig 2.8 and defined by

$$x = r \sin \theta \cos \phi, \quad y = r \sin \theta \sin \phi, \quad z = r \cos \theta \tag{2.42a}$$

where the ranges of r, θ and ϕ are

$$r \geqslant 0, \quad 0 \leqslant \theta \leqslant \pi, \quad 0 \leqslant \phi \leqslant 2\pi. \tag{2.42b}$$

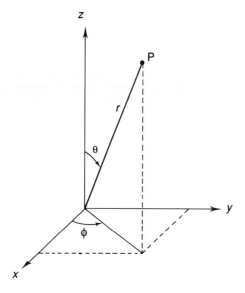

Fig. 2.8   The spherical polar coordinates r, θ, ϕ of the point P, given by Eqs. (2.42).

Expressed in spherical polar coordinates, the Cartesian components of $\hat{\mathbf{l}}$ are given by

$$\hat{l}_x = i\hbar\left(\sin\phi\,\frac{\partial}{\partial\theta} + \cot\theta\cos\phi\,\frac{\partial}{\partial\phi}\right) \tag{2.43a}$$

$$\hat{l}_y = i\hbar\left(-\cos\phi\,\frac{\partial}{\partial\theta} + \cot\theta\sin\phi\,\frac{\partial}{\partial\phi}\right) \tag{2.43b}$$

$$\hat{l}_z = -i\hbar\,\frac{\partial}{\partial\phi} \tag{2.43c}$$

and the square of the magnitude of the angular momentum operator by

$$\hat{l}^2 = \hat{l}_x^2 + \hat{l}_y^2 + \hat{l}_z^2$$
$$= -\hbar^2\left[\frac{1}{\sin\theta}\frac{\partial}{\partial\theta}\left(\sin\theta\,\frac{\partial}{\partial\theta}\right) + \frac{1}{\sin^2\theta}\frac{\partial^2}{\partial\phi^2}\right]. \tag{2.44}$$

Since the angular momentum operators (2.43) and (2.44) depend on the angles θ and ϕ only and are independent of the radial coordinate r, the eigenfunctions of these operators will have a specific dependence on θ and ϕ but will have an arbitrary dependence on r. Hence the radial dependence of these eigenfunctions is irrelevant to the discussion of angular momentum, and we shall take $Y(\theta, \phi)$ to be an eigenfunction of the angular momentum operator $\hat{l}^2$, belonging to the eigenvalue $\hbar^2\lambda$, i.e.

$$\hat{l}^2 Y(\theta, \phi) = \hbar^2\lambda Y(\theta, \phi). \tag{2.45}$$

It follows from Eq. (2.44) that $Y(\theta, \phi)$ is a solution of the 'angular equation'

$$-\left[\frac{1}{\sin\theta}\frac{\partial}{\partial\theta}\left(\sin\theta\,\frac{\partial}{\partial\theta}\right) + \frac{1}{\sin^2\theta}\frac{\partial^2}{\partial\phi^2}\right]Y(\theta, \phi) = \lambda Y(\theta, \phi), \tag{2.46}$$

Tc find the solutions of Eq. (2.46) which are finite single-valued functions of θ and ϕ is a standard problem of classical physics.* (It occurs, for example, in solving Laplace's equation in electrostatics.) The solutions are well known and it is not our purpose to rederive them. We shall only state the results and quote

---

* 'Single-valued' here means that the value of $Y(\theta, \phi)$ remains the same if θ or ϕ are replaced by $\theta + 2\pi$ or $\phi + 2\pi$ respectively. (Remember that (r, θ, ϕ) and $(r, \theta + 2\pi, \phi + 2\pi)$ label the same point in space.) The restriction to finite single-valued solutions seems reasonable in view of the probability interpretation of $|\psi|^2$. However, the full justification for this restriction needs further careful arguments. See, for example, Gottfried, pp. 85–86.

those properties most frequently needed. Eq. (2.46) possesses finite single-valued solutions for certain values of λ only. These eigenvalues are given by

$$\lambda = l(l + 1), \qquad l = 0, 1, 2, \ldots . \tag{2.47}$$

The eigenvalue $l(l + 1)$ is $(2l + 1)$-fold degenerate, i.e. there exist $(2l + 1)$ linearly independent functions which are eigenfunctions of $\hat{\mathbf{l}}^2$ belonging to the eigenvalue $\hbar^2 l(l + 1)$. Denoting these eigenfunctions by

$$Y_l^m(\theta, \phi), \qquad m = l, l - 1, \ldots, -l, \qquad l = 0, 1, 2, \ldots, \tag{2.48}$$

we can write the eigenvalue equation (2.45) as

$$\hat{\mathbf{l}}^2 Y_l^m(\theta, \phi) = \hbar^2 l(l + 1) Y_l^m(\theta, \phi), \qquad \left. \begin{array}{l} m = l, l - 1, \ldots, -l \\ l = 0, 1, \ldots . \end{array} \right\} \tag{2.49}$$

In the following, these ranges of values for m and l will always be understood, unless stated to the contrary.

The functions $Y_l^m(\theta, \phi)$ are the spherical harmonics and are given by

$$Y_l^m(\theta, \phi) = N_l^m P_l^{|m|}(\cos \theta)\, e^{im\phi} \tag{2.50}$$

where the $P_l^{|m|}(\cos \theta)$ are associated Legendre functions and the N_l^m are normalization constants. They will both be defined explicitly in section 2.3.1. The exponential ϕ-dependence of $Y_l^m(\theta, \phi)$ one verifies directly by substituting Eq. (2.50) in the angular equation (2.46). The latter equation in this way reduces to an equation in θ which depends on m through m^2 only; hence the occurrence of the superscript $|m|$ in Eq. (2.50).

The $(2l + 1)$-fold degeneracy of the eigenvalue $\hbar^2 l(l + 1)$ of $\hat{\mathbf{l}}^2$ means that the corresponding eigenfunctions $Y_l^m(\theta, \phi)$ are not uniquely determined. (See the discussion of degeneracy in section 1.2, p. 9.) The functions (2.50) were chosen because they are also eigenfunctions of the z-component of angular momentum. From Eqs. (2.43c) and (2.50) one sees directly that

$$\hat{l}_z Y_l^m(\theta, \phi) = -i\hbar \frac{\partial}{\partial \phi} Y_l^m(\theta, \phi) = \hbar m Y_l^m(\theta, \phi), \tag{2.51}$$

i.e. $Y_l^m(\theta, \phi)$ is an eigenfunction of $\hat{l}_z$ with the eigenvalue $\hbar m$.

Eqs. (2.49) and (2.51) express the quantization of angular momentum: the magnitude squared of the angular momentum, $\hat{\mathbf{l}}^2$, can only assume one of the discrete set of values

$$\hbar^2 l(l + 1), \qquad l = 0, 1, 2, \ldots \tag{2.52a}$$

and the z-component of angular momentum, $\hat{l}_z$, can only assume one of the discrete set of values

$$\hbar m, \qquad m = l, l - 1, \ldots, -l. \tag{2.52b}$$

Furthermore, a state

$$\psi(\mathbf{r}) = R(r)Y_l^m(\theta, \phi), \tag{2.53}$$

where $R(r)$ is an arbitrary radial function, is a *simultaneous eigenstate* of $\hat{\mathbf{l}}^2$ and $\hat{l}_z$. For a particle in this state, $\hat{\mathbf{l}}^2$ and $\hat{l}_z$ have the definite values (2.52a) and (2.52b) respectively.

l and m are called the angular momentum quantum number and the magnetic quantum number respectively. (We shall shortly see that m is very important in analysing the interaction of atoms, etc., with magnetic fields; hence its name.) One refers to a particle in a state with angular momentum quantum number l as having angular momentum l, rather than more cumbersomely as having an angular momentum of magnitude $\hbar\sqrt{[l(l + 1)]}$. For historical reasons, states with $l = 0, 1, 2, 3$ are called s-, p-, d-, f-states. For higher values $l = 4, 5, \ldots$ one then continues in alphabetical order, calling these states g-states, h-states, and so on. This nomenclature which originated in spectroscopy is known as the spectroscopic notation.

We have so far singled out the z-component of angular momentum l_z.* What can we say about components in other directions, say l_x or l_y? Suppose the particle is in the state (2.53), what are the values of l_x and l_y? It follows at once from Eq. (2.43a) for l_x that operating with l_x on the wave function (2.53) does not simply give a multiple of (2.53). A similar result holds for l_y. The state (2.53) is not an eigenstate of l_x or l_y. In this state neither l_x nor l_y have definite values; we only have probability distributions for predicting the results of measuring l_x or l_y for a particle in this state.[†] We must conclude that if a particle is in a state in which l_z has a definite value, we cannot ascribe definite values to l_x or l_y. This is a most remarkable and characteristic feature of quantum mechanics. It represents a limitation—not only in practice but in principle—of our possible knowledge about a system: in general, one cannot know the values of all the observables of a system simultaneously; one can only know some simultaneously. Observables for which simultaneous knowledge is always possible are called *compatible*. In the angular momentum case, l_z is compatible with $\mathbf{l}^2$, but not with l_x or l_y. In Chapter 3 we shall discuss fully the compatibility and incompatibility of observables.

In our discussion of angular momentum, the z-component l_z received preferential treatment: our results do not appear symmetric in x, y, z. This apparent

---

* Up to now I have written $\hat{l}_x, \hat{l}_y, \hat{l}_z$ and $\hat{\mathbf{l}}^2$, in order to stress to the reader that these are operators. Hereafter the reader should no longer need reminding of this and the circumflex accent will usually be omitted.

[†] The case $l = 0$ is an exception. It follows from Eqs. (2.43) that

$$Y_0^0(\theta, \phi) = \text{const.}$$

is a simultaneous eigenfunction of l_x, l_y, and l_z, with the eigenvalue 0 in each case. It is, of course, also an eigenfunction of $\mathbf{l}^2$ with eigenvalue 0.

asymmetry stems from our choice of spherical polar coordinates which singles out the z-axis as polar axis. But there is nothing special about this way of labelling axes; we could relabel them: x into y, y into z, z into x. With this relabelling, what had been labelled l_z becomes l_x.* Carrying the same relabelling out a second time, the original l_z goes over into l_y. Hence whatever we found for l_z must be equally true for l_x and l_y. It follows without further calculations that the eigenvalues of l_x or l_y are the same as those of l_z, Eq. (2.52b), and that l_x and l_y are compatible with $\mathbf{l}^2$, but not with each other. We conclude that we can only know simultaneously the value of $\mathbf{l}^2$ and of one of the components l_x, l_y, l_z. This knowledge then precludes ascribing definite values to either of the other two components. This situation is totally different from that of classical physics where $\mathbf{l}^2$ and all three components l_x, l_y, l_z have definite values which of course are not quantized and only subject to the restriction

$$l_x^2 + l_y^2 + l_z^2 = \mathbf{l}^2.$$

The quantum-mechanical properties of angular momentum are demonstrated particularly clearly in the Stern–Gerlach experiment which we shall consider in section 2.4, and this should help the reader to understand these concepts better and acquire a feeling for them.

2.3.1 Spherical harmonics

We summarize the most frequently used properties of the spherical harmonics.[†]

(a) *Definition*   The spherical harmonics $Y_l^m(\theta, \phi)$ are explicitly defined by

$$Y_l^m(\theta, \phi) = N_l^m P_l^{|m|}(\cos\theta)\, e^{im\phi}, \qquad \left.\begin{array}{l} l = 0, 1, 2, \ldots \\ m = l, l - 1, \ldots, -l \end{array}\right\} \qquad (2.50)$$

where the associated Legendre functions $P_l^{|m|}(\mu)$ are defined by

$$P_l^{|m|}(\mu) = \frac{1}{2^l l!}(1 - \mu^2)^{|m|/2}\frac{\mathrm{d}^{l+|m|}}{\mathrm{d}\mu^{l+|m|}}[(\mu^2 - 1)^l] \qquad (2.54)$$

and the normalization constant N_l^m by

$$N_l^m = (-1)^{(m+|m|)/2}\left[\frac{2l + 1}{4\pi}\frac{(l - |m|)!}{(l + |m|)!}\right]^{1/2}. \qquad (2.55)$$

---

* A cyclic permutation like this is needed, rather than just $z \to x$, $x \to z$, $y \to y$, so that the relabelled coordinate system is also right-handed.

† The reader should have a brief look at this section but may defer studying the details until they are required. For the derivation of these and other properties, see, for example, Merzbacher, sections 9.6–9.7, or Cohen-Tannoudji, Diu and Laloë, vol. I, pp. 678–689.

The first few spherical harmonics are given by

$$
\left.
\begin{aligned}
Y_0^0(\theta, \phi) &= \left(\frac{1}{4\pi}\right)^{1/2} \\[2mm]
Y_1^0(\theta, \phi) &= \left(\frac{3}{4\pi}\right)^{1/2} \cos\theta, \quad Y_1^{\pm 1}(\theta, \phi) = \mp\left(\frac{3}{8\pi}\right)^{1/2} \sin\theta\, e^{\pm i\phi} \\[2mm]
Y_2^0(\theta, \phi) &= \left(\frac{5}{16\pi}\right)^{1/2} (3\cos^2\theta - 1).
\end{aligned}
\right\}
\tag{2.56}
$$

For many purposes one needs only the general properties of the spherical harmonics and not the explicit definition. However, the reader should note the peculiar phase factor $(-1)^{(m+|m|)/2}$ in the normalization constant N_l^m which also manifests itself in Eqs. (2.56). This is the conventional phase factor of the Y_l^m in angular momentum theory. When adding angular momenta, one deals with linear combinations of spherical harmonics; then these phase factors become all-important in determining interference effects and must be chosen consistently. The definition (2.55) is consistent with the tabulated coefficients which one employs in the addition of angular momenta (see the last paragraph of section 5.6). So it is worth while—and just as easy—always to use these conventions.

(b) *Orthonormality*   The spherical harmonics (2.50) satisfy the ortho-normality relations

$$
\int d\Omega\, Y_l^{m*}(\theta, \phi) Y_{l'}^{m'}(\theta, \phi) = \delta_{ll'}\delta_{mm'},
\tag{2.57}
$$

where $d\Omega = \sin\theta\, d\theta\, d\phi$ is an element of solid angle, and the integration is over all solid angles, i.e. over the ranges

$$
0 \leqslant \theta \leqslant \pi, \quad 0 \leqslant \phi \leqslant 2\pi.
\tag{2.58}
$$

Integration over all solid angles will always be implied, unless stated otherwise. The orthogonality relation (2.57) with respect to m and m' follows easily from the ϕ-dependent part of the integral

$$
\int_0^{2\pi} d\phi\, e^{i(m'-m)\phi} = 2\pi\delta_{mm'}.
$$

Proofs of the orthogonality with respect to l and l' and of the normalization are longer. Of course, we must expect the two orthogonality relations from the general result of section 1.2, Eq. (1.23): the eigenfunctions belonging to different eigenvalues of an observable are orthogonal. Eq. (2.57) states this result for the observables $\mathbf{l}^2$ and l_z.

(c) *Completeness*   The spherical harmonics Y_l^m form a complete set, i.e. any 'reasonable' function $f(\theta, \phi)$ can be expanded in the form

$$f(\theta, \phi) = \sum_{l=0}^{\infty} \sum_{m=-l}^{l} C_{lm} Y_l^m(\theta, \phi). \tag{2.59a}$$

Using the orthonormality relation (2.57) one obtains the expansion coefficients

$$C_{lm} = \int d\Omega Y_l^{m*}(\theta, \phi) f(\theta, \phi). \tag{2.59b}$$

The ϕ-dependent part of this expansion should be familiar to the reader: it is the Fourier expansion of $f(\theta, \phi)$ in terms of the exponential functions $\exp(im\phi)$, with m taking on all integer values from $-\infty$ to $+\infty$.

(d) *Complex conjugation*   The associated Legendre function $P_l^{|m|}(\cos \theta)$, Eq. (2.54), is a real function of $\cos \theta$. The constant N_l^m, Eq. (2.55), is real and

$$N_l^{-m} = (-1)^m N_l^m.$$

Hence one has from the definition (2.50) of the Y_l^m that

$$Y_l^{m*}(\theta, \phi) = (-1)^m Y_l^{-m}(\theta, \phi). \tag{2.60}$$

(e) *Inversion*   One of the symmetry operations of importance, which we shall study later (section 4.1) is that of inversion, i.e. how a system behaves under the coordinate transformation

$$\mathbf{r} \to \mathbf{r}' = -\mathbf{r}. \tag{2.61a}$$

Expressed in Cartesian or spherical polar coordinates, Eq. (2.61a) becomes

$$(x, y, z) \to (x', y', z') = (-x, -y, -z) \tag{2.61b}$$

and

$$(r, \theta, \phi) \to (r', \theta', \phi') = (r, \pi - \theta, \phi + \pi). \tag{2.61c}$$

Under inversion $\cos \theta \to \cos (\pi - \theta) = -\cos \theta$ and from Eq. (2.54), with $\mu = \cos \theta$,

$$P_l^{|m|}(\mu) \to P_l^{|m|}(-\mu) = (-1)^{l+|m|} P_l^{|m|}(\mu). \tag{2.62}$$

We also have

$$e^{im\phi} \to e^{im(\phi + \pi)} = (-1)^m e^{im\phi}.$$

It follows from Eq. (2.50) that under inversion

$$Y_l^m(\theta, \phi) \to Y_l^m(\pi - \theta, \phi + \pi) = (-1)^l Y_l^m(\theta, \phi). \tag{2.63}$$

2.4   THE STERN–GERLACH EXPERIMENT

In the last section we derived the quantization of angular momentum: the component of angular momentum in any direction can only assume one of the values $\hbar m$, with $m = 0, \pm 1, \pm 2, \ldots$. This result goes against any intuitive classical way of picturing things. It is therefore very helpful that the Stern–Gerlach experiment seems to demonstrate this curious behaviour very directly.

The Stern–Gerlach experiment depends on the fact that a magnetic dipole, placed in an inhomogeneous magnetic field, experiences a force which depends on the orientation of the dipole relative to the field. To establish this result, we remind the reader that the interaction energy of a magnetic dipole with permanent moment μ placed in a magnetic field $\mathbf{B}$ is $-\mu \cdot \mathbf{B}$. If the field is inhomogeneous, work has to be done in displacing the dipole, i.e. a force $\mathbf{F}$ acts on the dipole which is the negative gradient of the interaction energy

$$\mathbf{F} = \nabla(\mu \cdot \mathbf{B}).$$

If we take the z-axis as the direction of the field, the force on the dipole in the z-direction is given by

$$F_z = \mu_z \frac{\partial B}{\partial z} = \mu \cos \theta \frac{\partial B}{\partial z} \tag{2.64}$$

where θ is the angle between μ and $\mathbf{B}$. It follows that if a beam of atoms, possessing a permanent magnetic moment μ, passes through an inhomogeneous magnetic field $\mathbf{B}$, it is deflected, with the deflections depending on the relative orientations of μ and $\mathbf{B}$.

In the Stern–Gerlach experiment an atomic beam, produced from an oven, is collimated by slits and passed between the poles of a magnet. The pole pieces are shaped so that the field possesses a strong gradient perpendicular to the direction of the beam. The deflected atoms are observed on a screen. Fig. 2.9 shows a schematic picture of this system.*

The atoms emerging from an oven have their magnetic moments pointing randomly in all directions. From Eq. (2.64), the force F_z varies continuously between the extremes $+\mu\, \partial B/\partial z$ (for $\theta = 0$) and $-\mu\, \partial B/\partial z$ (for $\theta = \pi$) depending on the orientation θ of the dipoles. The corresponding pattern which one would expect to see on the observation screen is a continuous band, spread along the z-axis and symmetric about the undeflected direction, $z = 0$. Of course, if the atoms have zero magnetic moment, they will not be deflected and will produce a single spot on the observation screen.

What does one find experimentally? One never finds a continuously spread pattern. Instead one obtains patterns of 1, 2, 3, ... spots, arranged symmetrically

---

* For a more realistic description, references and some interesting historical background, see French and Taylor, section 10-3.

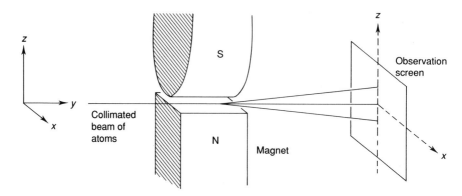

Fig. 2.9   Schematic diagram of the Stern–Gerlach apparatus.

about the undeflected direction, $z = 0$. The original beam has been split into several beams. The number of spots one observes depends on the atom; their spacings on the details of the magnetic field and the geometry of the apparatus. A single undeflected spot we can understand as atoms with zero magnetic moment, but patterns of 2, 3, ... spots are classically unintelligible. It suggests that θ cannot assume all values from 0 to π but only a certain discrete set of values, i.e. μ_z can only assume a certain discrete set of values. This is reminiscent of what we found for angular momentum in quantum mechanics.

The connection with angular momentum is readily established. Classically one easily shows that an electron, moving in an orbit with angular momentum **l**, possesses a magnetic moment

$$\mathbf{\mu} = \frac{-e}{2m_e}\mathbf{l} \tag{2.65}$$

where m_e and $(-e)$ are the mass and charge of the electron. Taking this equation over into quantum mechanics, it follows that the quantization of the angular momentum **l** implies that of the magnetic moment **μ**. If we think of this one-electron system as a model of an atom, we can 'understand' how in a Stern–Gerlach experiment a beam of such atoms produces $(2l + 1)$ spots, with $l = 0, 1, 2, \ldots$, depending on the angular momentum of the atomic electron.

These ideas can be generalized and applied to real many-electron atoms. The magnetic moment **μ** is now related to the resultant angular momentum of the atom, found by adding the contributions from the individual electrons. If one generalizes the above treatment in this way one finds that μ_z is restricted to sets of $(2L + 1)$ values, where L is the analogue for the many-electron atom of the angular momentum quantum number l for one electron. One can show that L can only have the values $L = 0, 1, 2, \ldots$. This implies that a Stern–Gerlach experiment always leads to patterns with an odd number of spots. In practice, patterns with an even number of spots are observed. For example, in their original experiment, Stern and Gerlach used a beam of silver atoms and obtained two spots. The explanation is, of course, that in addition to the *orbital* angular momentum $\mathbf{l}\,(= \mathbf{r} \wedge \mathbf{p}$, i.e. associated with its motion in an orbit) an electron possesses a *spin* angular momentum $\mathbf{s}$ (i.e. an intrinsic angular momentum like a spinning top). The z-component s_z of this spin angular momentum can assume the two values $\pm\hbar/2$ only, and it is this spin which leads to Stern–Gerlach patterns with an even number of spots whenever one deals with atoms with an odd number of electrons. We cannot derive these results at this stage since this requires knowledge of how to add angular momenta and calculate the magnetic moments of many-electron atoms.

Instead we shall consider further our model atom, i.e. a single electron, whose spin we ignore, moving in a central potential. We shall assume the electron is in a p-state, i.e. $l = 1$. This case allows us to analyse Stern–Gerlach experiments in detail using very simple mathematics and will provide much insight into the meaning of angular momentum quantization and, more generally, of the superposition of waves.

A beam of atoms in an $l = 1$ state is split into three beams by the Stern–Gerlach magnet of Fig. 2.9. Suppose we stop two of the beams. If the third beam is passed into a second Stern–Gerlach magnet, identical to the first, it will not be split but only deflected in the same sense as by the first magnet. For example, the top beam will be deflected upwards, as indicated schematically in Fig. 2.10. (In this figure, the bold arrows indicate the direction of the positive gradients of the magnetic fields. We assume that the deflections are small.) This conclusion is self-evident, since the addition of the second magnet is equivalent to making the first one longer. Similar results hold if the middle or the bottom beam only is allowed to enter the second magnet. Since the second magnet does not split the beams further, each beam must consist of atoms with a definite value of l_z. The corresponding atomic states are of the form

$$\psi_z^m(\mathbf{r}) = R(r)Y_1^m(\theta, \phi), \qquad m = 1, 0, -1. \tag{2.66}$$

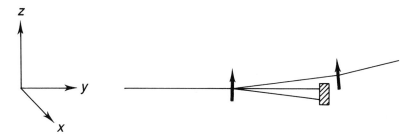

Fig. 2.10   Passing the top beam from a Stern-Gerlach magnet, whose field gradient is in the z-direction, through a second similar Stern–Gerlach magnet.

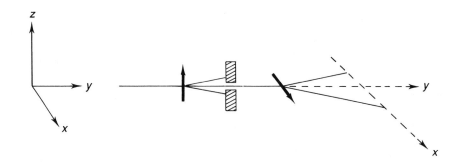

Fig. 2.11   Passing the middle beam from the first Stern-Gerlach magnet, whose field gradient is in the z-direction, through a second Stern–Gerlach magnet, whose field gradient is in the x-direction.

Let us next consider what happens if one of the beams emerging from the first Stern–Gerlach magnet is passed through a second Stern–Gerlach magnet oriented differently from the first. Suppose it has its field gradient in the x-direction and we analyse the middle beam, as indicated in Fig. 2.11. The atoms entering the second magnet are in the state $\psi_z^0(\mathbf{r})$. This is not an eigenstate of l_x as we know from section 2.3. Hence we expect the second magnet to split the beam into three beams, corresponding to the three possible eigenvalues $\hbar$, 0, $-\hbar$, of l_x in a p-state.

To check this quantitatively, we must expand the state $\psi_z^0(\mathbf{r})$ in terms of the eigenstates $\psi_x^m(\mathbf{r})$ of l_x, belonging to the eigenvalues $\hbar m$, $m = 1, 0, -1$. For p-states, these eigenfunctions are easily written down; that is why we chose this case. With $Y_1^m(\theta, \phi)$ given by Eqs. (2.56), we express Eqs. (2.66) in Cartesian coordinates as

$$\psi_z^1(\mathbf{r}) = \frac{-(x + iy)}{\sqrt{2}} f(r) \tag{2.67a}$$

$$\psi_z^0(\mathbf{r}) = z f(r) \tag{2.67b}$$

$$\psi_z^{-1}(\mathbf{r}) = \frac{(x - iy)}{\sqrt{2}} f(r) \tag{2.67c}$$

where $f(r) \equiv (3/4\pi)^{1/2} R(r)/r$. These states are orthogonal, since

$$\int_{\text{all space}} d^3\mathbf{r}\, xy |f(r)|^2 = 0, \quad \text{etc.,} \tag{2.68a}$$

(where etc. stands for two similar equations with xy replaced by yz or zx) and they are normalized to unity if we norm $f(r)$ so that

$$\int_{\text{all space}} d^3r z^2 |f(r)|^2 = 1, \quad \text{etc.} \tag{2.68b}$$

(where etc. stands for two similar equations with z^2 replaced by x^2 or y^2).

The eigenfunctions $\psi_x^m(\mathbf{r})$ are now easily written down. As explained on p. 42, we need only relabel axes: x into y, y into z, z into x. With this relabelling l_z goes over into l_x, and Eqs. (2.67) go over into the eigenfunctions of l_x:

$$\psi_x^1(\mathbf{r}) = \frac{-(y + iz)}{\sqrt{2}} f(r) \tag{2.69a}$$

$$\psi_x^0(\mathbf{r}) = x f(r) \tag{2.69b}$$

$$\psi_x^{-1}(\mathbf{r}) = \frac{(y - iz)}{\sqrt{2}} f(r). \tag{2.69c}$$

$\psi_z^0(\mathbf{r})$, Eq. (2.67b), is at once written as a superposition of the eigenfunctions $\psi_x^m(\mathbf{r})$:

$$\psi_z^0(\mathbf{r}) = \frac{i}{\sqrt{2}} [\psi_x^1(\mathbf{r}) + \psi_x^{-1}(\mathbf{r})]. \tag{2.70}$$

Using the results of section 1.2, Eq. (1.38), it follows from Eq. (2.70) that for atoms in the state $\psi_z^0(\mathbf{r})$ the probability of l_x having the value $\hbar$, 0 or $-\hbar$ is $\frac{1}{2}$, 0 or $\frac{1}{2}$ respectively. The intensities of the beams emerging from the second Stern–Gerlach magnet are, of course, proportional to these probabilities, i.e. two beams emerge from the second magnet, deflected in the $\pm x$-directions, as indicated in Fig. 2.11, and of equal intensity. Contrary to our expectations, the second Stern–Gerlach magnet in this case causes a splitting into two beams only. In general, a Stern–Gerlach magnet will split a beam of p-state atoms into three beams. It so happens that the expansion of $\psi_z^0(\mathbf{r})$ does not contain a component $\psi_x^0(\mathbf{r})$. For example, either of the other beams emerging from the first Stern–Gerlach magnet in Fig. 2.11 would be split into three beams, of unequal intensities, by the second magnet. The details are left as an exercise for the reader (problem 2.8).

The above analysis illustrates the incompatibility of the observables l_x and l_z which was already mentioned in section 2.3. Each beam entering the second Stern–Gerlach magnet consists of atoms with a definite value of l_z. The atoms in each beam emerging from the second magnet have a definite value of l_x, but our knowledge of the value of l_z has been lost. Each eigenstate of l_x is a superposition of eigenstates of l_z. A third Stern–Gerlach magnet, with field gradient in the z-direction, would again split each beam consisting of atoms with a definite value of l_x, and so on *ad infinitum*. We cannot know the values of both l_x and l_z. A measurement of one destroys our knowledge of the other.

So far we have not discussed how to specify the states of the atoms in the beam emerging from the oven, beyond saying that their magnetic moments μ are oriented at random. Classically, this means that a fraction $d\Omega/4\pi$ of the atoms will have μ pointing into any solid angle $d\Omega$. Quantum-mechanically it means (for our p-state model atoms) that for an *arbitrary* axis of quantization, specified by a unit vector $\mathbf{n}$, one third of the atoms are in the eigenstate of $\mathbf{n} \cdot \mathbf{l}$ with eigenvalue $\hbar$, one third in the eigenstate of $\mathbf{n} \cdot \mathbf{l}$ with eigenvalue 0, and one third in the eigenstate of $\mathbf{n} \cdot \mathbf{l}$ with eigenvalue $-\hbar$. The general wave function for our p-state model atom is a superposition of the form

$$\psi(\mathbf{r}) = \sum_{m=-1}^{1} c_m \psi_z^m(\mathbf{r}) \tag{2.71}$$

where the c_m are constants. This wave function has phase relationships between the three components ψ_z^m. For a beam of randomly oriented atoms no such phase relationships exist. Hence we cannot describe such a beam by a single wave function. To obliterate such phase relationships we must add the *intensities* of different angular momentum states; *not their amplitudes* as in Eq. (2.71). This situation is analogous to that for light where unpolarized light can be described as the *incoherent* sum of two polarization states (e.g. right- and left-circularly polarized light), i.e. one adds intensities. Adding amplitudes would only give another state of polarization. Correspondingly, one speaks of polarized and unpolarized atomic beams. For the former, the atoms are in a state with a definite wave function—one also calls this a *pure state*; for the latter, one deals with a *mixed state*, i.e. a mixture of pure states which are added incoherently. A simple procedure for dealing with an unpolarized atomic beam of intensity I is to consider in turn three fully polarized beams, with l_z having the eigenvalues $\hbar$, 0 and $-\hbar$, and of equal intensity $I/3$. (We are still considering our model atoms with $l = 1$; the generalization to other cases should be obvious.) One then calculates the intensities of deflected beams, etc., for each case separately and adds the intensities for the three cases.

2.5 CENTRAL POTENTIAL

We shall now study the motion of a particle in a central potential $V(r)$. The Coulomb potential is of this kind, and the hydrogen atom is an excellent example of such a system. More generally, the electrons in heavier atoms can to a good approximation be described as moving independently of each other in a central potential. So the problem we are now considering is of great importance in atomic physics, and there are many other systems where central forces pay a dominant role.

The Schrödinger equation for a particle of mass m in the central potential $V(r)$ is

$$\left[-\frac{\hbar^2}{2m} \nabla^2 + V(r) \right] \psi(\mathbf{r}) = E\psi(\mathbf{r}).$$ (2.72)

Because of the spherical symmetry of the potential $V(r)$, it is natural to introduce spherical polar coordinates r, θ, ϕ. With ∇^2 expressed in these coordinates, the Schrödinger equation becomes

$$-\frac{\hbar^2}{2m} \left[\frac{1}{r^2} \frac{\partial}{\partial r} \left(r^2 \frac{\partial}{\partial r} \right) + \frac{1}{r^2 \sin\theta} \frac{\partial}{\partial\theta} \left(\sin\theta \frac{\partial}{\partial\theta} \right) + \frac{1}{r^2 \sin^2\theta} \frac{\partial^2}{\partial\phi^2} \right] \psi(r, \theta, \phi)$$

$$+ [V(r) - E]\psi(r, \theta, \phi) = 0.$$ (2.73)

This equation is separable: substituting

$$\psi(r, \theta, \phi) = R(r) Y(\theta, \phi)$$ (2.74)

in it and dividing by ψ, we obtain

$$\frac{1}{R} \left\{ \left(\frac{-\hbar^2}{2m} \right) \frac{d}{dr} \left(r^2 \frac{dR}{dr} \right) + r^2 [V(r) - E]R \right\}$$

$$= \frac{1}{Y} \left(\frac{\hbar^2}{2m} \right) \left[\frac{1}{\sin\theta} \frac{\partial}{\partial\theta} \left(\sin\theta \frac{\partial}{\partial\theta} \right) + \frac{1}{\sin^2\theta} \frac{\partial^2}{\partial\phi^2} \right] Y.$$ (2.75)

The expression on the left-hand side of this equation is a function of r only, that on the right-hand side is a function of θ and ϕ only. Hence the two expressions must be equal to the same constant which we shall call $(-\hbar^2 \lambda / 2m)$. Eq. (2.75) then reduces to the 'radial equation'

$$-\frac{\hbar^2}{2m} \left[\frac{1}{r^2} \frac{d}{dr} \left(r^2 \frac{dR(r)}{dr} \right) \right] + \left[V(r) + \frac{\hbar^2 \lambda}{2mr^2} \right] R(r) = ER(r)$$ (2.76)

and the 'angular equation'

$$-\left[\frac{1}{\sin\theta} \frac{\partial}{\partial\theta} \left(\sin\theta \frac{\partial}{\partial\theta} \right) + \frac{1}{\sin^2\theta} \frac{\partial^2}{\partial\phi^2} \right] Y(\theta, \phi) = \lambda Y(\theta, \phi).$$ (2.77)

The angular equation we already met in section 2.3. Its solutions are the orbital angular momentum eigenfunctions $Y_l^m(\theta, \phi)$; for a given value of $l (= 0, 1, 2, \ldots)$, these belong to the $(2l + 1)$-fold degenerate eigenvalue $\lambda = l(l + 1)$, with $m = l, l - 1, \ldots, -l$. Since the angular equation is independent of the central potential $V(r)$, its solutions are the same for all central potentials, and a particle in such a potential will possess the angular momentum properties discussed in section 2.3.

The radial equation (2.76) depends on the potential $V(r)$ and must be solved separately for each potential. Analytic solutions are only possible in a few cases. These include the three-dimensional square-well potential and the Coulomb potential. Usually one will have to resort to computational methods. We shall start by discussing some general properties which hold for all $V(r)$, and conclude this section by considering the hydrogen atom.

2.5.1   Radial wave equation

With $\lambda = l(l + 1)$, Eq. (2.47), the radial equation (2.76) becomes

$$-\frac{\hbar^2}{2m}\left[\frac{1}{r^2}\frac{d}{dr}\left(r^2\frac{dR(r)}{dr}\right)\right] + \left[V(r) + \frac{\hbar^2 l(l + 1)}{2mr^2}\right]R(r) = ER(r). \qquad (2.78)$$

This equation does not contain the magnetic quantum number m.* Hence if we have found a solution $R(r)$ of this equation belonging to the energy E, then the three-dimensional Schrödinger equation (2.72) will have solutions

$$R(r)Y_l^m(\theta, \phi), \qquad m = l, l - 1, \ldots, -l, \qquad (2.79)$$

with this energy. In other words, E is a $(2l + 1)$-fold degenerate energy eigenvalue of the Schrödinger equation, i.e. of the central field Hamiltonian

$$H = -\frac{\hbar^2}{2m}\nabla^2 + V(r). \qquad (2.80)$$

This result is completely general. The degeneracy follows from the spherical symmetry of the potential $V(r)$. The value of l_z for a given state of the system depends on the orientation of the coordinate axes. But for a particle in a spherically symmetric potential, the energy cannot depend on this orientation, i.e. it cannot depend on m. The eigenfunctions (2.79) do however have an angular dependence in general; only for s-states ($l = 0$), are the eigenfunctions (2.79) isotropic. It also follows quite generally from the results about angular momentum in section 2.3 that the functions (2.79) are simultaneous eigenfunctions of H, $\mathbf{l}^2$ and l_z, with eigenvalues E, $\hbar^2 l(l + 1)$ and $\hbar m$ respectively.

The behaviour of the function $R(r)$ near the origin follows from the radial equation (2.78). We shall assume that $V(r)$ is finite at the origin or diverges no faster than $1/r$ as $r \to 0$, so that

$$\lim_{r \to 0} r^2 V(r) = 0. \qquad (2.81)$$

Suppose $R(r)$ is of the form

$$R(r) \sim Cr^s, \qquad \text{as } r \to 0. \qquad (2.82)$$

---

* No confusion should arise by the use of m for the mass of the particle and for the magnetic quantum number. In any case, this is the conventional notation, so the reader must get used to it.

Substituting this into the radial equation (2.78), we obtain

$$-\frac{\hbar^2}{2m}s(s+1)r^{s-2} + r^2V(r)r^{s-2} + \frac{\hbar^2 l(l+1)}{2m}r^{s-2} = Er^s, \qquad r \to 0. \quad (2.83)$$

The lowest power of r in this equation is r^{s-2}. Equating the coefficient of r^{s-2} to zero gives

$$s(s+1) = l(l+1),$$

i.e. $s = l$ or $s = -l - 1$. The solution which behaves like $1/r^{l+1}$ at the origin must be rejected as it does not lead to a solution of the full Schrödinger equation (2.72) at $r = 0$.* Hence the acceptable radial wave functions behave like

$$R(r) \sim Cr^l \qquad \text{as } r \to 0. \quad (2.84)$$

Except for $l = 0$, they vanish at the origin. For $l = 0$, $R(0) = $ const.

If we introduce the radial function $u(r)$, defined by

$$R(r) = \frac{1}{r}u(r), \quad (2.85)$$

into Eq. (2.78), it simplifies to

$$-\frac{\hbar^2}{2m}\frac{d^2u(r)}{dr^2} + \left[V(r) + \frac{\hbar^2 l(l+1)}{2mr^2}\right]u(r) = Eu(r). \quad (2.86)$$

This is like a one-dimensional Schrödinger equation with r restricted to the range $r \geqslant 0$. From Eq. (2.84) $u(r)$ behaves like

$$u(r) \sim Cr^{l+1} \qquad \text{as } r \to 0 \quad (2.87)$$

so that it satisfies the boundary condition

$$u_l(0) = 0, \qquad \text{all } l. \quad (2.88)$$

Eq. (2.86) contains the effective potential

$$V_{\text{eff}}(r) = \left[V(r) + \frac{\hbar^2 l(l+1)}{2mr^2}\right]. \quad (2.89)$$

---

* Just as $R(r) = 1/r$ is not a solution of $\nabla^2 R(r) = 0$ at the origin $r = 0$. Rather, $1/r$ is a solution of Poisson's equation

$$\nabla^2 \frac{1}{r} = -4\pi\rho(r)$$

where $\rho(r)$ is the charge density of a unit point charge situated at the origin, i.e. $1/(4\pi\varepsilon_0 r)$ is the Coulomb potential due to this charge.

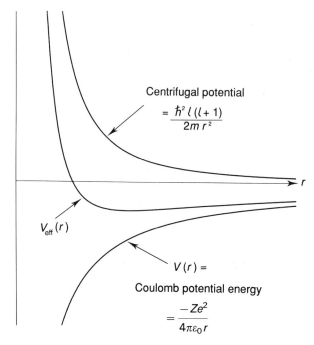

Fig. 2.12   The effective potential (2.89) for an attractive Coulomb potential and $l \geqslant 1$.

This is illustrated in Fig. 2.12 for the case of an attractive Coulomb potential. The second term on the right-hand side of Eq. (2.89) (which is only present for $l \neq 0$) is a purely repulsive potential: it increases monotonically as r decreases and becomes infinite as $r \to 0$. This explains the behaviour (2.87)–(2.88) of $u(r)$ near the origin. For $l \neq 0$, the particle can only penetrate into the classically forbidden region, where $V_{\text{eff}}(r) > E$, to a limited extent, and $V_{\text{eff}}(r)$ tending to infinity as $r \to 0$ implies $u(0) = 0$. The term $\hbar^2 l(l + 1)/2mr^2$ is called the centrifugal potential (or centrifugal barrier) as it has the form

$$\frac{(\text{angular momentum})^2}{2 \, (\text{moment of inertia})} \, ,$$

i.e it has the form of the centrifugal energy of a particle in an orbit.

Solving Eq. (2.86) for $u(r)$, with the boundary condition $u(0) = 0$, is very similar to solving the one-dimensional Schrödinger equation. If the potential $V(r)$ is sufficiently attractive to support bound states, we shall denote these solutions by $u_{nl}(r)$ and the corresponding energy eigenvalues by E_{nl}. We shall order the E_{nl}, for each value of l, in increasing energy:

$$E_{0l} < E_{1l} < \cdots, \qquad l = 0, 1, \ldots . \qquad (2.90)$$

Corresponding to each solution $u_{nl}(r)$, the radial equation (2.78) possesses a solution

$$R_{nl}(r) = \frac{1}{r} u_{nl}(r), \qquad (2.91)$$

belonging to the same energy eigenvalue E_{nl}, and the three-dimensional Schrödinger equation (2.72) possesses $(2l + 1)$ linearly independent eigenfunctions

$$\psi_{nlm}(\mathbf{r}) = R_{nl}(r) Y_l^m(\theta, \phi), \qquad m = l, l - 1, \ldots, -l, \qquad (2.92)$$

belonging to the same energy eigenvalue E_{nl}.

Since the $u_{nl}(r)$ are bound-state solutions of a one-dimensional eigenvalue problem and with the energies ordered as in Eq. (2.90), the node theorem of section 2.1 applies to $u_{nl}(r)$ and hence to $R_{nl}(r)$: for each value of l, $R_{0l}(r)$ has no internal node, $R_{1l}(r)$ has one internal node, and so on. Whether and how many bound states exist depends on $V(r)$. The situation is very similar to that in one dimension. Suppose $V(r) \to 0$ as $r \to \infty$. For $E > 0$, the energy spectrum is continuous: any positive energy can occur but these are not bound states. For an attractive potential and $E < 0$ the spectrum is discrete. In general, only a finite number of bound states occur for each value of $l = 0, 1, \ldots$; how many depends on how strongly attractive $V(r)$ is. As l increases, the effective potential $V_{\text{eff}}(r)$, Eq. (2.89), becomes less attractive and the number of bound states decreases. The bound s-states $(l = 0)$ of the three-dimensional attractive square well potential are derived in problem 2.4. The bound states for $l > 0$ can be written down in terms of spherical Bessel functions.

2.5.2 The hydrogen atom

We conclude this section with a brief discussion of the hydrogenic system, i.e. an electron moving in the Coulomb field of a positive point charge Ze; for $Z = 1$ we are dealing with the hydrogen atom, for $Z = 2, 3, \ldots$ with the one-electron ions He^+, $Li^{++}, \ldots$. Being a two-body system, the hydrogenic system allows exact or highly accurate solutions which are free of the complexities and approximations typical of many-body problems. For this reason, it has played a key role in establishing non-relativistic quantum mechanics, the relativistic theory of the electron (the Dirac equation) and quantum electrodynamics (the quantum theory of the electromagnetic field—the famous Lamb shift experiment). Furthermore, the hydrogenic wave functions often form the basis of the approximation methods for many-electron atoms, and the conventional notation used for labelling the hydrogenic states and energy levels is also used for labelling the states of many-electron atoms. For these reasons we shall explain these conventions and results although we shall not derive them.

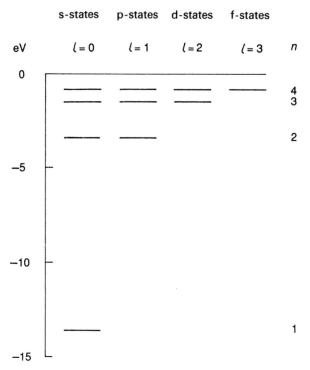

Fig. 2.13   The first few energy levels E_n of the hydrogen atom.

The Schrödinger equation for the hydrogenic system is

$$\left(-\frac{\hbar^2}{2m}\nabla^2 + \frac{-Ze^2}{4\pi\varepsilon_0 r} \right)\psi(\mathbf{r}) = E\psi(\mathbf{r}). \qquad * \tag{2.93}$$

The attractive Coulomb potential goes to zero rather slowly as $r \to \infty$. As a result there exist infinitely many bound-state solutions for each value of the angular momentum quantum number l, which are of the form (2.92). For the hydrogenic system, the radial functions $R_{nl}(r)$, corresponding to a given value of l, are conventionally labelled $n = l + 1, l + 2, \ldots$, and not $n = 0, 1, \ldots$ as we did for the general case. Thus, for a given value of l, the magnetic quantum number m takes on the $(2l + 1)$ values $m = l, l - 1, \ldots, -l$, and n takes on

---

* For m we must take the reduced mass of the electron to allow for the finite nuclear mass. The derivation of this result is similar to that in classical mechanics. (See problem 1.5.)

the infinitely many values $n = l + 1, l + 2, \ldots$. One can state this more conveniently by saying that the quantum numbers (n, l, m) assume the values

$$n = 1, 2, \ldots \qquad (2.94a)$$

$$l = 0, 1, \ldots, (n - 1) \qquad (2.94b)$$

$$m = l, l - 1, \ldots, -l. \qquad (2.94c)$$

n is called the principal quantum number.
The hydrogenic energy levels are given by

$$E_n = -\frac{1}{2}\left(\frac{e^2}{4\pi\varepsilon_0 a_0}\right)\frac{Z^2}{n^2} \equiv -\mathrm{Ry}\,\frac{Z^2}{n^2} = -13.6\frac{Z^2}{n^2}\,\mathrm{eV}, \qquad n = 1, 2, \ldots, \quad (2.95)$$

with the Bohr radius a_0 defined by

$$a_0 = \frac{4\pi\varepsilon_0\hbar^2}{me^2} = 0.53 \times 10^{-10}\,\mathrm{m}, \qquad (2.96)$$

and where Ry is the Rydberg unit of energy with the value $1\,\mathrm{Ry} = 13.6\,\mathrm{eV}$. (1 Ry is the ionization energy of atomic hydrogen in its ground state.)
    This is an unexpected result: the energy levels (2.95) depend on the principal quantum number n only and are degenerate with respect to l and m. For a given level E_n, l and m take on the range of values given by Eqs. (2.94b–c); hence

$$\text{degeneracy of } E_n = \sum_{l=0}^{n-1}(2l + 1) = n^2. \quad * \qquad (2.97)$$

    In spectroscopic notation, one denotes states with angular momentum quantum number $l = 0, 1, 2, \ldots$ and with principal quantum number n as (ns), (np), (nd), $\ldots$ states. We see that the hydrogenic ground state is the non-degenerate (1s) state with energy E_1, that the (2s) and (2p) states are degenerate with energy E_2, etc. The first few energy levels of atomic hydrogen are shown in Fig. 2.13. The degeneracy of these levels with respect to l is a peculiarity of the non-relativistic hydrogenic system, i.e. of the point Coulomb potential, and is removed by any deviations from this potential. In contrast, the degeneracy with respect to m is a general feature of any central potential $V(r)$, as we have seen.
    The hydrogenic radial wave functions are of the form

$$R_{nl}(r) = \mathrm{e}^{-Zr/na_0}r^l(c_0 + c_1 r + \cdots + c_{n-l-1}r^{n-l-1}) \qquad (2.98)$$

where $c_0, c_1, \ldots, c_{n-l-1}$ are real constants. These functions have the expected r^l behaviour at the origin. $R_{nl}(r)$ tends to zero as $r \to \infty$, more slowly for

---

* For the present we are ignoring the electron's internal degree of freedom, its spin. When this is taken into account, the degeneracy of each level E_n is doubled.

larger n, consistent with the fact that the states with larger n are less tightly bound: $|E_n|$ is smaller. The polynomial factor in (2.98) ensures that the node theorem holds. The radial functions $R_{nl}(r)$ are orthogonal for a given value of l and different values of the principal quantum number. Conventionally, they are normalized so that the complete hydrogenic wave functions $\psi_{nlm}(\mathbf{r})$, Eq. (2.92), are orthonormal

$$\int d^3r \psi^*_{nlm}(\mathbf{r})\psi_{n'l'm'}(\mathbf{r}) = \delta_{nn'}\delta_{ll'}\delta_{mm'}. \tag{2.99}$$

The first three radial functions are given by

$$\left.\begin{aligned}
R_{10}(r) &= 2\left(\frac{Z}{a_0}\right)^{3/2} e^{-Zr/a_0} \\
R_{20}(r) &= 2\left(\frac{Z}{2a_0}\right)^{3/2}\left(1 - \frac{Zr}{2a_0}\right)e^{-Zr/2a_0} \\
R_{21}(r) &= \frac{1}{\sqrt{3}}\left(\frac{Z}{2a_0}\right)^{3/2}\frac{Zr}{a_0}e^{-Zr/2a_0}.
\end{aligned}\right\} \tag{2.100}$$

The reader should sketch these functions and see that they possess the general properties discussed above.

2.6 MOMENTUM EIGENSTATES

The momentum eigenstates are the solutions of the eigenvalue problem

$$\hat{\mathbf{p}}\psi(\mathbf{r}) \equiv -i\hbar\nabla\psi(\mathbf{r}) = \mathbf{p}\psi(\mathbf{r}). \tag{2.101}$$

One sees at once that for any momentum vector $\mathbf{p}$

$$\psi(\mathbf{r}) = \text{const. } e^{i\mathbf{p}\cdot\mathbf{r}/\hbar} \tag{2.102}$$

is a momentum eigenstate with $\mathbf{p}$ as eigenvalue. Thus the momentum operator $\hat{\mathbf{p}}$ has a continuous eigenvalue spectrum. Unfortunately, the normalization integral

$$\int |\psi(\mathbf{r})|^2 \, d^3r$$

of the eigenstate (2.102), taken over all space, is infinite. This difficulty is typical of eigenfunctions belonging to a continuous spectrum. It occurs whenever the state of a system is not localized. For the plane wave (2.102) the probability density $|\psi(\mathbf{r})|^2$ is constant and the particle is equally likely to be found at any point in space. In reality, a wave never extends throughout all space. Describing it by a plane wave is an approximation which leads to non-normalizable states. In the following, I shall show two ways of coping with these.

2.6.1   Box normalization

It is possible to avoid the continuous eigenvalue spectrum altogether. This is achieved by enclosing the system in a large but finite box, imposing suitable boundary conditions on the wave functions at the walls of the box. The system is now localized and has a discrete eigenvalue spectrum, as we shall see. Since any actual system is always to some extent localized, the walls of a sufficiently large box and the boundary conditions imposed at the walls cannot affect the system significantly, and the modified localized system will simulate the non-localized system as closely as we like.

We shall enclose the system in a cubic box of side L, defined as the region

$$-\frac{L}{2} \leqslant x \leqslant \frac{L}{2}, \quad -\frac{L}{2} \leqslant y \leqslant \frac{L}{2}, \quad -\frac{L}{2} \leqslant z \leqslant \frac{L}{2}. \tag{2.103}$$

At the surfaces of this cubic enclosure we subject the wave functions $\psi(\mathbf{r}) = \psi(x, y, z)$ to the periodic boundary conditions

$$\psi\left(-\frac{L}{2}, y, z\right) = \psi\left(\frac{L}{2}, y, z\right), \quad \text{etc.} \tag{2.104}$$

The momentum eigenstates are the solutions of Eq. (2.101) which satisfy the boundary conditions (2.104). They are given by

$$u_{\mathbf{p}}(\mathbf{r}) = \frac{1}{L^{3/2}} e^{i\mathbf{p} \cdot \mathbf{r}/\hbar}, \tag{2.105}$$

where the momentum eigenvalues $\mathbf{p}$ must be of the form

$$\mathbf{p} = \frac{2\pi\hbar}{L}(n_1, n_2, n_3), \quad n_1, n_2, n_3 = 0, \pm 1, \pm 2, \ldots, \tag{2.106}$$

so that the boundary conditions (2.104) are satisfied. We see that by enclosing the system in the finite-size box we have converted the continuum of momentum eigenvalues into the discrete spectrum (2.106) which for sufficiently large L approximates the continuous spectrum arbitrarily closely.

We easily verify that the momentum eigenstates (2.105) satisfy the ortho-normality relation (1.29), i.e.

$$\int d^3r u_{\mathbf{p}}^*(\mathbf{r}) u_{\mathbf{p}}(\mathbf{r}) = \frac{1}{L^3} \int d^3r \exp[i(\mathbf{p}' - \mathbf{p}) \cdot \mathbf{r}/\hbar] = \delta_{\mathbf{p}\mathbf{p}'}. \tag{2.107}$$

Here—as throughout this subsection—the integration is over the volume of the cubic enclosure, and $\mathbf{p}$ and $\mathbf{p}'$ are two of the eigenvalues (2.106).

An arbitrary function $\psi(\mathbf{r})$, defined within the cubic enclosure, can be expanded as a Fourier series of exponential terms:

$$\psi(\mathbf{r}) = \sum c_{\mathbf{p}} u_{\mathbf{p}}(\mathbf{r}), \tag{2.108}$$

where the summation is over all momenta (2.106) and the expansion coefficients c_p are given by

$$c_p = \int d^3 r u_p^*(\mathbf{r})\psi(\mathbf{r}). \tag{2.109}$$

Eqs. (2.108) and (2.109) illustrate the general expansion theorem, Eqs. (1.31) and (1.32), for the complete set of momentum eigenstates.

For a particle in the normed state $\psi(\mathbf{r})$, the expectation value of the momentum is given by

$$\langle \hat{\mathbf{p}} \rangle = \int d^3 r \psi^*(\mathbf{r})(-i\hbar\nabla)\psi(\mathbf{r}). \tag{2.110}$$

If in this equation we substitute the Fourier expansion (2.108) for ψ, its complex conjugate for ψ^*, and use the orthonormality relation (2.107) we obtain

$$\langle \hat{\mathbf{p}} \rangle = \sum |c_p|^2 \mathbf{p}. \tag{2.111}$$

From this equation follows the interpretation of

$$P(\mathbf{p}) = |c_p|^2 = \left| \int d^3 r u_p^*(\mathbf{r})\psi(\mathbf{r}) \right|^2 \tag{2.112}$$

as the probability that for a particle in the state ψ a measurement of the momentum gives the value $\mathbf{p}$.

★2.6.2   The continuous spectrum

It is often convenient to force a system to have a discrete momentum spectrum by enclosing it in a box. But it is more natural to consider the infinite domain, as we shall do now. For simplicity we shall study one-dimensional motion along the x-axis, generalizing the results to three dimensions at the end.

The momentum eigenfunctions are now given by

$$u_p(x) = \frac{1}{\sqrt{(2\pi\hbar)}} e^{ipx/\hbar}, \tag{2.113}$$

with the continuous eigenvalue spectrum $-\infty < p < \infty$. (The reason for the choice of normalization constant will become clear shortly.) Unlike a plane wave, an actual state is always localized to some degree. It corresponds to a superposition of plane waves, called a wave packet. If $\psi(x)$ is a normalizable state, i.e.

$$\int_{-\infty}^{\infty} dx |\psi(x)|^2 < \infty, \tag{2.114}$$

its representation as a superposition of plane waves is given by the Fourier integral

$$\psi(x) = \int dp\phi(p) \frac{e^{ipx/\hbar}}{\sqrt{(2\pi\hbar)}} = \int dp\phi(p)u_p(x). \tag{2.115}$$

The amplitude $\phi(p)$ is given by the inverse Fourier transform

$$\phi(p) = \int dx \frac{e^{-ipx/\hbar}}{\sqrt{(2\pi\hbar)}} \psi(x) = \int dx u_p^*(x)\psi(x). \tag{2.116}$$

In Eqs. (2.115) and (2.116), as throughout this subsection, integrals with limits of integration not shown are over the range $(-\infty, \infty)$.

Eqs. (2.115) and (2.116) are our basic result. They are the analogues for the continuous spectrum of Eqs. (2.108) and (2.109) for the discrete case. If we wish to take the analogy with the discrete case further, we must introduce an orthonormality relation for the momentum eigenstates. For this purpose we substitute Eq. (2.115) in Eq. (2.116)

$$\phi(p) = \int dx u_p^*(x) \int dp' \phi(p')u_{p'}(x) \tag{2.117}$$

and interchange the order of integration, obtaining

$$\phi(p) = \int dp' \phi(p')\left[\int dx u_p^*(x)u_{p'}(x) \right]. \tag{2.118}$$

We shall see that the integral in square brackets is a function of the difference $(p' - p)$ only and we shall denote it by $\delta(p' - p)$:

$$\delta(p' - p) = \int dx u_p^*(x)u_{p'}(x). \tag{2.119}$$

Eq. (2.118) then becomes

$$\phi(p) = \int dp' \phi(p')\delta(p' - p). \tag{2.120}$$

The integral in Eq. (2.120) picks out the value of $\phi(p')$ at $p' = p$. Since the amplitude $\phi(p')$ is essentially an arbitrary function, Eq. (2.120) can only hold if the function $\delta(p' - p)$ has the following properties:

(i)
$$\delta(p' - p) = 0, \quad \text{if } p' \neq p; \tag{2.121a}$$

(ii) at $p' = p$, $\delta(p' - p)$ diverges, i.e. $\delta(0) = \infty$, in such a manner that

$$\int_{-\infty}^{\infty} dp' \delta(p' - p) = 1. \tag{2.121b}$$

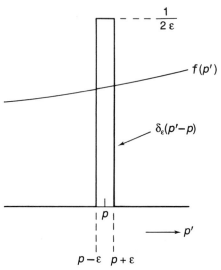

Fig. 2.14   An approximate representation of the Dirac δ-function by the function $\delta_\varepsilon(p' - p)$. For sufficiently small ε:

$$f(p')\delta_\varepsilon(p' - p) = \begin{cases} f(p)/2\varepsilon, & \text{if } |p' - p| < \varepsilon, \\ 0, & \text{if } |p' - p| > \varepsilon, \end{cases} \text{ so that } \int_{-\infty}^{\infty} dp' f(p')\delta_\varepsilon(p' - p) = f(p).$$

Eqs. (2.121) define the Dirac delta function $\delta(p' - p)$. From these equations follows the alternative definition: for any function $f(p')$ which is continuous at $p' = p$ and any interval $p_1 < p' < p_2$

$$\int_{p_1}^{p_2} dp' f(p')\delta(p' - p) = \begin{cases} f(p), & \text{if } p_1 < p < p_2, \\ 0, & \text{if } p < p_1 \text{ or } p > p_2. \end{cases} \tag{2.122}$$

We can visualize the δ-function $\delta(p' - p)$ as an extremely high and extremely narrow peak, centred at $p' = p$, the area under the curve being unity. In Fig. 2.14 we show the function

$$\delta_\varepsilon(p' - p) = \begin{cases} \dfrac{1}{2\varepsilon}, & \text{if } |p' - p| < \varepsilon, \\ 0, & \text{if } |p' - p| > \varepsilon. \end{cases}$$

This function approximates the properties (2.121) and (2.122) of the Dirac δ-function $\delta(p' - p)$ very closely for sufficiently small values of ε. Hence we can consider the Dirac δ-function formally as the limiting function

$$\delta(p' - p) = \lim_{\varepsilon \to 0} \delta_\varepsilon(p' - p).$$

A very useful explicit expression for the δ-function is obtained from Eq. (2.119). Substitution of Eq. (2.113) for the plane wave states leads to

$$\delta(p' - p) = \frac{1}{2\pi\hbar} \int_{-\infty}^{\infty} dx \, e^{i(p' - p)x/\hbar}$$

and writing k for $x/\hbar$ and x for $(p' - p)$ in this equation we obtain

$$\delta(x) = \frac{1}{2\pi} \int_{-\infty}^{\infty} dk \, e^{ikx}. \tag{2.123}$$

Eq. (2.119) is the desired orthonormality relation for continuum eigenfunctions. Instead of the Kronecker delta, it contains the Dirac δ-function. Clearly no ordinary mathematical function with the properties (2.121) exists. It is a highly pathological function. That we ended up with this improper function is due to the fact that the change of order of integration in going from Eq. (2.117) to (2.118) is not permissible. In applications, the Dirac δ-function always occurs under an integral sign. Carrying out this integration, using the formal properties of the δ-function, in effect amounts to inverting the order of integration once more, thereby getting back to a mathematically correct expression. In this way, the purely formal manipulation of δ-functions can be justified. In practice, the δ-function simplifies the analysis and enables us to formulate the cases of continuous and discrete spectra in very similar ways.

Using the δ-function, we easily calculate the expectation value of the momentum. For any normed state ψ, it is given by

$$\langle \hat{p}_x \rangle = \int dx \psi^*(x) \left(-i\hbar \frac{\partial}{\partial x} \right) \psi(x). \tag{2.124}$$

Substituting the Fourier transform (2.115) and its complex conjugate for ψ and ψ^* in Eq. (2.124) and using the orthonormality relation (2.119) leads to

$$\langle \hat{p}_x \rangle = \int dp |\phi(p)|^2 p. \tag{2.125}$$

From this equation follows the probability interpretation for a continuous eigenvalue spectrum analogous to our earlier result (2.112) for the discrete case:

$$|\phi(p)|^2 \, dp = \left| \int dx u_p^*(x) \psi(x) \right|^2 dp \tag{2.126}$$

is the probability that for a particle in the normed state $\psi(x)$ a measurement of the momentum produces a value in the range p to $p + dp$. More concisely, one refers to expression (2.126) as the momentum distribution of a particle in the state $\psi(x)$. In the same way in which we transformed Eq. (2.124) into (2.125), we can transform the normalization condition

$$\int dx |\psi(x)|^2 = 1 \tag{2.127}$$

into

$$\int dp |\phi(p)|^2 = 1. \tag{2.128}$$

This relation shows that the probability distribution (2.126) is correctly normalized to unity.

The extension of these results to three dimensions is straightforward. The three-dimensional δ-function is defined by

$$\delta^{(3)}(\mathbf{r}) = \delta(x)\delta(y)\delta(z), \tag{2.129}$$

leading to the obvious generalizations of the properties (2.121) and (2.122). In particular, for any function $f(\mathbf{r})$, continuous at $\mathbf{r} = \mathbf{r}_0$, and any three-dimensional region of integration R, we have

$$\int_R d^3r f(\mathbf{r})\delta^{(3)}(\mathbf{r} - \mathbf{r}_0) = \begin{cases} f(\mathbf{r}_0), & \text{if } \mathbf{r}_0 \text{ lies inside the region } R, \\ 0, & \text{if } \mathbf{r}_0 \text{ lies outside the region } R. \end{cases} \tag{2.130}$$

In three dimensions, the momentum eigenstates are given by

$$u_\mathbf{p}(\mathbf{r}) = \frac{1}{(2\pi\hbar)^{3/2}} e^{i\mathbf{p}\cdot\mathbf{r}/\hbar}, \tag{2.131}$$

with $\mathbf{p}$ taking on any value. We leave it to the reader to write down for these eigenstates the results analogous to the Eqs. (2.115–116), (2.119) and (2.124)–(2.128) of the one-dimensional case.

★2.7 HARMONIC OSCILLATOR

Harmonic oscillations are of fundamental importance in both classical and quantum physics. For a system in stable equilibrium, small displacements from equilibrium will bring into play forces which aim to restore the equilibrium. These forces are usually proportional to the displacement: this is the hallmark of simple harmonic motion. If a particle of mass m, which is displaced a distance x from its equilibrium position, experiences a restoring force $-kx$ (where k is a constant), its equation of motion

$$m\frac{d^2x}{dt^2} = -kx \tag{2.132}$$

leads to simple harmonic motion of angular frequency

$$\omega = \sqrt{(k/m)}. \tag{2.133}$$

This result generalizes to systems with more than one degree of freedom. In this case small displacements from stable equilibrium lead to complicated coupled oscillations of the whole system. By suitable choice of coordinates (known as normal coordinates) one can describe the oscillations of the whole system in terms of these normal modes, i.e. as a superposition of simple harmonic oscillations, each with its own angular frequency. For example, the vibrations of the atoms in a polyatomic molecule or in a solid can be analysed

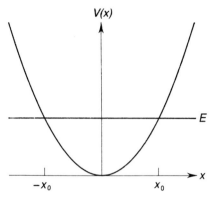

Fig. 2.15   The harmonic oscillator potential $V(x) = \frac{1}{2}m\omega^2 x^2$. $\pm x_0$ are the classical turning points for a particle of energy E, i.e. classically only the region $|x| < x_0$ is accessible to the particle.

in this way, and this forms the basis for understanding their heat capacities and the vibrational spectra of polyatomic molecules.*

To treat the harmonic oscillator quantum-mechanically, we need its Schrödinger equation. The classical equation (2.132) represents the one-dimensional motion of a particle of mass m in the parabolic potential

$$V(x) = \tfrac{1}{2}kx^2 = \tfrac{1}{2}m\omega^2 x^2, \tag{2.134}$$

shown in Fig. 2.15. The corresponding Schrödinger equation is

$$\left(-\frac{\hbar^2}{2m}\frac{d^2}{dx^2} + \tfrac{1}{2}m\omega^2 x^2\right)\psi(x) = E\psi(x). \tag{2.135}$$

Classically, a particle subject to the oscillator potential (2.134) is always in a bound state; if it has energy E, it is confined to the region where $V(x) \leqslant E$, i.e. to the region between the classical turning points $-x_0$ and x_0, where

$$x_0 = \left(\frac{2E}{m\omega^2}\right)^{1/2}. \tag{2.136}$$

Quantum-mechanically, a particle in the oscillator potential (2.134) is also always in a bound state. We know from the discussion in section 2.1 that the wave function $\psi(x)$ can penetrate into the classically forbidden regions $x < -x_0$ and $x > x_0$, in which $V(x) > E$, but in these regions $\psi(x)$ is non-oscillatory and tends to zero as $x \to \pm \infty$.

---

* In particular, the observed temperature dependence of the heat capacities can only be understood in terms of the quantum-mechanical treatment of the harmonic oscillator, to be given in this section. See, for example, Mandl, chapter 6.

For the finite square-well potential, we saw in section 2.1 that the bound state wave functions decay exponentially as $x \to \pm\infty$ [compare Eqs. (2.8a–b)]. For the oscillator potential (2.134) which becomes infinite as $x \to \pm\infty$ we expect a much more rapid decay. One verifies by direct substitution that

$$\psi_0(x) = e^{-\alpha^2 x^2/2}, \tag{2.137}$$

with

$$\alpha = (m\omega/\hbar)^{1/2}, \tag{2.138}$$

is an eigenfunction of the Schrödinger equation (2.135) belonging to the energy eigenvalue

$$E_0 = \tfrac{1}{2}\hbar\omega. \tag{2.139}$$

The wave function (2.137) possesses no nodes. It follows from the node theorem, stated in section 2.1, that (2.137) represents the state of lowest energy, the ground state, of the particle in the oscillator potential. For the probability interpretation, the wave function (2.137) still must be multiplied by an appropriate constant, so that it is normalized to unity. For the present, we shall work with the unnormalized wave functions.

To simplify the writing in what follows, we introduce the dimensionless variables

$$\xi = \alpha x, \quad \lambda = E/(\tfrac{1}{2}\hbar\omega) \tag{2.140}$$

and the short-hand notation

$$D \equiv d/d\xi. \tag{2.141}$$

With

$$\psi(x) = \psi(\xi/\alpha) = \phi(\xi), \tag{2.142}$$

the Schrödinger equation (2.135) becomes

$$(D^2 + \lambda - \xi^2)\phi(\xi) = 0, \tag{2.143}$$

and the wave function (2.137) and energy eigenvalue (2.139) of the ground state go over into

$$\phi_0(\xi) = e^{-\xi^2/2}, \quad \lambda_0 = 1. \tag{2.144}$$

To obtain the other harmonic oscillator eigenfunctions and eigenvalues, we start from the following identity for the differential operators $(D^2 + \lambda - \xi^2)$ and $(\xi - D)$. For any function $\phi(\xi)$ and any parameter λ

$$(D^2 + \lambda - \xi^2)(\xi - D)\phi - (\xi - D)(D^2 + \lambda - \xi^2)\phi = -2(\xi - D)\phi. \tag{2.145}$$

This identity is derived by carrying out the differentiations on the left-hand side of this equation. In this identity we take $\phi = \phi_n(\xi)$ and $\lambda = \lambda_n$, where $\phi_n(\xi)$ is an oscillator eigenfunction belonging to the eigenvalue λ_n, i.e.

$$(D^2 + \lambda_n - \xi^2)\phi_n(\xi) = 0. \tag{2.146}$$

The identity (2.145) then reduces to

$$[D^2 + (\lambda_n + 2) - \xi^2](\xi - D)\phi_n(\xi) = 0. \tag{2.147}$$

Comparing this equation with the Schrödinger equation (2.143), we see that the function $(\xi - D)\phi_n(\xi)$ is an oscillator eigenstate belonging to the eigenvalue $(\lambda_n + 2)$. Denoting these by

$$\phi_{n+1}(\xi) = (\xi - D)\phi_n(\xi), \quad \lambda_{n+1} = \lambda_n + 2, \tag{2.148}$$

we can write Eq. (2.147)

$$(D^2 + \lambda_{n+1} - \xi^2)\phi_{n+1}(\xi) = 0. \tag{2.149}$$

Eqs. (2.148) give a prescription for generating an eigenfunction ϕ_{n+1} and its eigenvalue λ_{n+1} from ϕ_n and λ_n. Repeated application of (2.148) starting from the ground state wave function and the ground state energy, Eqs. (2.144), generates a whole sequence of oscillator eigenstates

$$\phi_n(\xi) = (\xi - D)^n e^{-\xi^2/2}, \quad n = 0, 1, 2, \dots, \tag{2.150}$$

belonging to the eigenvalues

$$\lambda_n = (2n + 1), \quad n = 0, 1, 2, \dots. \tag{2.151}$$

The corresponding energy eigenvalues are, from Eq. (2.140), given by

$$E_n = \hbar\omega(n + \tfrac{1}{2}), \quad n = 0, 1, \dots. \tag{2.152}$$

This is an infinite sequence of equally spaced levels, with separation $\hbar\omega$ between adjacent levels. It can be shown that the eigenfunctions (2.150) form the complete set of all oscillator eigenfunctions.

Operating with $(\xi - D)^n$ on $e^{-\xi^2/2}$, to generate the functions $\phi_n(\xi)$ in Eq. (2.150), has the effect of multiplying this exponential by a polynomial of degree n in ξ, i.e. we can rewrite Eq. (2.150) in the form

$$\phi_n(\xi) = (\xi - D)^n e^{-\xi^2/2} = e^{-\xi^2/2} H_n(\xi), \quad n = 0, 1, 2, \dots, \tag{2.153}$$

which defines the Hermite polynomials $H_n(\xi)$. The first few of these are

$$\left.\begin{array}{ll} H_0(\xi) = 1, & H_1(\xi) = 2\xi \\ H_2(\xi) = 4\xi^2 - 2, & H_3(\xi) = 8\xi^3 - 12\xi. \end{array}\right\} \tag{2.154}$$

The first three oscillator eigenfunctions (2.153) are shown in Fig. 2.16. These curves illustrate the node theorem and the fact that the eigenfunctions divide into even and odd functions. This can be seen more generally from Eq. (2.153). Changing ξ to $-\xi$ (and therefore D to $-$D) in this equation, it follows that

$$\phi_n(-\xi) = (-1)^n\phi_n(\xi), \quad H_n(-\xi) = (-1)^n H_n(\xi), \tag{2.155}$$

i.e. both $\phi_n(\xi)$ and $H_n(\xi)$ are even and odd functions of ξ for $n = 0, 2, 4, \dots$ and $n = 1, 3, 5, \dots$ respectively. That the oscillator states (2.153) divide into even

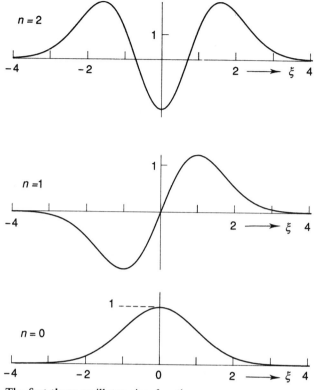

Fig. 2.16   The first three oscillator eigenfunctions

$$\phi_n(\xi) = e^{-\xi^2/2} H_n(\xi), \qquad n = 0, 1, 2.$$

and odd functions in this way is a consequence of the fact that the oscillator potential (2.134) is an even function of x (and therefore of ξ), as already mentioned when discussing the bound states of the one-dimensional square-well potential in section 2.1.

It follows from the general theory about eigenfunctions (section 1.2) that the oscillator eigenfunctions (2.153) are orthogonal. However, they are not normed to unity. I shall not derive the normalization and only state the result.* One can show that

$$\int_{-\infty}^{\infty} dx\, \psi_m(x)\psi_n(x) = \frac{1}{\alpha} \int_{-\infty}^{\infty} d\xi\, \phi_m(\xi)\phi_n(\xi) = \delta_{mn} \frac{2^n n! \sqrt{\pi}}{\alpha}, \qquad (2.156)$$

---

* For a derivation of this result and other properties of Hermite polynomials see, for example, Arfken, pp 712–717, or Merzbacher, section 5.3.

so that the normalized eigenfunctions are given by $C_n\psi_n(x)$, with

$$C_n = \left(\frac{\alpha}{2^n n! \sqrt{\pi}}\right)^{1/2}. \tag{2.157}$$

We conclude this section with some useful mathematical properties of Hermite polynomials. By substituting the oscillator eigenvalues and eigenfunctions, Eqs. (2.151) and (2.153), into the Schrödinger equation (2.143), we obtain the differential equation which the Hermite polynomials satisfy:

$$(D^2 - 2\xi D + 2n)H_n(\xi) = 0. \tag{2.158}$$

We note that the definition (2.153) of $H_n(\xi)$ can be shown to be equivalent to the more usual definition (see problem 2.10)

$$H_n(\xi) = (-1)^n e^{\xi^2} D^n e^{-\xi^2}. \tag{2.159}$$

Finally, we state two recurrence relations which are useful in manipulating Hermite polynomials:

$$\xi H_n(\xi) = \tfrac{1}{2} H_{n+1}(\xi) + n H_{n-1}(\xi), \tag{2.160}$$

$$D H_n(\xi) = 2n H_{n-1}(\xi). \tag{2.161}$$

PROBLEMS 2

2.1 A particle moving in the one-dimensional square-well potential

$$V(x) = 0 \quad \text{for } |x| < a, \qquad V(x) = \infty \quad \text{for } |x| > a, \tag{2.162}$$

is in the state

$$\psi(x) = [\phi_1(x) + \phi_2(x)]/\sqrt{2}$$

at time $t = 0$, with $\phi_n(x)$ given by Eqs. (2.25). What is its wave function at time t?

Calculate the probabilities $P_+(t)$ and $P_-(t)$ that at time t the particle is in the intervals $0 < x < a$ and $-a < x < 0$ respectively. Interpret the time-dependence of these probabilities.

2.2 Discuss qualitatively how the results of the last problem are affected if a potential barrier of height V_0 and width $2b$ is inserted in the middle of the well, i.e. the potential (2.162) is modified to

$$V(x) = \begin{cases} V_0, & |x| < b, \\ 0, & -a < x < -b \text{ and } b < x < a, \\ \infty, & |x| > a. \end{cases} \tag{2.163}$$

Assume the barrier height V_0 is much greater than the two lowest-lying energy levels E_1 and E_2 in the potential (2.163). Consider, in particular, the limit $V_0 \to \infty$ for which the exact solutions are easily written down.

2.3   In section 2.1 a graphical method was developed for finding the energies of the odd bound states of a particle in the square-well potential (2.1). Derive a similar procedure for the even bound states of this potential. Show that there always exists at least one even bound state, in contrast to what we found for the odd bound states.

2.4   Obtain the energy eigenfunctions and eigenvalues of the bound states of a particle with zero angular momentum in the three-dimensional square-well potential

$$V(r) = -V_0(<0) \quad \text{for } r < a, \qquad V(r) = 0 \quad \text{for } r > a. \tag{2.164}$$

[The wave function must satisfy the continuity conditions: ψ and $\partial\psi/\partial r$ are continuous at $r = a$; see Eqs. (2.11a–b).]

2.5   The nuclear force responsible for the binding of the neutron and the proton in the deuteron can, very roughly, be represented by the square-well potential (2.164). The range a of the force is approximately the Compton wavelength $\hbar/\mu c$ of the pion (pion mass $\mu \approx M/7$, with the proton mass $M \approx 1000$ MeV). The deuteron ground state is an s-state with binding energy $\mathscr{E} = 2.2$ MeV. Assuming that $\mathscr{E} \ll V_0$, calculate V_0. (This assumption will have been justified, if you find $V_0 \gg \mathscr{E}_0$.) Hence show that $V_0 a^2 \approx \pi^2 \hbar^2/4M$.

2.6   For the one-dimensional potential barrier (2.27), derive the ratio of the amplitude C of the transmitted wave to the amplitude A of the incident wave, Eq. (2.37), and hence the transmission and reflection coefficients (2.38), for the case in which the particle energy E is less than the barrier height V_0.

2.7   A beam of particles of mass m and energy E is incident in the positive x-direction on the potential step

$$V(x) = 0, \ x < 0; \quad V(x) = V_0(>0), \ x > 0.$$

Calculate the transmission and reflection coefficients for the cases: (a) $E > V_0$, (b) $E < V_0$.

2.8   For a particle in the normalized state $\psi(\mathbf{r}) = (x + iy)f(r)/\sqrt{2}$, calculate the probabilities $P(\hbar m)$, $m = 1, 0, -1$, that a measurement of its x-component of angular momentum yields the results $\hbar$, 0 and $-\hbar$ respectively.

2.9   The energy levels of a particle of mass m moving in the potential $V(r) = \frac{1}{2}m\omega^2 r^2$ (known as a three-dimensional oscillator) can be derived from the levels of the one-dimensional oscillator, since the Schrödinger equation for the potential $V(r)$ is separable when expressed in Cartesian coordinates. Obtain the energy level spectrum of the three-dimensional oscillator and the degeneracies of the three lowest-lying levels.

Since $V(r)$ is a central potential, the energy eigenstates of the three-dimensional oscillator can be chosen as angular momentum eigenstates. Derive the values of the angular momentum quantum number l associated with the three lowest-lying levels of the three-dimensional oscillator, and obtain the wave functions of the corresponding states.

2.10 Derive the definition (2.159) of the Hermite polynomial $H_n(\xi)$ from Eq. (2.153). Hint: for any function $f = f(\xi)$, one easily shows that

$$D(f\,e^{-\xi^2/2}) = e^{-\xi^2/2}(D - \xi)f. \tag{2.165}$$

2.11 Use the results of section 2.6.2 to calculate the momentum distribution of the electron in the ground state of the hydrogen atom. What is the probability that the magnitude of the momentum lies in the range p to $p + dp$? The ground state wave function of the hydrogen atom is

$$\psi(r) = e^{-r/a_0}/\sqrt{(\pi a_0^3)}$$

where a_0 is the Bohr radius.

CHAPTER

Several observables

In the classical description of a system, it is in principle always possible to have a complete knowledge of all observables. For example, for a particle moving in a central potential we can know at any instant of time its position and momentum coordinates $\mathbf{r}$ and $\mathbf{p}$, all components of its angular momentum $\mathbf{l} = \mathbf{r} \wedge \mathbf{p}$, etc. In discussing angular momentum in section 2.3, we found that in quantum mechanics this is no longer the case: we may have simultaneous knowledge of $\mathbf{l}^2$ and l_z, say, but not of l_z and l_x. A measurement involves an interaction with, i.e. a disturbance of, a system. Classically, this disturbance can (in principle) be made as small as we please. Quantum-mechanically, i.e. on the atomic scale, a measurement in general represents a severe disturbance. Measuring l_x with a Stern–Gerbach magnet forces the atom into an eigenstate of l_x, and this obliterates any prior knowledge one might have had of l_z. Quantum-mechanically, the observables of a system divide into compatible observables for which simultaneous exact knowledge is possible and incompatible observables for which it is not. We shall consider compatible observables in section 3.1. Observables compatible with the Hamiltonian of a system are of special interest, as we shall see in section 3.2. For incompatible observables the question arises: how accurately can one know them simultaneously? The answer is provided by the uncertainty principle. This is the topic of section 3.3.

3.1 COMPATIBLE OBSERVABLES

When discussing angular momentum in section 2.3 we saw that the spherical harmonics

$$Y_l^m(\theta, \phi), \qquad m = l, l - 1, \ldots, -l, \quad l = 0, 1, \ldots$$

form a complete set of simultaneous eigenfunctions of the angular momentum operators l^2 and l_z. In such a mutual eigenstate both l^2 and l_z have definite values. In contrast, l_z and l_x are incompatible, i.e. they do not possess a complete set of common eigenfunctions.

We now generalize these ideas and define two observables A and B of a system as compatible if they possess a complete orthonormal set of simultaneous eigenfunctions ψ_{abk}:

$$A\psi_{abk} = a\psi_{abk}, \quad B\psi_{abk} = b\psi_{abk}. \tag{3.1}$$

Here the eigenvalues of A are denoted by a (which takes on different values a', a'', ...) and those of B by b (with values b', b'', ...). The index k allows for degeneracy: there may be several linearly independent eigenfunctions for which A and B have the values a and b respectively; $k = 1, 2, \ldots$ labels these different eigenfunctions. If the state with the eigenvalues a and b is unique, the label k ($= 1$ only) is redundant and we shall write the eigenfunction ψ_{ab}. In case of degeneracy, we shall always choose the eigenfunctions $\psi_{abk}, k = 1, 2, \ldots$, belonging to particular values of a and b, orthonormal. The orthogonality of eigenfunctions belonging to different eigenvalues of A or of B follows from the general theory [see Eq. (1.23)].

For a system in the state ψ_{abk}, the observables A and B have the exact values a and b. We shall now show that from the formalism of section 1.2 a more general statement follows: for the system in any state ψ, a measurement of A, giving the value a', followed immediately by a measurement of B, giving the value b', leaves the system in a state where an immediate remeasurement of A or B gives the same values a' or b' again, i.e. after the first two measurements we have a simultaneous knowledge of the values of A and of B. The proviso that the two measurements and the remeasurements be carried out immediately ensures that between the measurements the values of the observables have not altered due to the state of the system changing according to its equations of motion.

To demonstrate this result, we expand the general state ψ of the system in terms of the complete set ψ_{abk}:*

$$\psi = \sum_a \sum_b \sum_k c_{abk}\psi_{abk}. \tag{3.2a}$$

After a measurement of the observable A of the system in the state ψ, giving the value a' (this must be one of the eigenvalues of A), the system is in the state

$$\psi' \propto \sum_b \sum_k c_{a'bk}\psi_{a'bk} \tag{3.2b}$$

---

* I am assuming that we are dealing with a discrete set of eigenfunctions, i.e. the eigenvalues of A and B, as well as the values which the index k assumes, form discrete sets.

where the constant of proportionality is fixed by the normalization of ψ'. The result (3.2b) follows from our postulate (1.45)–(1.46). Similarly, measuring B on the system in the state ψ', with the result b', leaves the system in the state

$$\psi'' \propto \sum_k c_{a'b'k} \psi_{a'b'k}.$$ (3.2c)

ψ'' is a simultaneous eigenfunction of A and B with the eigenvalues a' and b' respectively. Remeasuring A or B on the system in the state of ψ'' necessarily gives the values a' or b' respectively. Thus for compatible observables, simultaneous knowledge of their values is always possible.

Our criterion for the compatibility of two observables, to ascertain whether they possess a complete set of common eigenfunctions, is not a very practical one. Life is made much simpler by a general theorem about observables. Quite generally, we define the *commutator* of two operators A and B by

$$[A, B] \equiv AB - BA.$$ (3.3)

Two operators for which the commutator vanishes,

$$[A, B] = 0,$$ (3.4a)

are said to be commuting operators. One can then prove the following theorem: two observables A and B possess a complete orthonormal set of simultaneous eigenfunctions if and only if they commute.

I shall relegate the proof (which the reader may omit) to the end of this section. Instead, we shall discuss its consequences. It states that two observables A and B are compatible if they are commuting observables, and vice versa. If A and B do not commute,

$$[A, B] \neq 0,$$ (3.4b)

they are incompatible. Furthermore, the theorem allows extension to more than two observables: if $A_1, A_2, \ldots, A_n$ are commuting observables, i.e.

$$[A_r, A_s] = 0, \qquad r, s = 1, \ldots, n,$$

they possess a complete orthonormal set of simultaneous eigenfunctions, i.e. they are compatible, and vice versa. This is a very powerful result: it is easy to evaluate commutators even in very complicated situations where one has no hope of obtaining the eigenfunctions.

To exploit this result, one needs some simple properties of commutators. If A, B and C are any operators, the following identities are easily verified:

$$[A, B] = -[B, A]$$ (3.5a)

$$[A, B + C] = [A, B] + [A, C]$$ (3.5b)

$$[A, BC] = [A, B]C + B[A, C]$$ (3.5c)

$$[A, [B, C]] + [B, [C, A]] + [C, [A, B]] = 0.$$ (3.5d)

Eq. (3.5d) is known as Jacobi's identity. In Eq. (3.5c) the order of the factors must be preserved, since in general we are dealing with non-commuting operators. From Eq. (3.5c), A commutes with $A^2 = AA$ and, more generally, with A^n. It follows that corresponding to any function $f(x)$ which can be expanded in a power series we can define an operator

$$f(A) = c_0 + c_1 A + c_2 A^2 + \cdots$$

and that $f(A)$ commutes with A:

$$[f(A), A] = 0. \qquad (3.5e)$$

Now for some examples. The Hamiltonian for a free particle

$$H = \frac{1}{2m}\,\hat{\mathbf{p}}^2 = -\frac{\hbar^2}{2m}\,\nabla^2$$

commutes with $\hat{\mathbf{p}}$. Hence the momentum eigenstates

$$u_{\mathbf{p}}(\mathbf{r}) \propto e^{i\mathbf{p}\cdot\mathbf{r}/\hbar}$$

are also energy eigenstates with energy $E = \mathbf{p}^2/2m$, as can be seen directly from the Schrödinger equation.

The position and momentum coordinates $\mathbf{r}$ (with components $x_1 = x$, $x_2 = y$, $x_3 = z$) and $\mathbf{p}$ (with components $p_1 = p_x$, $p_2 = p_y$, $p_3 = p_z$) satisfy the commutation relations

$$[x_r, x_s] = 0, \quad [p_r, p_s] = 0, \qquad r, s = 1, 2, 3 \qquad (3.6a)$$

$$[x_r, p_s] = i\hbar\delta_{rs}, \qquad r, s = 1, 2, 3. \qquad (3.6b)$$

These are easily derived. For example, with $p_r = -i\hbar\,\partial/\partial x_r$ and any function $\psi = \psi(\mathbf{r})$

$$[x_r, p_s]\psi = x_r\left(-i\hbar\,\frac{\partial\psi}{\partial x_s}\right) + i\hbar\,\frac{\partial}{\partial x_s}(x_r\psi)$$

$$= i\hbar\left(\frac{\partial x_r}{\partial x_s}\right)\psi = i\hbar\delta_{rs}\psi.$$

Since this relation holds for any function $\psi(\mathbf{r})$, the operator identity (3.6b) follows. The generalization of the commutation relations (3.6) to the case of N particles follows trivially from Eqs. (1.68).

Next consider the angular momentum of a particle $\mathbf{l} = \mathbf{r} \wedge \mathbf{p}$. From Eqs. (2.43c) and (2.44), $l_z = -i\hbar\,\partial/\partial\phi$ and $\mathbf{l}^2$ depends on ϕ through $\partial^2/\partial\phi^2$ only. Hence for any function $\psi(\mathbf{r})$

$$\{(\mathbf{l}^2)l_z - l_z(\mathbf{l}^2)\}\psi(\mathbf{r}) = 0$$

and consequently

$$[\mathbf{l}^2, l_z] = 0. \qquad (3.7)$$

Since $l^2 = l_x^2 + l_y^2 + l_z^2$, it follows from symmetry that (3.7) also implies

$$[l^2, l_x] = [l^2, l_y] = 0, \tag{3.7a}$$

and more generally the component of l in any direction commutes with l^2. In contrast, the components l_x, l_y, l_z do not commute with each other, e.g.

$$[l_x, l_y] \neq 0 \tag{3.8}$$

(see problem 3.1). Thus l^2 and any one component only of l are compatible. In section 2.3 we obtained this result from the explicit eigenfunctions.

For a particle in a central potential $V(r)$ the Hamiltonian is given by

$$H = \frac{\mathbf{p}^2}{2m} + V(r). \tag{3.9}$$

It follows from the commutation relations for $\mathbf{r}$ and $\mathbf{p}$, Eqs. (3.6), that neither $\mathbf{r}$ nor $\mathbf{p}$ is compatible with H. This is really obvious: the energy eigenfunctions are wave functions $\psi(\mathbf{r})$ which give a probability distribution of finding the particle in any region of space; except for $V(r) \equiv 0$, the eigenfunctions of H are not plane waves. On account of Eqs. (2.73) and (2.44), we can write the Hamiltonian (3.9) as

$$H = -\frac{\hbar^2}{2mr^2} \frac{\partial}{\partial r}\left(r^2 \frac{\partial}{\partial r}\right) + V(r) + \frac{l^2}{2mr^2}. \tag{3.10}$$

Since l^2, Eq. (2.44), depends on θ and ϕ but is independent of r we have for any radial function $f(r)$

$$[f(r), l^2] = 0 \tag{3.11}$$

and hence from Eq. (3.10) that

$$[H, l^2] = 0. \tag{3.12}$$

It follows from Eq. (3.10) and $[l^2, l_z] = 0$ that $l_z = -i\hbar \, \partial/\partial \phi$ also commutes with H:

$$[H, l_z] = 0. \tag{3.13}$$

Thus H, l^2 and l_z are compatible. l_z could again be replaced by any other component of l.* These results hold for any central potential. In section 2.5.1 we derived them from the explicit form of the angular momentum eigenfunctions.

---

* This follows most simply from the fact that for a central potential $V(r)$ the Hamiltonian $H = \mathbf{p}^2/2m + V(r)$ is spherically symmetric, i.e. no direction is distinguished; hence Eq. (3.13) implies that H and any component of l commute.

We can now answer the question how to determine the state of a system. We saw above how the measurement of an observable A, with the result a', restricts the system to be in a state ψ' of the form (3.2b). If a' is non-degenerate, the state is fully determined. If a' is a degenerate eigenvalue, measurement of a second observable B, compatible with A, with the result b', limits the system to a state ψ'' of the more restricted form (3.2c). If corresponding to the pair of eigenvalues a' and b' there exist several eigenfunctions, we continue this process and measure further observables $C, D, \ldots$, compatible with A, B and each other. At each stage the number of terms in the wavefunctions (3.2a), (3.2b), etc., is reduced. Eventually we end up with a set of commuting operators $A, B, \ldots, Q$, such that for any given set of eigenvalues $a', b', \ldots, q'$ of these observables there exists only one simultaneous eigenfunction. Measuring such a set of observables in quick succession then determines the state of the system completely. Such a set is called a complete (or maximal) set of commuting observables. For example, for the hydrogen atom, H, $\mathbf{l}^2$ and l_z represent such a set. (See section 2.5.2. We have neglected electron spin. Taking this into account, we would have to add a spin component as a further observable to have a maximal set.) To a given complete set of commuting observables $A, B, \ldots, Q$ we can always add further commuting operators $f_1(A)$, $f_2(B)$, $f_3(A, B), \ldots$, where $f_1, f_2, f_3, \ldots$ are arbitrary functions, but these provide no additional information about the state of the system and are really redundant. On the other hand, a maximal set only allows us to determine observables compatible with this set. Furthermore, for a given system there exist in general several complete sets of commuting observables, incompatible with each other. In the case of the hydrogen atom, discussed above, replacing l_z by l_x or l_y in $H, \mathbf{l}^2, l_z$ would give alternative complete sets of compatible observables.

★3.1.1   Compatibility and commutativity of observables

We shall now derive the theorem, stated on p. 73, that two observables A and B are compatible if and only if they commute.

(i) Assume the existence of a complete set of simultaneous eigenfunctions $\psi_1, \psi_2, \ldots$ of A and B with eigenvalues $a_1, a_2, \ldots$ and $b_1, b_2, \ldots$ respectively. Hence

$$[A, B]\psi_n = (AB - BA)\psi_n = (a_n b_n - b_n a_n)\psi_n = 0. \tag{3.14}$$

Since the ψ_n form a complete set, any state ψ of the system can be expanded as

$$\psi = \sum_n c_n \psi_n.$$

From Eq. (3.14)

$$[A, B]\psi = 0$$

follows, and since the last equation holds for any state ψ it implies

$$[A, B] = 0.$$

(ii) Assume that A and B commute. If ψ_a is an eigenfunction of A with the eigenvalue a, then

$$AB\psi_a = BA\psi_a = aB\psi_a. \tag{3.15}$$

Hence $B\psi_a$ is also an eigenfunction of A belonging to the same eigenvalue a. If a is a non-degenerate eigenvalue, $B\psi_a$ must be a multiple of ψ_a:

$$B\psi_a = b\psi_a \tag{3.16}$$

where b is a constant; i.e. ψ_a is an eigenfunction of B, with eigenvalue b, and is therefore a simultaneous eigenfunction of A and B. This proves the existence of a complete set of mutual eigenfunctions, provided all eigenvalues of A are non-degenerate, since the corresponding eigenfunctions of A form a complete set.

If the eigenvalue a is degenerate, Eq. (3.15) still holds but (3.16) no longer follows. Suppose a is v-fold degenerate and $\psi_{ak}, k = 1, 2, \ldots, v$, are v orthonormal eigenfunctions of A belonging to this eigenvalue. Any linear combination

$$\psi_a = \sum_{k=1}^{v} c_k \psi_{ak} \tag{3.17}$$

is also such an eigenfunction. We shall now show that one can use this freedom to choose the coefficients $c_1, c_2, \ldots, c_v$ in (3.17) so that ψ_a is also an eigenfunction of B:

$$B \sum_{k=1}^{v} c_k \psi_{ak} = b \sum_{k=1}^{v} c_k \psi_{ak}.$$

Taking the scalar product of this equation with $\psi_{al}, l = 1, \ldots, v$, gives

$$\sum_{k=1}^{v} c_k \int d^3r \psi_{al}^* B \psi_{ak} = b \sum_{k=1}^{v} c_k \int d^3r \psi_{al}^* \psi_{ak}, \quad * \quad v = 1, \ldots, l. \tag{3.18}$$

---

* For simplicity, we have written the scalar product $\int d^3r \ldots$, i.e. for a one-particle wave function. The argument easily generalizes. For a system of particles without spin one merely replaces the scalar products by the generalizations

$$\int d^3r_1 \ldots d^3r_N \psi_{al}^*(r_1, \ldots r_N)\psi_{ak}(r_1, \ldots r_N),$$

etc., which follow from section 1.4. For particles with spin, the appropriate generalization of the scalar product will be given later.

The integral on the right-hand side of (3.18) equals the Kronecker delta δ_{lk} since the ψ_{ak} are orthonormal. We introduce the notation

$$B_{lk} = \int d^3r \psi_{al}^* B \psi_{ak}.$$

Eqs. (3.18) then become

$$\sum_{k=1}^{v} (B_{lk} - b\delta_{lk})c_k = 0, \qquad l = 1, \ldots, v. \tag{3.19}$$

This set of v homogeneous equations in the unknowns $c_1, \ldots, c_v$ possesses non-zero solutions only if the coefficient determinant vanishes:

$$\det (B_{lk} - b\delta_{lk}) = 0. \tag{3.20}$$

This is a polynomial equation of degree v in b. Its solutions are eigenvalues of B. The solution of Eqs. (3.19) for each of these eigenvalues gives the corresponding eigenfunction (3.17). These are mutual eigenfunctions of A and B, establishing the above theorem for the case of degeneracy.

3.2   CONSTANTS OF THE MOTION

A fundamental feature of classical physics is the existence of conservation laws. For example, for a particle moving in a central potential, the energy and the angular momentum are conserved. Similar results hold in quantum physics, as we shall now see. We define an observable A as a *constant of the motion* of a system if

(i) A is compatible with the Hamiltonian H of the system:

$$[A, H] = 0, \tag{3.21}$$

and

(ii) $$\frac{\partial A}{\partial t} = 0. \tag{3.22}$$

This condition states that the observable A does not have an explicit time dependence. Most observables one deals with are of this kind; examples are the position and momentum coordinates, $\mathbf{r}$ and $\mathbf{p}$, of a particle, its angular momentum $\mathbf{r} \wedge \mathbf{p}$, and all observables which are functions of $\mathbf{r}$ and $\mathbf{p}$ only. It may seem surprising that one can describe the changing position and momentum of a particle by means of time-independent operators $\mathbf{r}$ and $\mathbf{p}$. The explanation is that in our formulation of quantum mechanics the time development of a system is described by time-dependent wave functions. For example, the time dependence of $\langle \mathbf{r} \rangle$ or $\langle \mathbf{p} \rangle$ has its origin in the time dependence of the wave functions which enter these expectation values, as we shall see in a moment. On the other hand, the Hamiltonian of a particle possessing a magnetic moment $\boldsymbol{\mu}$ and moving in a time-varying magnetic field $\mathbf{B}(t)$ contains the interaction energy $-\boldsymbol{\mu} \cdot \mathbf{B}(t)$ and so depends explicitly on the time: $H = H(t)$.

An observable A which satisfies the two conditions (3.21) and (3.22) has the important property that for *any* state of the system the expectation value $\langle A \rangle$ is constant in time:

$$\frac{\mathrm{d}}{\mathrm{d}t} \langle A \rangle = 0; \tag{3.23}$$

hence the name 'constant of the motion'.

To derive this result we calculate $\mathrm{d}\langle A \rangle / \mathrm{d}t$ for an observable $A = A(t)$, i.e. it may be explicitly time-dependent. The evolution in time of the state of the system, $\psi = \psi(\mathbf{r}, t)$, is given by the Schrödinger equation

$$i\hbar \frac{\partial \psi}{\partial t} = H(t)\psi. \tag{3.24}$$

As indicated, the Hamiltonian of the system may also be explicitly time-dependent. The time dependence of the expectation value

$$\langle A \rangle = \int \mathrm{d}^3\mathbf{r}\,\psi^*(\mathbf{r}, t)A(t)\psi(\mathbf{r}, t) \tag{3.25}$$

arises from that of the wave functions and from any explicit time dependence which A may possess. Differentiating Eq. (3.25) with respect to t and substituting in the resulting equation for $\partial\psi/\partial t$ and $\partial\psi^*/\partial t$ from Eq. (3.24) and its complex conjugate gives

$$\frac{\mathrm{d}}{\mathrm{d}t} \langle A \rangle = \int \mathrm{d}^3\mathbf{r} \left\{ \psi^* \frac{\partial A(t)}{\partial t} \psi \right.$$
$$\left. + \left[\frac{1}{i\hbar} H(t)\psi \right]^* A(t)\psi + \psi^* A(t) \left[\frac{1}{i\hbar} H(t)\psi \right] \right\}. \tag{3.26}$$

The middle term on the right-hand side of this equation can be written

$$\frac{-1}{i\hbar} \int \mathrm{d}^3\mathbf{r}[H(t)\psi]^* A(t)\psi = \frac{-1}{i\hbar} \int \mathrm{d}^3\mathbf{r}\,\psi^* H(t)A(t)\psi \tag{3.27}$$

since $H(t)$ is a Hermitian operator.* Substituting this equation in (3.26) and rearranging terms, we obtain

$$i\hbar \frac{\mathrm{d}}{\mathrm{d}t} \langle A \rangle = i\hbar \left\langle \frac{\partial A(t)}{\partial t} \right\rangle + \langle [A(t), H(t)] \rangle \tag{3.28}$$

---

* With $\psi_1 = \psi$ and $\psi_2 = A\psi$, Eq. (3.27) is just the Hermiticity condition (1.13) for H.

where $\langle O \rangle$ denotes the expectation value of the operator O in the state $\psi(\mathbf{r}, t)$. Although we derived Eq. (3.28) for a single-particle system, it holds in complete generality for more complicated systems with the expectation values interpreted accordingly, and its proof follows along the above lines. (Eq. (1.67) defines the expectation value for a system of particles without spin. The generalization to particles with spin will be considered later.) Eq. (3.28) gives the desired result: provided conditions (3.21) and (3.22) hold, $\langle A \rangle$ is constant in time for any state ψ of the system. In particular, if the Hamiltonian H of the system is not explicitly time-dependent, the energy is a constant of the motion. The analogous classical result states that for a system whose Hamiltonian does not depend explicitly on the time (known as a conservative system) energy is conserved.

A second important property of constants of the motion follows from the theorem in the last section. If A is a constant of the motion of a system, A and the Hamiltonian H of the system possess a complete set of simultaneous eigenfunctions. If H has no explicit time dependence—as we shall assume from now on—we see that if

$$H\phi(\mathbf{r}) = E\phi(\mathbf{r}) \quad \text{and} \quad A\phi(\mathbf{r}) = a\phi(\mathbf{r})$$

(i.e. $\phi(\mathbf{r})$ is a mutual eigenfunction of H and A with eigenvalues E and a), then

$$\psi(\mathbf{r}, t) = \phi(\mathbf{r}) \, e^{-iEt/\hbar} \tag{3.29}$$

is a simultaneous eigenfunction of H and A with the same eigenvalues E and a for all t.* If the system is in the state $\phi(\mathbf{r})$ at $t = 0$, it will be in the state (3.29) subsequently (unless disturbed), i.e. it will continue in an eigenstate of H and A with the eigenvalues E and a. For this reason, the eigenvalues of a constant of the motion are known as *good quantum numbers*. More generally, if $A, B, \ldots, Q$ is a complete set of commuting observables of a system which includes its Hamiltonian H, measurement of these observables determines a unique stationary state ψ which is a simultaneous eigenstate of these observables. Subsequent to the measurement, the system will continue in the state ψ and the observables will retain their original values.

In the last section we had several examples of observables which commute with the Hamiltonian of a system. We saw that for a particle moving in a central potential $V(r)$, the Hamiltonian H and the angular momentum operators $\mathbf{l}^2$ and l_z form a complete set of commuting observables; hence $\mathbf{l}^2$ and l_z are constants of the motion, and the angular momentum and magnetic quantum numbers l and m are good quantum numbers for this system. Classically, angular

---

* The time-dependent Schrödinger equation (3.24) reduces to the eigenvalue problem $H\phi = E\phi$ and the time dependence (3.29) only if H has no explicit time dependence.

momentum is, of course, also conserved in this case: we have

$$\frac{d}{dt}\mathbf{l} = \frac{d}{dt}(\mathbf{r} \wedge \mathbf{p}) = \frac{d\mathbf{r}}{dt} \wedge \mathbf{p} + \mathbf{r} \wedge \frac{d\mathbf{p}}{dt} = 0,$$

since $\mathbf{p} = m\,d\mathbf{r}/dt$ and

$$\frac{d\mathbf{p}}{dt} = -\nabla V(r) = -\frac{dV(r)}{dr}\frac{1}{r}\mathbf{r}.$$

Hence $\mathbf{l}$ is constant: classically, all components of $\mathbf{l}$ are conserved. Quantum-mechanically, we can make a statement about only one component of $\mathbf{l}$ (and $\mathbf{l}^2$), since different components of $\mathbf{l}$ are not compatible with each other.

Another example of a constant of the motion is the momentum $\mathbf{p}$ of a free particle. For a particle moving in a potential $V(\mathbf{r})$

$$[\mathbf{p}, H]\psi = [\mathbf{p}, V(\mathbf{r})]\psi = \{-i\hbar\nabla V(\mathbf{r})\}\psi$$

for *any* state ψ. Hence the operator identities

$$[\mathbf{p}, H] = [\mathbf{p}, V(\mathbf{r})] = -i\hbar\nabla V(\mathbf{r}) \tag{3.30}$$

follow. If the potential is constant in space, i.e. for a free particle, $[\mathbf{p}, H] = 0$ and momentum is conserved. The same result holds classically, since a particle moving in a constant potential experiences no force.

For an observable which is not a constant of the motion Eq. (3.28) gives the time development of its expectation value. We apply this equation to the position and momentum coordinates, $\mathbf{r}$ and $\mathbf{p}$, of a particle moving in a potential $V(\mathbf{r})$. From the commutation relations (3.6) and the commutator identities (3.5) one finds

$$[\mathbf{r}, H] = \frac{1}{2m}[\mathbf{r}, \mathbf{p}^2] = \frac{1}{2m}[\mathbf{r}, \mathbf{p}]\cdot\mathbf{p} + \frac{1}{2m}\mathbf{p}\cdot[\mathbf{r}, \mathbf{p}] = \frac{i\hbar}{m}\mathbf{p}.$$

From this equation, Eq. (3.30) and (3.28) one obtains

$$\frac{d}{dt}\langle\mathbf{r}\rangle = \frac{1}{m}\langle\mathbf{p}\rangle, \quad \frac{d}{dt}\langle\mathbf{p}\rangle = -\langle\nabla V(\mathbf{r})\rangle. \tag{3.31}$$

These equations, known as Ehrenfest's theorem, state that the expectation values obey the classical equations of motion. However, these equations are not the classical limit of the quantal equations. Eqs. (3.31) hold for any state ψ of the particle, and they are equations for expectation values about which dispersion occurs, i.e. the standard deviations $\Delta\mathbf{r}$ and $\Delta\mathbf{p}$ are not negligible in general. The classical limit only applies for states in which this dispersion is negligible, so that particles travel along well-defined trajectories, etc.*

---

* For a discussion of the classical limit see, for example, Gottfried, section 8.1.

3.3 THE UNCERTAINTY PRINCIPLE

We next consider incompatible observables. If the observables A and B do not commute, they do not possess a complete set of mutual eigenfunctions, and we cannot, in general, know simultaneously the values of A and B. The limitation imposed on our simultaneous knowledge of these values is given by Heisenberg's uncertainty principle.

Suppose the observables A and B satisfy the commutation relation

$$[A, B] = ic \tag{3.32}$$

where c is a constant. (From the Hermiticity of A and B, it follows that c is real. See problem 3.4.) For any state ψ of a system, we can take the standard deviations

$$\Delta A = \sqrt{\langle (A - \langle A \rangle)^2 \rangle}, \quad \Delta B = \sqrt{\langle (B - \langle B \rangle)^2 \rangle},$$

as measuring the uncertainties of the values of A and B in the state ψ. The uncertainty principle states that for any state ψ of the system

$$\Delta A \Delta B \geqslant \tfrac{1}{2}|c|. \tag{3.33}$$

Eq. (3.33) provides a lower limit on the product of the 'uncertainties' ΔA and ΔB. We see from Eq. (3.33) that the more sharply we know A, the less accurately we know B, and vice versa. In general the inequality holds. Only for certain very special states does (3.33) become an equality so that the product of the uncertainties is minimized. For $c = 0$, A and B are compatible, and $\Delta A = \Delta B = 0$ as required.

Eq. (3.33) follows trivially from the following more general inequality: for any state ψ of a system the product of the standard deviations of two observables A and B satisfies the inequality

$$\Delta A \Delta B \geqslant \tfrac{1}{2}|\langle [A, B] \rangle|. \tag{3.33a}$$

We defer the proof of this relation, which is straightforward and which the reader may omit, to the end of this section. Instead, we shall briefly discuss the uncertainty principle. It is essentially a negative statement, expressing the limitations imposed on our knowledge of the values of incompatible observables. In specifying systems and in calculating their properties, compatible observables are of much greater use. Nevertheless, the uncertainty principle is important, not only historically, but also in pinpointing the differences between classical and quantum physics, and in bringing out the essential limitations in our knowledge about a system. In section 2.4 we gave a purely qualitative analysis of successive Stern–Gerlach experiments and saw how a measurement of the component of angular momentum l_x obliterates any knowledge of l_z gained in an earlier experiment. The quantitative analysis of idealized experiments demonstrates how successive measurements of incompatible observables disturbs a system in such a way that knowledge about it is lost to a degree which ensures that the uncertainty principle is satisfied. This is so, however 'clever' the measuring apparatus one employs: one cannot beat the uncertainty principle. In essence this comes about because the measuring apparatus are themselves subject to the limitations of the uncertainty principle.

The most famous and most important examples of the uncertainty relations are those for the position and momentum coordinates, x_r and p_r, of a particle. These satisfy the commutation relations (3.6), and Eq. (3.33) becomes

$$\Delta x_r \Delta p_s \geqslant \tfrac{1}{2}\hbar \delta_{rs}, \qquad r, s = 1, 2, 3. \tag{3.34}$$

We can have exact knowledge of x and p_y, but not of x and p_x, etc. The uncertainty relations (3.34) allow a simple interpretation in terms of deBroglie waves. A momentum eigenstate is a plane monochromatic wave, i.e. it has constant amplitude throughout space, and the position of a particle in this state is completely undetermined. A localized state of the particle is described by a wave packet; this is a superposition of plane waves with different wave numbers, introducing a spread in momentum (see problem 3.7).

Another example of the position–momentum uncertainty relations (3.34) is provided by the deflection of particles which have passed through a small aperture in a screen. Again, this can be understood in terms of deBroglie waves; the smaller the aperture, the wider the diffraction pattern.*

An interesting point arises from the generalized uncertainty relation (3.33a). The orbital angular momentum operators **l** for a particle satisfy the commutation relations

$$[l_x, l_y] = i\hbar l_z \tag{3.35}$$

(see problem 3.1), whence

$$\Delta l_x \Delta l_y \geqslant \tfrac{1}{2}\hbar |\langle l_z \rangle|. \tag{3.36}$$

By cyclic permutations of x, y and z (i.e. $x \to y, y \to z, z \to x$) we obtain two more commutation relations like (3.35) and two more inequalities like (3.36). Although l_x and l_y do not commute, they can have simultaneous sharp values (i.e. $\Delta l_x = \Delta l_y = 0$) provided the right-hand side of (3.36) is zero. In particular, all three components have sharp values provided

$$\langle l_x \rangle = \langle l_y \rangle = \langle l_z \rangle = 0.$$

These equations hold for a particle in an s-state, and in such a state any component of **l** has the sharp value zero.

Although our examples of the uncertainty relations (3.33) and (3.33a) have been for single-particle systems, these relations are completely general. They

---

* For a detailed analysis of such experiments see, for example, Messiah, vol. I, pp. 139–149, or Heisenberg, Chapters 2 and 3.

also hold for systems of any number of particles which may possess internal degrees of freedom such as spin. One need only interpret the expectation values in the appropriately generalized fashion.

There exists also a relation known as the time–energy uncertainty relation. This name is unfortunate as *the origin and meaning of this relation is completely different* from what we have been discussing so far, namely incompatible observables represented by non-commuting operators. However, time is not an operator; it is an ordinary parameter which commutes with *all* observables of a system; in particular with the energy operator, i.e. the Hamiltonian of the system. Energy, momentum, etc., are properties, i.e. observables, of a system; time is not. We can measure the position or momentum of a particle; we do not measure the time of a particle. Rather, time is a parameter which identifies the instant at which we specify a property of a system.

The time–energy uncertainty relation relates the *rate at which the state of a system changes* to the *uncertainty δE of its energy*. If the state of the system changes appreciably during a time interval δt, then the time–energy uncertainty relation states that

$$\delta t \, \delta E \gtrsim \hbar. \tag{3.37}$$

I have written δt and δE, rather than Δt and ΔE, to emphasize that these are not standard deviations.

For a stationary state, the relation (3.37) is satisfied. A stationary state is an energy eigenstate, so that $\delta E = 0$. From section 1.3 we know that the physical properties of a system in a stationary state do not change with time, i.e. $\delta t = \infty$, which is consistent with (3.37).

We next show how the uncertainty relation (3.37) follows for the case, already considered in section 1.3, of a state $\psi(\mathbf{r}, t)$ which is a superposition of two stationary states

$$\psi(\mathbf{r}, t) = c_1 \phi_1(\mathbf{r}) \, e^{-iE_1 t/\hbar} + c_2 \phi_2(\mathbf{r}) \, e^{-iE_2 t/\hbar}. \tag{1.59}$$

In this case we know from Eq. (1.63) that the physical properties of the system depend on the time through the exponential function

$$\exp\left(i \frac{E_1 - E_2}{\hbar} t \right).$$

For time intervals $\delta t \ll \hbar/|E_1 - E_2|$, the properties do not change significantly. For appreciable changes to occur we require

$$\delta t \gtrsim \hbar/|E_1 - E_2|. \tag{3.38a}$$

A measurement of the energy in the state (1.59) gives either the value E_1 or E_2. Hence

$$\delta E = |E_1 - E_2| \tag{3.38b}$$

measures the energy uncertainty in this state. Combining the last two equations leads to the time–energy relation (3.37). The generalization of this result to the superposition of more than two stationary states is obvious.

I want to stress again that the nature of the time–energy uncertainty relation (3.37) is utterly different from that of the uncertainty relations (3.33)–(3.33a) for non-commuting observables. I hope the above discussion has highlighted this difference. It should also be noted that in a fully relativistic quantum field theory, where the position coordinate $\mathbf{r}$ and the time t stand on an equal footing (they form a four-vector), it is not the time coordinate t which becomes an operator; rather the position coordinate $\mathbf{r}$ is demoted to become an ordinary parameter, like t. $\mathbf{r}$ and t together label points in space–time, and it is the fields (e.g. the electromagnetic field) at these points which are the observables.

The most important example of the relation (3.37) occurs for the decay of a metastable state. Consider an excited state of an atom emitting a photon and decaying to the atomic ground state. The latter is a stable stationary state. The former is only approximately stable insofar as we neglect the possibility of photon emission, but strictly speaking it is unstable with a finite lifetime τ, i.e. it is not a stationary state with a sharp energy but a superposition of stationary states of different energies with an energy spread δE. The detailed calculation* shows that

$$\tau \delta E \sim \hbar.$$

As a consequence of this uncertainty δE in the energy of the excited state, the emitted photon has an energy spread δE, i.e. the spectral line has a non-zero width $\delta E \sim \hbar/\tau$: the shorter the lifetime of the excited state, the broader the spectral line. $\hbar/\tau$ is known as the natural width of the line. It will be considered further in section 9.5.3.

★3.3.1   Proof of the uncertainty principle

We derive the generalized uncertainty relation (3.33a) for a single-particle system. Let $\psi = \psi(\mathbf{r}, t)$ be any state of this system. If the operators A and B are Hermitian, so are the operators

$$\alpha = A - \langle A \rangle, \quad \beta = B - \langle B \rangle.$$

For any state $\phi = \phi(\mathbf{r}, t)$

$$\int d^3 r \, \phi^* \phi \geqslant 0. \tag{3.39}$$

---

* See, for example, Davydov, section 80.

Let

$$\phi = (\alpha + i\lambda\beta)\psi$$

where λ is a real parameter. Eq. (3.39) then becomes

$$\int d^3\mathbf{r}[(\alpha + i\lambda\beta)\psi]^*[(\alpha + i\lambda\beta)\psi] = \int d^3\mathbf{r}\psi^*(\alpha - i\lambda\beta)(\alpha + i\lambda\beta)\psi \geqslant 0$$

where the middle expression follows from the Hermiticity of α and β. The last inequality can be written

$$\lambda^2\langle\beta^2\rangle - \lambda\langle\gamma\rangle + \langle\alpha^2\rangle \geqslant 0 \qquad (3.40)$$

where the operator γ is defined by

$$\gamma = -i[\alpha, \beta] = -i[A, B]. \qquad (3.41)$$

We rewrite Eq. (3.40) in the form

$$\langle\beta^2\rangle\left[\lambda - \frac{\langle\gamma\rangle}{2\langle\beta^2\rangle}\right]^2 + \frac{1}{4\langle\beta^2\rangle}[4\langle\alpha^2\rangle\langle\beta^2\rangle - \langle\gamma\rangle^2] \geqslant 0. \qquad (3.42)$$

Since the operator γ is Hermitian (see problem 3.4), $\langle\gamma\rangle$ is real. For the inequality (3.42) to hold for all real values of λ, we require

$$4\langle\alpha^2\rangle\langle\beta^2\rangle \geqslant \langle\gamma\rangle^2. \qquad (3.43)$$

The standard deviations ΔA and ΔB are given by

$$\Delta A = \sqrt{\langle\alpha^2\rangle}, \quad \Delta B = \sqrt{\langle\beta^2\rangle}.$$

Hence, taking the square root of Eq. (3.43) we obtain

$$\Delta A \Delta B \geqslant \tfrac{1}{2}|\langle\gamma\rangle|,$$

and substituting expression (3.41) for γ in this inequality gives the generalized uncertainty relation

$$\Delta A \Delta B \geqslant \tfrac{1}{2}|\langle[A, B]\rangle|. \qquad (3.33)$$

Eq. (3.33a) was derived for a single-particle system. The extension to more complicated systems is trivial. Eq. (3.33a), like Eq. (3.40) from which it was derived, contains expectation values only. For more complicated systems, one need only interpret these expectation values in the appropriately generalized manner.

PROBLEMS 3

3.1 From the commutation relations (3.6) for the position and momentum coordinates of a particle, derive the following commutation relations for the angular momentum operators $\mathbf{l} = \mathbf{r} \wedge \mathbf{p}$:

$$[l_x, x] = 0, \quad [l_x, y] = i\hbar z, \quad [l_x, z] = -i\hbar y, \qquad (3.44a)$$

$$[l_x, p_x] = 0, \quad [l_x, p_y] = i\hbar p_z, \quad [l_x, p_z] = -i\hbar p_y \qquad (3.44b)$$

$$[l_x, l_y] = i\hbar l_z \qquad (3.44c)$$

$$[l_x, \mathbf{l}^2] = [l_x, \mathbf{r}^2] = [l_x, \mathbf{p}^2] = 0. \qquad (3.44d)$$

All other angular momentum commutation relations follow from these equations by cyclic permutations of x, y and z, i.e. $x \to y$, $y \to z$, $z \to x$.

3.2 Derive constants of the motion for a particle moving in three dimensions in each of the following potentials:

   (i) $V_1(\mathbf{r}) = V(\rho)$,   where $\rho = (x^2 + y^2)^{1/2}$,

   (ii) $V_2(\mathbf{r}) = Az$,   where A is a constant,

   (iii) $V_3(\mathbf{r}) = V(r) + Bl_x$,   where B is a constant and l_x is the x-component of the angular momentum $\mathbf{r} \wedge \mathbf{p}$.

3.3 The Hamiltonian H of two interacting particles is given by

$$H = \frac{1}{2m_1} \mathbf{p}_1^2 + \frac{1}{2m_2} \mathbf{p}_2^2 + V(|\mathbf{r}_1 - \mathbf{r}_2|). \qquad (3.45)$$

From the commutation relations of the operators $\mathbf{r}_n$ and $\mathbf{p}_n = -i\hbar\nabla_n$, $n = 1, 2$, show that the total linear momentum $\mathbf{P} = \mathbf{p}_1 + \mathbf{p}_2$ is a constant of the motion.

Derive the commutation relations of the centre-of-mass and relative position and momentum operators $\mathbf{R}$, $\mathbf{r}$, $\mathbf{P}$ and $\mathbf{p}$, defined in problem 1.5 and its solution. Use the result of problem 1.5 to show that $[\mathbf{P}, H] = 0$.

3.4 If A and B are Hermitian operators, show that the operator $C = i[A, B]$ is also Hermitian.

3.5 $\psi(x)$ and E are the energy eigenfunction and eigenvalue of the ground state of the one-dimensional harmonic oscillator, i.e.

$$H\psi(x) = \left[\frac{1}{2m} p_x^2 + \tfrac{1}{2}m\omega^2 x^2 \right] \psi(x) = E\psi(x), \qquad -\infty < x < \infty,$$

where $p_x = -i\hbar\, d/dx$. $\psi(x)$ is an even function: $\psi(-x) = \psi(x)$. Obtain an expression for E in terms of the standard deviations of x and p_x. Estimate E by minimizing this expression in a way consistent with the uncertainty principle.

3.6 The normalized Gaussian wave function

$$\psi(x) = \left(\frac{a}{\pi} \right)^{1/4} e^{-ax^2/2}, \qquad -\infty < x < \infty, \qquad (3.46)$$

describes a one-dimensional wave packet, centred at $x = 0$. Calculate the standard deviation of x in this state.

Calculate the mean value and the standard deviation of the momentum p_x of the particle in this state and relate your results to the uncertainty principle.

$$\left(\int_{-\infty}^{\infty} dx\, e^{-ax^2} = \left(\frac{\pi}{a} \right)^{1/2}, \quad \int_{-\infty}^{\infty} dx\, x^2\, e^{-ax^2} = \frac{1}{2a} \left(\frac{\pi}{a} \right)^{1/2}. \right)$$

3.7 Use the results of section 2.6.2 to expand the Gaussian wave function $\psi(x)$, Eq. (3.46), as a superposition of plane waves, and hence obtain the momentum distribution for a particle in the state $\psi(x)$, i.e. calculate the probability $|\phi(p)|^2 \, \mathrm{d}p$ that the momentum p_x has a value in the range p to $p + \mathrm{d}p$.

$$\left(\int_{-\infty}^{\infty} \mathrm{d}x \, \mathrm{e}^{-Ax^2 - iBx} = \left(\frac{\pi}{A} \right)^{1/2} \mathrm{e}^{-B^2/4A}, \qquad A > 0. \right)$$

Symmetries

In the last chapter we saw that conserved quantities are characterized by observables which commute with the Hamiltonian of a system. In this chapter we shall gain an understanding of the origin of conservation laws. We shall show that they stem from the symmetries, i.e. the invariance properties, of a system. This connection is also of great practical importance. The generality of the arguments permits definitive deductions to be made about systems which are too complicated for exact calculations or for which the forces are not even known precisely; it suffices to know the symmetry properties of the forces without knowing the exact law of force. To exploit these ideas fully for complicated systems one requires group theory. We shall see that even without the use of group theory these ideas allow important results of great generality to be derived by means of simple arguments without elaborate calculations.

4.1 INVERSION

As a first example of a symmetry operation we shall consider inversion. Although this is not the most familiar symmetry in classical physics, it is very important in quantum physics and its mathematics is particularly simple. The operation of inversion is defined as the transformation

$$\mathbf{r} \to \mathbf{r}' = -\mathbf{r}. \qquad (4.1)$$

If the vectors $\mathbf{r}$ and $\mathbf{r}'$ have Cartesian components (x, y, z) and (x', y', z') respectively, we can write (4.1)

$$(x, y, z) \to (x', y', z') = (-x, -y, -z). \qquad (4.1a)$$

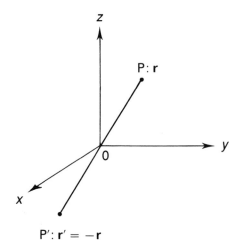

Fig. 4.1   Inversion through the origin transforms the point P into the point P'.

We can interpret this transformation in two ways. We can think of **r** and **r'** as labelling two different points P and P' in the same coordinate system (Fig. 4.1), or as labelling the same point P in two different coordinate systems (Fig. 4.2). Inversion turns a right-handed coordinate system (as in Fig. 4.2a) into a left-handed one (as in Fig. 4.2b). It is a discontinuous transformation: a system is either right- or left-handed, and we cannot go from the one to the other in a continuous way, as would be the case if we were considering rotations.

To study the symmetry properties of a system under inversion, we consider its Hamiltonian, which we shall call $H(\mathbf{r})$ when it is expressed in terms of the coordinates **r**. Since from Eq. (4.1) $\mathbf{r} = -\mathbf{r'}$, it follows that

$$H(\mathbf{r}) = H(-\mathbf{r'}). \qquad (4.2a)$$

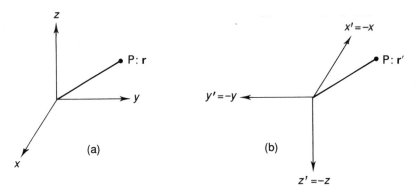

Fig. 4.2   Inversion of the coordinate system. **r** and **r'** label the same point P in the two systems obtained from each other by inversion.

We shall call the Hamiltonian of the system, when expressed in terms of the coordinates $\mathbf{r}'$, $H'(\mathbf{r}')$:

$$H'(\mathbf{r}') = H(-\mathbf{r}').   \tag{4.2b}$$

This equation *defines* the function $H'(\mathbf{r}')$. Combining the last two equations, we can also write

$$H'(\mathbf{r}') = H(\mathbf{r}).   \tag{4.2c}$$

Eqs. (4.2) always hold. They represent a transformation of variables. In particular, Eq. (4.2c) merely expresses the Hamiltonian in terms of two different sets of variables. The Hamiltonians H and H' are, of course, in general *different functions*,

$$H'(\mathbf{r}) \neq H(\mathbf{r}),$$

and the equality sign will hold only in special circumstances. A system is called invariant under inversion if its Hamiltonian is invariant under the transformation (4.1), i.e. if

$$H'(\mathbf{r}) = H(\mathbf{r}).   \tag{4.3a}$$

Eq. (4.3a) states that the transformed function $H'(\mathbf{r})$ has the same functional form as the original Hamiltonian $H(\mathbf{r})$. (We shall have an example in a moment.) Substituting the definition (4.2b) of H', with $\mathbf{r}'$ replaced by $\mathbf{r}$, in Eq. (4.3a) it follows that for a Hamiltonian which is invariant under inversion

$$H(\mathbf{r}) = H(-\mathbf{r}).   \tag{4.3b}$$

Eqs. (4.3a–b) are alternative forms of the invariance criterion.

We illustrate these ideas for two simple examples. First, consider a particle of mass m moving in a potential $V(\mathbf{r})$. The Hamiltonian of this system is given by

$$H(\mathbf{r}) = -\frac{\hbar^2}{2m} \nabla^2 + V(\mathbf{r}).   \tag{4.4}$$

The kinetic energy operator

$$-\frac{\hbar^2}{2m}\left(\frac{\partial^2}{\partial x^2} + \frac{\partial^2}{\partial y^2} + \frac{\partial^2}{\partial z^2} \right)$$

is obviously invariant if (x, y, z) is replaced by $(-x, -y, -z)$. Hence $H(\mathbf{r})$, Eq. (4.4), is invariant if $V(\mathbf{r})$ is, i.e. if

$$V(-\mathbf{r}) = V(\mathbf{r}).$$

A central potential $V(r) = V(|\mathbf{r}|)$ is an important example of a potential invariant under inversion.

Next, consider a potential of the form

$$V(\mathbf{r}) = V_1(|\mathbf{r} + \mathbf{a}|) + V_2(|\mathbf{r} - \mathbf{a}|)   \tag{4.5a}$$

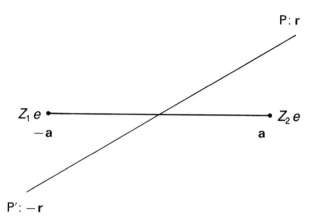

Fig. 4.3   Inversion for a two-centre system. With point charges Z_1e and Z_2e at $\mp\mathbf{a}$, an electron will experience the same potential at the points P and P' if $Z_1 = Z_2$.

i.e. a particle subject to forces from two different centres of force situated at different positions $\pm\mathbf{a}$. Under inversion, Eq. (4.5a) becomes

$$V(-\mathbf{r}) = V_1(|\mathbf{r} - \mathbf{a}|) + V_2(|\mathbf{r} + \mathbf{a}|) \qquad (4.5b)$$

and $V(-\mathbf{r}) = V(\mathbf{r})$ only if $V_1 = V_2$. An example of the potential (4.5a) is provided by an electron moving in the Coulomb fields of two point charges Z_1e and Z_2e, positioned at $-\mathbf{a}$ and $\mathbf{a}$ respectively (Fig. 4.3). The electron situated at P and P' would experience the same forces only if $Z_1 = Z_2$.

We next derive the special properties of a system which is invariant under inversion. Suppose $\psi(\mathbf{r})$ is an energy eigenfunction with the energy eigenvalue E of a Hamiltonian $H(\mathbf{r})$ which is invariant under inversion:

$$H(\mathbf{r})\psi(\mathbf{r}) = E\psi(\mathbf{r}). \qquad (4.6)$$

With $\mathbf{r}' = -\mathbf{r}$, Eq. (4.6) becomes

$$H(-\mathbf{r}')\psi(-\mathbf{r}') = E\psi(-\mathbf{r}'),$$

and, since $H(-\mathbf{r}') = H(\mathbf{r}')$, we can write the last equation

$$H(\mathbf{r}')\psi(-\mathbf{r}') = E\psi(-\mathbf{r}').$$

In this equation we can again write $\mathbf{r}$ instead of $\mathbf{r}'$, since this is merely a label for a general point, giving

$$H(\mathbf{r})\psi(-\mathbf{r}) = E\psi(-\mathbf{r}). \qquad (4.7)$$

Eqs. (4.6) and (4.7) show that if $H(\mathbf{r})$ is invariant under inversion, then if $\psi(\mathbf{r})$ is an eigenfunction of H with the eigenvalue E, so is $\psi(-\mathbf{r})$. Two cases arise:

(i) The eigenvalue E is non-degenerate. In this case $\psi(\mathbf{r})$ and $\psi(-\mathbf{r})$ are essentially the same functions; they can differ at most by a constant factor which we shall call Π:

$$\psi(-\mathbf{r}) = \Pi\psi(\mathbf{r}). \tag{4.8}$$

Replacing $\mathbf{r}$ by $-\mathbf{r}$ in this equation gives

$$\psi(\mathbf{r}) = \Pi\psi(-\mathbf{r}),$$

and combining the last two equations we have

$$\psi(\mathbf{r}) = \Pi^2\psi(\mathbf{r}).$$

Hence $\Pi^2 = 1$, Π can only take on one of the values

$$\Pi = \pm 1, \tag{4.9}$$

and

$$\psi(-\mathbf{r}) = \pm\psi(\mathbf{r}). \tag{4.10}$$

Thus, if the eigenvalue E is non-degenerate, the eigenfunction $\psi(\mathbf{r})$ belonging to E is an even or an odd function of $\mathbf{r}$. One refers to the even wave functions as having even parity (or as having parity $+1$ or positive parity), and to the odd wave functions as having odd parity (or as having parity -1 or negative parity).

(ii) If the eigenvalue E is degenerate, Eqs. (4.6) and (4.7) still hold, but $\psi(\mathbf{r})$ and $\psi(-\mathbf{r})$ can differ significantly, i.e. not only by a multiplicative factor. In this case, any linear combination

$$a\psi(\mathbf{r}) + b\psi(-\mathbf{r})$$

(with a and b constants) is also an eigenfunction of $H(\mathbf{r})$, belonging to the same eigenvalue E. One can use this freedom to choose linear combinations which are even and odd parity states, namely

$$\psi(\mathbf{r}) \pm \psi(-\mathbf{r}). \tag{4.11}$$

We therefore conclude that in the case of degeneracy too we can make the energy eigenfunctions have definite parity by choosing appropriate linear combinations of the degenerate energy eigenfunctions.

There exists an extremely useful alternative formulation of these ideas. We define the parity operator P as follows: operating on any wave function $f(\mathbf{r})$, P transforms it into a new wave function

$$f'(\mathbf{r}) = Pf(\mathbf{r}) \tag{4.12a}$$

which is defined by

$$f'(\mathbf{r}) = Pf(\mathbf{r}) = f(-\mathbf{r}). \tag{4.12b}$$

Substituting $\mathbf{r} = -\mathbf{r}'$ in $f(\mathbf{r})$, it follows from Eq. (4.12b) that

$$f(\mathbf{r}) = f(-\mathbf{r}') = f'(\mathbf{r}'), \tag{4.12c}$$

i.e. $f'(\mathbf{r}')$ is the function of $\mathbf{r}'$ into which $f(\mathbf{r})$ transforms when one substitutes $\mathbf{r} = -\mathbf{r}'$ in $f(\mathbf{r})$. It follows from Eq. (4.12b) that

$$P^2 f(\mathbf{r}) = P f(-\mathbf{r}) = f(\mathbf{r}),$$

and since this equation holds for *any* wave function $f(\mathbf{r})$, we must have

$$P^2 = 1: \tag{4.13}$$

operating on *any* wave function $f(\mathbf{r})$ with P^2 is the same as multiplying it by 1, i.e. leaving it alone. (This is what we would expect: carrying out inversion twice gets us back to where we started from.) If Π is an eigenvalue of the operator P, then P^2 has the eigenvalue Π^2. But from Eq. (4.13) $\Pi^2 = 1$; hence $\Pi = \pm 1$, i.e. the only eigenvalues of the parity operator P are $+1$ and -1. If $\psi(\mathbf{r})$ is a corresponding eigenfunction, i.e.

$$P\psi(\mathbf{r}) = \pm \psi(\mathbf{r}), \tag{4.14}$$

it follows from $P\psi(\mathbf{r}) = \psi(-\mathbf{r})$ that

$$\psi(-\mathbf{r}) = \pm \psi(\mathbf{r}), \tag{4.15}$$

i.e. parity eigenstates with eigenvalue $+1$ (-1) are even (odd) functions of $\mathbf{r}$.

Generally, for any operator A, we define the inverse operator A^{-1} by

$$A^{-1}A = AA^{-1} = 1; \tag{4.16}$$

thus if $\psi = A\phi$, then

$$A^{-1}\psi = A^{-1}A\phi = \phi. \tag{4.17}$$

For the parity operator we have $P^2 = 1$, Eq. (4.13), and hence

$$P^{-1} = P. \tag{4.18}$$

Let us next operate with the parity operator P on the wave function

$$f(\mathbf{r}) = H(\mathbf{r})\psi(\mathbf{r}),$$

where $\psi(\mathbf{r})$ is *any* wave function and we are *not* assuming that $H(\mathbf{r})$ is invariant under inversion. From the definition (4.12b) of P we have

$$PH(\mathbf{r})\psi(\mathbf{r}) = H(-\mathbf{r})\psi(-\mathbf{r}). \tag{4.19a}$$

Since $P^{-1}P = 1$, we can also write the left-hand side of this equation

$$PH(\mathbf{r})P^{-1}P\psi(\mathbf{r}) = PH(\mathbf{r})P^{-1}\psi(-\mathbf{r}), \tag{4.19b}$$

and combining Eqs. (4.19a–b) we obtain

$$PH(\mathbf{r})P^{-1}\psi(-\mathbf{r}) = H(-\mathbf{r})\psi(-\mathbf{r}).$$

This equation holds for *any* wave function $\psi(\mathbf{r})$; hence we must have

$$PH(\mathbf{r})P^{-1} = H(-\mathbf{r}). \qquad (4.20)$$

In deriving Eq. (4.20) we did not assume invariance of $H(\mathbf{r})$ under inversion. If $H(\mathbf{r})$ is invariant, i.e. if $H(-\mathbf{r}) = H(\mathbf{r})$, Eq. (4.20) becomes

$$PH(\mathbf{r})P^{-1} = H(\mathbf{r}). \qquad (4.21a)$$

Postmultiplying this equation by P, we can write it

$$PH(\mathbf{r}) = H(\mathbf{r})P$$

or

$$[P, H(\mathbf{r})] = 0. \qquad (4.21b)$$

We have thus shown that if $H(\mathbf{r})$ is invariant under inversion, it commutes with the parity operator P. We know from the last chapter that this is precisely the condition for H and P to possess a complete set of mutual eigenfunctions, i.e. the energy eigenfunctions can be chosen to be states of definite parity. This is the result which we derived from first principles earlier in this section. It also follows that if $H(\mathbf{r})$ is invariant under inversion, parity is a constant of the motion.

The extension of the above results to a system of N particles is straightforward. In the above treatment we need only interpret $\mathbf{r}$ to stand for the set of coordinates $\mathbf{r}_1, \mathbf{r}_2, \ldots, \mathbf{r}_N$ of the N particles, i.e. we make the replacements

$$H(\mathbf{r}) \to H(\mathbf{r}_1, \ldots, \mathbf{r}_N), \quad \psi(\mathbf{r}) \to \psi(\mathbf{r}_1, \ldots, \mathbf{r}_N).$$

Under inversion we now have

$$\mathbf{r}_i \to \mathbf{r}'_i = -\mathbf{r}_i, \quad i = 1, \ldots, N. \qquad (4.22)$$

The parity operator P, defined analogously to Eq. (4.12b) by

$$Pf(\mathbf{r}_1, \ldots, \mathbf{r}_N) = f(-\mathbf{r}_1, \ldots, -\mathbf{r}_N)$$

for all wave functions f, has the eigenvalues $+1$ and -1 as before. If H is invariant under inversion, i.e.

$$H(-\mathbf{r}_1, \ldots, -\mathbf{r}_N) = H(\mathbf{r}_1, \ldots, \mathbf{r}_N),$$

one proves, as before, that H and P commute so that the energy eigenfunctions can be chosen as parity eigenstates.

These ideas have wide and important applications which we shall now illustrate.

The Hamiltonian of an atom with atomic number Z is given by

$$H(\mathbf{r}_1, \ldots, \mathbf{r}_Z) = -\frac{\hbar^2}{2m} \sum_{i=1}^{Z} \nabla_i^2 + \sum_{i=1}^{Z} \frac{-Ze^2}{4\pi\varepsilon_0 r_i} + \sum_{\substack{i,j=1 \\ i<j}}^{Z} \frac{e^2}{4\pi\varepsilon_0 |\mathbf{r}_i - \mathbf{r}_j|}. \quad (4.23)$$

This is obviously invariant under inversion, so that the energy eigenstates can be classified into positive and negative parity states.

Classically, the atomic electric dipole moment is defined by

$$\sum_{i=1}^{Z} -e\mathbf{r}_i. \quad (4.24)$$

Quantum-mechanically, (4.24) is interpreted as an operator, and the electric dipole moment $\mathbf{D}$ of the atom in the state $\psi(\mathbf{r}_1, \ldots, \mathbf{r}_Z)$ is defined as the expectation value of (4.24) in this state:

$$\mathbf{D} = \int d^3 r_1 \ldots d^3 r_Z \psi^*(\mathbf{r}_1, \ldots, \mathbf{r}_Z) \left[\sum_{i=1}^{Z} -e\mathbf{r}_i \right] \psi(\mathbf{r}_1, \ldots, \mathbf{r}_Z). \quad * \quad (4.25)$$

If the atomic state ψ has definite parity, the dipole moment (4.25) vanishes: $\mathbf{D} = 0$. To prove this, we introduce new variables of integration

$$\mathbf{r}_i' = -\mathbf{r}_i, \quad i = 1, \ldots, Z, \quad (4.26)$$

in Eq. (4.25) which becomes

$$\mathbf{D} = \int d^3 r_1' \ldots d^3 r_Z' \psi^*(-\mathbf{r}_1', \ldots, -\mathbf{r}_Z') \left[\sum_{i=1}^{Z} e\mathbf{r}_i' \right] \psi(-\mathbf{r}_1', \ldots, -\mathbf{r}_Z'). \quad (4.27)$$

If the state ψ has the parity Π (equal to $+1$ or -1), i.e.

$$\psi(-\mathbf{r}_1, \ldots, -\mathbf{r}_Z) = \Pi\psi(\mathbf{r}_1, \ldots, \mathbf{r}_Z), \quad \Pi = \pm 1,$$

then the right-hand side of (4.27) becomes

$$(-1)\Pi^2\mathbf{D} = -\mathbf{D}$$

and hence $\mathbf{D} = 0$. It is clear from the derivation of this result that it applies equally to nuclei. The nuclear electric dipole moment operator is given by $\Sigma e\mathbf{r}_i$, where the summation is over the proton coordinates only, and in a nuclear state of definite parity the nuclear electric dipole moment vanishes.

Next, let us look at diatomic molecules. In a molecule, the nuclei as well as the electrons move. However, because the nuclear masses are many thousand times larger than the electron's mass, the motion of the nuclei is very slow

---

* The atomic wave functions are not only functions of the space coordinates; they also depend on the electron spins. The spins are not relevant here. We shall be considering inversion, and inversion affects only the space coordinates but not the spins.

compared to that of the electrons, and as a first approximation one can treat the electrons as moving in the fields of stationary nuclei, fixed at their equilibrium positions. In this approximation, the Hamiltonian for the electrons of a diatomic molecule is given by

$$
H(\mathbf{r}_1, \ldots, \mathbf{r}_N) = -\frac{\hbar^2}{2m} \sum_{i=1}^{N} \nabla_i^2 + \sum_{i=1}^{N} \frac{-Z_1 e^2}{4\pi\varepsilon_0 |\mathbf{r}_i + \mathbf{a}|}
$$

$$
+ \sum_{i=1}^{N} \frac{-Z_2 e^2}{4\pi\varepsilon_0 |\mathbf{r}_i - \mathbf{a}|} + \sum_{\substack{i,j=1 \\ i<j}}^{N} \frac{e^2}{4\pi\varepsilon_0 |\mathbf{r}_i - \mathbf{r}_j|} \tag{4.28}
$$

where Z_1 and Z_2 are the atomic numbers of the two atoms of the molecule, $N = Z_1 + Z_2$, and $\mp\mathbf{a}$ are the positions of the two nuclei. Eq. (4.28) is called the electronic Hamiltonian of the molecule. The electron–electron repulsion term in Eq. (4.28) is clearly invariant under the inversion operation (4.22), and Eq. (4.28) has the same symmetry as Eq. (4.5a). It is invariant under inversion if $Z_1 = Z_2$, i.e. for homonuclear diatomic molecules such as N_2 or O_2, but not if $Z_1 \neq Z_2$, i.e. for heteronuclear diatomic molecules such as NO. Thus for homonuclear diatomic molecules the energy eigenstates can be chosen as parity eigenstates and these have zero electric dipole moment.

We have seen that for heteronuclear diatomic molecules parity is not a good quantum number. This is due to the lopsidedness of our system: on inversion an electron experiences a different environment: if originally it was close to the nucleus with charge $Z_1 e$, it will be close to the other nucleus, with charge $Z_2 e$, after inversion. It is the geometry of our system which is asymmetric. The Coulomb potential between two point charges depends only on the magnitude of the distance between them and is invariant under inversion. More generally, all electromagnetic interactions are invariant under inversion, as are the strong interactions which are responsible for nuclear forces. (In fact, we implied this above when we said that nuclei in states of definite parity have zero electric dipole moments.) In contrast, the weak interactions, which are responsible for beta-decay, are themselves asymmetric: under inversion, the law of force itself changes. The asymmetry is at a much more fundamental level here.*

An interesting and important application of parity, involving the dipole operator (4.24), occurs when studying the emission or absorption of electromagnetic radiation by an atom making a transition from an initial state i, with wave function $\psi_i(\mathbf{r}_1, \ldots, \mathbf{r}_Z)$, to a final state f, with wave function $\psi_f(\mathbf{r}_1, \ldots, \mathbf{r}_Z)$. The optical line spectra of atoms arise in this way. In the

---

* A splendid discussion of these questions is given in Feynman I, Chapter 52.

quantum-mechanical analysis of this transition* one shows that the dominant contribution arises from an expression like the electric dipole moment (4.25) with the wave functions ψ^* and ψ replaced by ψ_f^* and ψ_i:

$$\mathbf{D}_{fi} = \int d^3\mathbf{r}_1 \ldots d^3\mathbf{r}_Z \psi_f^*(\mathbf{r}_1, \ldots, \mathbf{r}_Z) \left[\sum_{j=1}^{Z} -e\mathbf{r}_j \right] \psi_i(\mathbf{r}_1, \ldots, \mathbf{r}_Z). \qquad (4.29)$$

$\mathbf{D}_{fi}$ is called the electric dipole matrix element for the transition $i \to f$ and the transition is called an electric dipole transition. It follows from the analysis that the intensity of the radiation emitted in the transition is proportional to $|\mathbf{D}_{fi}|^2$; if $\mathbf{D}_{fi} = 0$, the transition cannot occur.[†] The line of argument used above to obtain $\mathbf{D} = 0$ for a state of definite parity can at once be applied to Eq. (4.29). If the states ψ_i and ψ_f have parities Π_i and Π_f respectively, changing integration variables in Eq. (4.29) to $\mathbf{r}_j' = -\mathbf{r}_j, j = 1, \ldots, Z$, gives

$$\mathbf{D}_{fi} = (-1)\Pi_f \Pi_i \mathbf{D}_{fi};$$

i.e. $\mathbf{D}_{fi} \neq 0$ only if

$$\Pi_f = -\Pi_i: \qquad (4.30)$$

the electric dipole transition $i \to f$ can only occur if the initial and final atomic states have opposite parity. This is known as Laporte's rule. It is an example of a selection rule, i.e. a rule stating which transitions may occur. Laporte's rule is an exact and completely general result: it only depends on the atomic Hamiltonian being invariant under inversion and ψ_i and ψ_f having definite parities.

Eq. (4.30) states that in the transition $i \to f$ the parity of the atom has changed. So, the reader may wonder, what has happened to parity conservation? The explanation is that the emitted or absorbed photon possesses parity. Conservation of parity applies to the whole system: atom plus electromagnetic field; just as conservation of energy or momentum only applies to the whole system.

We illustrate the parity selection rule (4.30) for atomic hydrogen. We found in section 2.5 that the energy eigenfunctions of a particle in a central potential are of the form

$$\psi_{nlm}(\mathbf{r}) = R_{nl}(r) Y_l^m(\theta, \phi). \qquad (2.92)$$

Under inversion, the spherical polar coordinates (r, θ, ϕ) transform into

$$(r', \theta', \phi') = (r, \pi - \theta, \phi + \pi), \qquad (4.31)$$

---

* This result will be derived in Chapter 9, section 9.5.

† To be more accurate, it cannot occur as an electric dipole transition, i.e. by the mechanism envisaged here. Usually this is the dominant mechanism but if $\mathbf{D}_{fi} = 0$ the same atomic transition may still occur involving other mechanisms.

and it follows from

$$Y_l^m(\pi - \theta, \phi + \pi) = (-1)^l Y_l^m(\theta, \phi) \qquad (2.63)$$

that

$$\psi_{nlm}(-\mathbf{r}) = (-1)^l \psi_{nlm}(\mathbf{r}). \qquad (4.32)$$

Thus a hydrogenic state with orbital angular momentum l has parity $(-1)^l$. And quite generally, a particle with orbital angular momentum l has parity $(-1)^l$: s-, d-, g-, ... states have even parity; p-, f-, h-, ... states have odd parity. Consequently, electric dipole transitions cannot occur between hydrogenic states both with even or both with odd orbital angular momentum quantum numbers: we must have

$$\Delta l = l_f - l_i = \pm(\text{odd integer}) \qquad (4.33)$$

where l_i and l_f are the orbital angular momentum quantum numbers of the initial and final states. Thus the parity selection rule (4.30) forbids the transitions

$$(n\text{s}) \to (1\text{s}), n \geqslant 2, \quad (n\text{p}) \to (2\text{p}), n \geqslant 3, \quad (n\text{d}) \to (1\text{s}), n \geqslant 3, \qquad (4.34)$$

but allows the transitions

$$(n\text{p}) \to (1\text{s}), \qquad n \geqslant 2, \qquad (4.35)$$

to occur. In addition to the parity selection rule, conservation of angular momentum also leads to angular momentum selection rules* which, for electric dipole transitions, restrict Δl to

$$\Delta l = l_f - l_i = \pm 1. \qquad (4.36)$$

For example, the transitions $(n\text{f}) \to (1\text{s})$, $n \geqslant 4$, have $\Delta l = 3$. They satisfy the parity selection rule (4.33) but violate the angular momentum selection rule (4.36) and so cannot occur as electric dipole transitions.

The parity selection rule (4.30) is equally important in the spectroscopy of many-electron atoms. We shall return to these applications later, after we have learned to construct the wave functions for such atoms.

The considerations and results of this section can easily be taken over to one-dimensional problems, such as the square-well potential which we treated in section 2.1. We need only interpret $\mathbf{r}$ to stand for the one space coordinate x. Inversion now is replaced by reflection of the x-axis in the origin

$$x \to x' = -x, \qquad (4.37)$$

and a one-dimensional Hamiltonian

$$H(x) = -\frac{\hbar^2}{2m} \frac{d^2}{dx^2} + V(x) \qquad (4.38)$$

---

* The selection rule (4.36) will be derived in section 6.2.2, Eq. (6.15).

is invariant under this reflection if $V(-x) = V(x)$. We shall show at the end of this section that the energy eigenvalues of the bound states of the one-dimensional Hamiltonian (4.38) are non-degenerate. It follows that if $V(-x) = V(x)$, then the bound-state wave functions of the Hamiltonian (4.38) are necessarily even or odd functions of x. This is precisely the result which we obtained 'by brute force' in section 2.1 for the one-dimensional square-well potentials of finite or infinite depth; in both cases we had chosen the potential symmetric about $x = 0$.

To show that the bound-state energy eigenvalues of the one-dimensional Hamiltonian (4.38) are non-degenerate, assume that $u(x)$ and $v(x)$ are two bound-state eigenfunctions belonging to the same energy E. From the Schrödinger equations for u and v

$$u''/u = v''/v = -(2m/\hbar^2)[E - V(x)],$$

where primes denote differentiation with respect to x. Hence $vu'' - uv'' = 0$ which on integration gives

$$v(x)u'(x) - u(x)v'(x) = \text{const.} \tag{4.39}$$

For a particle restricted to a region $a \leqslant x \leqslant b$ (where we could have $a = -\infty$ and/or $b = +\infty$) a bound-state wave function vanishes at the end points; thus $u(a) = v(a) = 0$, the constant on the right-hand side of (4.39) vanishes and $v'/v = u'/u$. Integrating this equation gives $v(x) = \text{const.} \times u(x)$, i.e. apart from at most a constant of proportionality $u(x)$ and $v(x)$ are the same function, and the energy eigenvalue E is non-degenerate.

4.2 TRANSLATIONS

Next we consider translations. These are the transformations

$$\mathbf{r} \to \mathbf{r}' = \mathbf{r} + \mathbf{a} \tag{4.40}$$

where $\mathbf{a}$ is an arbitrary vector. The translation (4.40) displaces the point P, labelled by the vector $\mathbf{r}$, to the point P′, labelled by $\mathbf{r}'$ (see Fig. 4.4). Alternatively, we can interpret Eq. (4.40) as a translation of the coordinate system through $-\mathbf{a}$, with $\mathbf{r}$ and $\mathbf{r}'$ labelling the same point P. Unlike reflections or inversion, translations are continuous transformations; the displacement $\mathbf{a}$ can be varied continuously and we can effect the displacement from P to P′, in Fig. 4.4, along a continuous path.

We consider a system whose Hamiltonian we shall denote by $H(\mathbf{r})$, when expressed in terms of the coordinates $\mathbf{r}$, and by $H'(\mathbf{r}')$, when expressed in terms of the coordinates $\mathbf{r}'$. It follows from Eq. (4.40) that

$$H(\mathbf{r}) = H(\mathbf{r}' - \mathbf{a}) = H'(\mathbf{r}'), \tag{4.41}$$

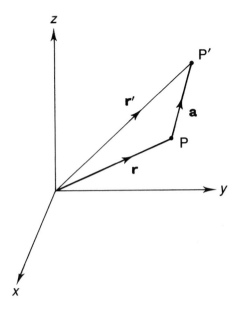

Fig. 4.4   Translation through **a** moves the point P to P'.

where the last step *defines* the function $H'(\mathbf{r}')$. (Eq. (4.41) is analogous to Eqs. (4.2) for inversion.) In general, $H(\mathbf{r})$ and $H'(\mathbf{r})$ will be different functions. If

$$H(\mathbf{r}) = H'(\mathbf{r}), \qquad (4.42a)$$

we call $H(\mathbf{r})$ invariant under the translation $\mathbf{r} \to \mathbf{r}' = \mathbf{r} + \mathbf{a}$. Combining Eq. (4.42a) with Eqs. (4.41), which always hold, we obtain

$$H(\mathbf{r}) = H'(\mathbf{r}') = H(\mathbf{r}') = H(\mathbf{r} + \mathbf{a}),$$

i.e.

$$H(\mathbf{r}) = H(\mathbf{r} + \mathbf{a}) \qquad (4.42b)$$

is an alternative form of the invariance criterion (4.42a).

If a system is invariant under translations, its Hamiltonian must possess this invariance, i.e. Eq. (4.42b) must hold for arbitrary constant vectors **a**. For an arbitrary infinitesimal displacement **a**, Eq. (4.42b) becomes

$$0 = H(\mathbf{r} + \mathbf{a}) - H(\mathbf{r}) = \mathbf{a} \cdot \nabla H(\mathbf{r}). \qquad (4.43)$$

Now for any operator $F(\mathbf{r})$ and any wave function $\psi(\mathbf{r})$

$$[\mathbf{p}, F(\mathbf{r})]\psi(\mathbf{r}) = [-i\hbar\nabla, F(\mathbf{r})]\psi(\mathbf{r})$$

$$= -i\hbar\nabla\{F(\mathbf{r})\psi(\mathbf{r})\} - F(\mathbf{r})\{-i\hbar\nabla\psi(\mathbf{r})\}$$

$$= -i\hbar\{\nabla F(\mathbf{r})\}\psi(\mathbf{r}).$$

Since this equation holds for any wave function $\psi(\mathbf{r})$, it must hold for the operators alone; i.e. for any operator $F(\mathbf{r})$ we have the identity

$$[\mathbf{p}, F(\mathbf{r})] = -i\hbar \nabla F(\mathbf{r}). \tag{4.44}$$

With $F(\mathbf{r}) = H(\mathbf{r})$, this equation becomes

$$\frac{1}{-i\hbar} [\mathbf{p}, H(\mathbf{r})] = \nabla H(\mathbf{r}).$$

Combining this equation with Eq. (4.43), we obtain

$$\mathbf{a} \cdot [\mathbf{p}, H(\mathbf{r})] = 0$$

and since $\mathbf{a}$ is an arbitrary infinitesimal vector this implies

$$[\mathbf{p}, H(\mathbf{r})] = 0. \tag{4.45}$$

Thus the momentum $\mathbf{p}$ is a constant of the motion if the Hamiltonian is invariant under translations. For the Hamiltonian

$$H(\mathbf{r}) = -\frac{\hbar^2}{2m} \nabla^2 + V(\mathbf{r})$$

to be translationally invariant, the potential $V(\mathbf{r})$ must have a constant value. In this case we are dealing with a free particle and its momentum is constant, as expected and already derived in section 3.2.

The generalization to several particles is trivial. In the above treatment we interpret $\mathbf{r}$ to stand for $\mathbf{r}_1, \ldots, \mathbf{r}_N$. For a system invariant under translations

$$H(\mathbf{r}_1 + \mathbf{a}, \ldots, \mathbf{r}_N + \mathbf{a}) - H(\mathbf{r}_1, \ldots, \mathbf{r}_N) = 0,$$

and for such a system the same procedure which led to Eq. (4.45) now gives

$$[\mathbf{P}, H(\mathbf{r}_1, \ldots, \mathbf{r}_N)] = 0 \tag{4.45a}$$

where $\mathbf{P} = -i\hbar \sum_{j=1}^{N} \nabla_j$; i.e. the total momentum of the system of N particles is conserved.

A simple but important example of a translationally invariant system is given by two particles interacting through a central potential and not subject to any other forces. The Hamiltonian for this system is

$$H(\mathbf{r}_1, \mathbf{r}_2) = \sum_{j=1}^{2} \left(\frac{-\hbar^2}{2m_j} \nabla_j^2 \right) + V(|\mathbf{r}_1 - \mathbf{r}_2|). \tag{4.46}$$

This is obviously invariant under translations, and it follows without further calculations that the total momentum $\mathbf{P} = \mathbf{p}_1 + \mathbf{p}_2$ is conserved. In problem 3.3, this result was derived by explicit evaluation of the commutator $[\mathbf{P}, H]$.

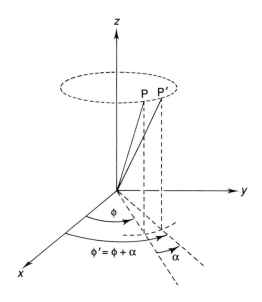

Fig. 4.5 A rotation about the z-axis through the angle α transforms the point P into the point P'. For $\alpha > 0$, the sense of the rotation is that of a right-handed screw advancing in the direction of the positive z-axis.

4.3 ROTATIONS

Rotational invariance of a system leads to conservation of angular momentum of that system. For this reason rotations are an important symmetry operation. They are much more complicated than the symmetries we have considered so far and we shall give a very incomplete account only.

A rotation is specified by an axis and an angle of rotation about this axis. We shall always take the axis as passing through the origin of coordinates. Using spherical polar coordinates (r, θ, ϕ) to specify a vector $\mathbf{r}$, consider the transformation

$$\mathbf{r} = (r, \theta, \phi) \to \mathbf{r}' = (r', \theta', \phi') = (r, \theta, \phi + \alpha) \qquad (4.47)$$

i.e. r and θ stay the same, ϕ changes to $\phi' = \phi + \alpha$. If we think of P and P' as two points having the coordinates $\mathbf{r}$ and $\mathbf{r}'$, referred to the same coordinate system, then (4.47) represents a rotation through an angle α about the z-axis (Fig. 4.5). The angle of rotation α is positive if the sense of the rotation is that of a right-handed screw advancing in the direction of the positive z-axis. Alternatively, we can interpret Eq. (4.47) as a rotation of the coordinate system through an angle $-\alpha$ about the positive z-axis, with $\mathbf{r}$ and $\mathbf{r}'$ labelling the same point P.

We shall start by studying the effect of the transformation (4.47), i.e. of rotations about the z-axis, on a general function of $\mathbf{r}$. We shall find that this is closely related to the orbital angular momentum operator l_z. Substituting

$$r = r', \quad \theta = \theta', \quad \phi = \phi' - \alpha \qquad (4.47a)$$

in any function $f(\mathbf{r}) = f(r, \theta, \phi)$ gives

$$f(\mathbf{r}) = f(r, \theta, \phi) = f(r', \theta', \phi' - \alpha) = f'(r', \theta', \phi') = f'(\mathbf{r}'). \qquad (4.48)$$

This equation *defines* the function $f'(\mathbf{r}') = f'(r', \theta', \phi')$ as the function of $\mathbf{r}' = (r', \theta', \phi')$ into which $f(\mathbf{r}) = f(r, \theta, \phi)$ transforms when one substitutes (4.47a) in $f(\mathbf{r}) = f(r, \theta, \phi)$.* For example, if

$$f(\mathbf{r}) = f(r, \theta, \phi) = r \sin \theta \cos m\phi$$

then

$$f(\mathbf{r}) = f(r, \theta, \phi) = r \sin \theta \cos m\phi = r' \sin \theta' \cos m(\phi' - \alpha) = f'(r', \theta', \phi') = f'(\mathbf{r}'),$$

i.e. the function $f(\mathbf{r}) = r \sin \theta \cos m\phi$ has been transformed into the different function

$$f'(\mathbf{r}) = r \sin \theta \cos m(\phi - \alpha).$$

Similar conclusions hold for the general case of Eq. (4.48). The functions $f(\mathbf{r})$ and $f'(\mathbf{r})$ will usually be different functions, and the change induced in $f(\mathbf{r})$ by the transformation (4.47) is given by

$$f'(\mathbf{r}) - f(\mathbf{r}) = f'(r, \theta, \phi) - f(r, \theta, \phi)$$
$$= f(r, \theta, \phi - \alpha) - f(r, \theta, \phi)$$

where the last line follows from Eq. (4.48). For infinitesimal α, i.e. for an infinitesimal rotation about the z-axis, the last equation becomes

$$f'(\mathbf{r}) - f(\mathbf{r}) = f(r, \theta, \phi - \alpha) - f(r, \theta, \phi) = -\alpha \frac{\partial}{\partial \phi} f(\mathbf{r}). \qquad (4.49)$$

If we define the operator

$$l_z = -i\hbar \frac{\partial}{\partial \phi} \qquad (4.50)$$

we can rewrite Eq. (4.49) as

$$f'(\mathbf{r}) - f(\mathbf{r}) = -\alpha \frac{i}{\hbar} l_z f(\mathbf{r}). \qquad (4.51)$$

The operator l_z arose in a natural way when considering the infinitesimal rotations about the z-axis. It is the operator for the z-component of the orbital angular momentum of a particle. [We met it previously in Eq. (2.43c).]

In general, the function $f(\mathbf{r})$ and the transformed function $f'(\mathbf{r})$ will differ. If $f'(\mathbf{r}) = f(\mathbf{r})$, i.e. if $f(\mathbf{r})$ and $f'(\mathbf{r})$ are identical, we call $f(\mathbf{r})$ invariant under the

---

* Eq. (4.48) is closely analogous to Eq. (4.12c) for inversion and to Eq. (4.41) for translations.

transformation (4.47), i.e. under rotations about the z-axis.* In this case, the left-hand side of Eq. (4.51) vanishes and

$$l_z f(\mathbf{r}) = 0. \tag{4.52a}$$

A similar result holds for rotations about any other axis. In particular, if $f(\mathbf{r})$ is invariant under rotations about the x-axis or the y-axis Eq. (4.52a) is replaced by

$$l_x f(\mathbf{r}) = 0 \quad \text{or} \quad l_y f(\mathbf{r}) = 0, \tag{4.52b}$$

where

$$l_j = -i\hbar \, \partial/\partial\phi_j, \quad j = x, y, z. \tag{4.53}$$

For $j = z$, Eq. (4.53) is identical with Eq. (4.50) for l_z if we define $\phi_z = \phi$. ϕ_x and ϕ_y are the angles analogous to $\phi_z = \phi$ for rotations about the x- and y-axes respectively, and l_x and l_y are the x- and y-components of orbital angular momentum. If the function $f(\mathbf{r})$ is invariant under *all* rotations—such a quantity is called a scalar—it follows that for all three components l_j

$$l_j f(\mathbf{r}) = 0, \quad j = x, y, z. \tag{4.54}$$

The converse statement also is true. An arbitrary rotation can be built up out of suitable successive rotations about the x-, y- and z-axes. Hence if Eqs. (4.54) hold, $f(\mathbf{r})$ is invariant under all rotations, i.e. it is a scalar.

We next extend these results to a general operator $F = F(\mathbf{r}, -i\hbar\nabla)$. For any function $\psi = \psi(\mathbf{r})$

$$[l_j, F]\psi = -i\hbar\left[\frac{\partial}{\partial\phi_j}, F\right]\psi$$

$$= -i\hbar\left\{\frac{\partial}{\partial\phi_j}(F\psi) - F\frac{\partial\psi}{\partial\phi_j}\right\} = -i\hbar\left(\frac{\partial F}{\partial\phi_j}\right)\psi,$$

and since $\psi(\mathbf{r})$ is an arbitrary function the last equation implies the operator identity

$$[l_j, F] = -i\hbar\left(\frac{\partial F}{\partial\phi_j}\right). \tag{4.55}$$

It follows at once that

$$[l_j, F(\mathbf{r}, -i\hbar\nabla)] = 0, \quad j = x, y, z, \tag{4.56}$$

---

* We see from Eq. (4.48) that the invariance criterion $f'(\mathbf{r}) = f(\mathbf{r})$ can be written $f(\mathbf{r}) = f(\mathbf{r}')$ or $f(r, \theta, \phi) = f(r, \theta, \phi + \alpha)$. In the last section, we derived similar relations for translational invariance.

if and only if $F(\mathbf{r}, -i\hbar\nabla)$ is a scalar quantity. This is an important result: any scalar operator $F(\mathbf{r}, -i\hbar\nabla)$ commutes with the angular momentum operators l_j. Obvious examples of scalars are the magnitude of a vector, the scalar product of two vectors, or functions of these. The angular momentum operator

$$\mathbf{l}^2 = l_x^2 + l_y^2 + l_z^2 \tag{4.57}$$

is a scalar, as are $r = |\mathbf{r}|$, any function of r only, and ∇^2. It follows that

$$[l_j, \mathbf{l}^2] = 0, \qquad j = x, y, z, \tag{4.58}$$

and from Eqs. (4.56) that for any scalar operator $F(\mathbf{r}, -i\hbar\nabla)$

$$[F(\mathbf{r}, -i\hbar\nabla), \mathbf{l}^2] = 0. \tag{4.59}$$

The central field Hamiltonian

$$H = -\frac{\hbar^2}{2m}\nabla^2 + V(r) \tag{4.60}$$

is a scalar, and hence

$$[l_j, H] = 0, \quad [\mathbf{l}^2, H] = 0, \qquad j = x, y, z. \tag{4.61}$$

All these results we have derived before. Here they come out trivially from our rotational considerations. To complete the picture, we need the angular momentum commutation relations

$$[l_x, l_y] = i\hbar l_z, \quad \text{etc.} \tag{4.62}$$

where etc. stands for two similar equations obtained by cyclic permutations of x, y, z (i.e. $x \to y, y \to z, z \to x$) carried out once or twice. The commutation relations (4.62) follow directly from the operators l_j, Eqs. (4.53), when expressed in terms of Cartesian coordinates:

$$l_x = yp_z - zp_y, \quad \text{etc.} \tag{4.63}$$

(see problem 3.1).

I state, without proof,* that these ideas can be generalized to systems of particles, which may have spin.† By considering infinitesimal rotations for such systems one obtains operators J_x, J_y, J_z which always satisfy the commutation relations

$$[J_x, J_y] = i\hbar J_z, \quad \text{etc.} \tag{4.64}$$

---

* The proofs are quite advanced, well beyond the level of this book. For many purposes it is more important to understand the results and their implications. Proofs are, for example, found in Gottfried, pp. 264–267, and Sakurai, section 3.1.

† Spin will be discussed later. Here we only note that some particles, such as electrons or protons, possess a non-zero angular momentum in their own rest-frame. So this angular momentum is not of the familiar $\mathbf{r} \wedge \mathbf{p}$ kind, associated with the orbital motion of the particle. Rather it must be thought of as an intrinsic internal angular momentum which is called the spin of the particle.

(These commutation relations have their origin in the geometry of rotations in space. They do not depend on the detailed structure of the system; hence their generality.) We *define* these operators to be the angular momentum operators of the system. Eqs. (4.64) are known as the angular momentum commutation relations. For spinless particles, these definitions reduce to the expressions in terms of **rs** and **ps** which one would expect. For one spinless particle, we saw that $\mathbf{J} = \mathbf{l}$ and Eqs. (4.64) are just the angular momentum, commutation relations (4.62) for **l**. For a system of N spinless particles, one easily shows that $\mathbf{J} = \mathbf{L}$, where **L** is defined by

$$\mathbf{L} = \sum_{s=1}^{N} \mathbf{l}_s = \sum_{s=1}^{N} \mathbf{r}_s \wedge \mathbf{p}_s, \tag{4.65}$$

(see problem 4.1). From the commutation relations (4.62) for $\mathbf{l}_s$, those for **L** follow:

$$[L_x, L_y] = i\hbar L_z, \quad \text{etc.} \tag{4.66}$$

(see problem 4.2). **L** is clearly the total orbital angular momentum of the system of particles. On the other hand, spin angular momentum is not related to **rs** and **ps** and cannot be expressed in terms of them. There is no classical analogue and we must be content with the more abstract definition derived from infinitesimal rotations.

The earlier discussion of rotational invariance, etc., for **l** at once generalizes to a system of particles with or without spin. Any operator F invariant under rotations commutes with the components of the *total* angular momentum operator **J** of the system:

$$[J_j, F] = 0, \quad j = x, y, z. \tag{4.56a}$$

With

$$\mathbf{J}^2 = J_x^2 + J_y^2 + J_z^2 \tag{4.57a}$$

it follows that

$$[J_j, \mathbf{J}^2] = 0, \quad j = x, y, z, \tag{4.58a}$$

and for a scalar operator F

$$[F, \mathbf{J}^2] = 0. \tag{4.59a}$$

In particular, if the Hamiltonian of a system is a scalar one has

$$[J_j, H] = 0, \quad [\mathbf{J}^2, H] = 0, \quad j = x, y, z, \tag{4.61a}$$

i.e. $\mathbf{J}^2$ and all components J_j are constants of the motion. However, since the components of **J** do not commute with each other, one can in general specify the values of $\mathbf{J}^2$ and only one component, say J_z. Hence one often speaks of $\mathbf{J}^2$ and only one component, say J_z, as constants of the motion.

We give some applications of these results. An important example of a Hamiltonian invariant under rotations is the atomic Hamiltonian H, Eq. (4.23). Hence if $\mathbf{J}$ is the total angular momentum of the electrons of the atom, then $\mathbf{J}^2$ and J_z, say, are conserved. Actually further deductions are possible in this case. The Hamiltonian H depends only on the space coordinates of the electrons but not on their spins. Looking at Eq. (4.23), one cannot tell whether it refers to particles with or without spin. Consequently H also commutes with the orbital angular momentum operators $\mathbf{L}^2$ and L_z, and these are also constants of the motion.*

The atomic Hamiltonian (4.23) is only an approximation, albeit a very good one. It takes into account the Coulomb interaction between charged particles but it neglects the much weaker spin–orbit interaction. On account of its spin, the electron possesses an intrinsic magnetic moment. This intrinsic magnetic moment of an atomic electron interacts with the magnetic field which the electron experiences due to its motion relative to the rest of the atom. This interaction is the spin–orbit interaction. For each atomic electron it contributes a term of the form $f(r)\,\mathbf{l}\cdot\mathbf{s}$ to the atomic Hamiltonian. $\mathbf{s}$ is the electron's spin angular momentum operator, analogous to $\mathbf{l}$, and will be dealt with more fully later. The scalar product $\mathbf{l}\cdot\mathbf{s}$ is invariant under rotations, as is $f(r)$. Hence the spin–orbit interaction terms are invariant under rotations, and so is the modified atomic Hamiltonian which includes these terms. Hence $\mathbf{J}^2$ and J_z continue to be constants of the motion. On the other hand, one can show that $\mathbf{L}^2$ and L_z do not commute with the modified Hamiltonian and so lose their status as constants of the motion. Note the generality of this type of argument. It depends on symmetries, not details; for example, it does not depend on the form of the function $f(r)$ in the spin–orbit interaction. Considerations such as these are of great importance in assigning angular momentum quantum numbers to atomic energy levels, in atomic spectroscopy and in constructing atomic wave functions. They are of course equally applicable to molecules and nuclei.

We conclude this section with a molecular example. In contrast to the atomic case, the electronic Hamiltonian H, Eq. (4.28), of a diatomic molecule is not invariant under *all* rotations. It only possesses axial symmetry about the internuclear axis and is invariant under rotations about this axis only. (See Fig. 4.6.) Calling this the z-axis and $\mathbf{J}$ the total angular momentum of the electrons in the molecule, it follows that H commutes with J_z, but not with the other components of $\mathbf{J}$ or with $\mathbf{J}^2$. Hence J_z is conserved but not $\mathbf{J}^2$ or the other components of $\mathbf{J}$. Since the Hamiltonian (4.28) does not depend on the electron spins, we can, in the above argument, replace $\mathbf{J}$ by the total orbital

---

* However, it will be shown in section 5.4 (see Eq. (5.68a) and the comment following it) that L_z does not commute with $\mathbf{J}^2$, so that the values of $\mathbf{J}^2$ and L_z cannot be specified simultaneously. The question of how to choose complete sets of commuting observables from the angular momentum operators is discussed in section 5.4.

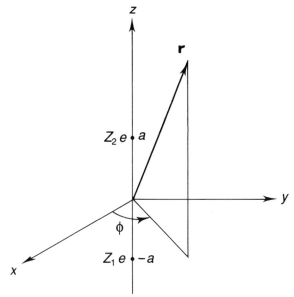

Fig. 4.6 The coordinate system and symmetries for a diatomic molecule. The internuclear axis is taken as z-axis with the nuclei positioned at $z = \pm a$. The system has axial symmetry about the z-axis. It is also invariant under reflections in the (x, z) plane. These reflections transform ϕ into $\phi' = -\phi$.

angular momentum $\mathbf{L}$ of the electrons and conclude that L_z is a constant of the motion, but $\mathbf{L}^2$ and the other components of $\mathbf{L}$ are not. Consequently, we can choose the eigenstates of H to be simultaneously eigenstates of L_z.

The Hamiltonian (4.28) possesses a further symmetry. It is invariant under reflections in any plane which passes through the internuclear axis. We continue to call this axis the z-axis and shall take the (x, z) plane as the reflection plane. (See Fig. 4.6.) Let ψ_M be a simultaneous eigenfunction of H and L_z with eigenvalues E and $\hbar M$ respectively.* Since H is unaltered under the reflection in the (x, z) plane, the state $\tilde{\psi}_M$ into which ψ_M transforms under the reflection is also an eigenstate of H belonging to the same energy E. Reflection in the (x, z) plane reverses the sense of rotation about the z-axis, and an eigenfunction of L_z with the eigenvalue $\hbar M$ transforms into an eigenfunction of L_z with the eigenvalue $-\hbar M$. (To illustrate this for a very simple case. An eigenfunction of l_z with the eigenvalue $\hbar m$ has the ϕ-dependence $e^{im\phi}$. Under reflection in the (x, z) plane, $\phi \to -\phi$ (see Fig. 4.6) and $e^{im\phi} \to e^{-im\phi}$, which is an eigenfunction of l_z with the eigenvalue $-\hbar m$.) Thus the state $\tilde{\psi}_M$ into which ψ_M transforms is an eigenstate of L_z with eigenvalue $-\hbar M$. It follows that for $M \neq 0$, ψ_M and $\tilde{\psi}_M$ are different states, and the energy E to which they belong is doubly degenerate.

---

* The possible eigenvalues of the operator L_z are $\hbar M$, with $M = 0, \pm 1, \pm 2, \ldots$, since from Eq. (4.65) $L_z = \Sigma l_{sz}$ and since the possible eigenvalues of l_{sz} are $0, \pm \hbar, \pm 2\hbar, \ldots$.

For $M = 0$, both ψ_0 and $\tilde{\psi}_0$ belong to the same eigenvalue 0 of L_z, and E is non-degenerate, i.e.

$$\tilde{\psi}_0 = \text{const. } \psi_0.$$

In section 4.1 we showed that for a system invariant under inversion a non-degenerate energy eigenstate has a definite parity (either $+1$ or -1). By essentially the same argument it follows here that under reflection in the (x, z) plane ψ_0 transforms into either $+\psi_0$ or $-\psi_0$.

Finally, we remind the reader that we saw in section 4.1 that for homonuclear diatomic molecules the electronic Hamiltonian H, Eq. (4.28), is invariant under inversion of the electron coordinates and so commutes with the parity operator P. It follows from the definition (4.12b) of P, that P also commutes with any angular momentum operator l and therefore with L. (Essentially this is so since under inversion $r \to -r$, so that $\nabla \to -\nabla$ and $p = -i\hbar\nabla \to -p$; hence $l = r \wedge p \to +l$.) Hence we can classify the energy eigenstates of a homonuclear diatomic molecule according to the absolute value of L_z (remember the states with $L_z = \pm\hbar M$ have the same energy) and according to parity. In the case of molecules the even and odd parity states are conventionally called g- and u-states (from gerade and ungerade, which is the German for even and odd). In addition, an $L_z = 0$ state has a definite reflection symmetry ($+$ or $-$) under reflection in a plane containing the internuclear axis.

4.4 IDENTICAL PARTICLES

The Hamiltonian of the helium atom

$$H_0(\mathbf{r}_1, \mathbf{r}_2) = \sum_{i=1}^{2} \left(-\frac{\hbar^2}{2m} \nabla_i^2 + \frac{-2e^2}{4\pi\varepsilon_0 r_i} \right) + \frac{e^2}{4\pi\varepsilon_0 r_{12}} \tag{4.67}$$

is symmetric in the electron coordinates, i.e. it is invariant under the permutation $1 \leftrightarrow 2$ of the indices labelling the two electrons:

$$H_0(\mathbf{r}_1, \mathbf{r}_2) = H_0(\mathbf{r}_2, \mathbf{r}_1). \tag{4.68}$$

This is no more than a statement that the two electrons are identical. If H_0 did not possess this symmetry, the two electrons would possess some property which could be used to distinguish them and label them 1, 2 rather than 2, 1. No such distinguishing marks exist, and it is, for example, meaningless to ask for the probability that in the helium ground state electron no. 1 is in the volume element dV_a at $\mathbf{a}$ and electron no. 2 is in the volume element dV_b at $\mathbf{b}$. It only makes sense to ask for the probability that *one* of the electrons is in dV_a and *the other* in dV_b.

The above argument is completely general. *All* observables for our two-electron system must be symmetric in the electron coordinates. The same argument holds for identical particles other than electrons, and it can obviously be generalized to systems of more than two identical particles.

Eq. (4.67) is an approximate Hamiltonian for the helium atom. The exact Hamiltonian—whatever it looks like—must also be symmetric in the electron labels 1 and 2. This leads to an essential refinement of the above reasoning. As discussed in the last section, electrons possess spin which enters the exact Hamiltonian H through the spin–orbit interaction. Hence H is of the form $H(\mathbf{r}_1, \mathbf{s}_1; \mathbf{r}_2, \mathbf{s}_2)$, where $\mathbf{s}_1$ and $\mathbf{s}_2$ are the spin angular momentum operators of the electrons analogous to their orbital angular momentum operators $\mathbf{l}_1$ and $\mathbf{l}_2$. (A proper discussion of spin will be given in Chapter 5.) Interchanging particle labels means interchanging the labels of the space and the spin variables, and our symmetry requirement for H becomes

$$H(\mathbf{r}_1, \mathbf{s}_1; \mathbf{r}_2, \mathbf{s}_2) = H(\mathbf{r}_2, \mathbf{s}_2; \mathbf{r}_1, \mathbf{s}_1). \tag{4.69}$$

Since we are only interested in the particle labels at present, we shall write the last equation more concisely as

$$H(1, 2) = H(2, 1). \tag{4.70a}$$

Similar arguments apply, of course, to particles with other kinds of internal degrees of freedom: we must always permute all particle labels.

The result (4.70a) at once generalizes. For a system of N identical particles, the Hamiltonian $H(1, 2, \ldots, N)$ must be invariant under any permutation of the particle labels $1, 2, \ldots, N$; in particular under the interchange of one pair of labels only, $i \leftrightarrow j$, leaving all other labels unaltered:

$$H(1, \ldots, i, \ldots, j, \ldots, N) = H(1, \ldots, j, \ldots, i, \ldots, N). \tag{4.70b}$$

It is an immediate consequence of the symmetry (4.70) that the eigenvalues of the Hamiltonian H are degenerate. (This is known as exchange degeneracy.) Consider two identical particles. Suppose

$$H(1, 2)\psi(1, 2) = E\psi(1, 2).$$

Interchanging the labels 1 and 2 in this equation and using Eq. (4.70a), we obtain

$$H(1, 2)\psi(2, 1) = E\psi(2, 1)$$

i.e. E is a two-fold degenerate eigenvalue with $\psi(1, 2)$, $\psi(2, 1)$ or any linear combination of these as eigenfunctions. In particular we can choose the combinations

$$\psi(1, 2) \pm \psi(2, 1) \tag{4.71}$$

which are respectively symmetric and antisymmetric under the permutation $1 \leftrightarrow 2$.

It follows from the Schrödinger equation for two identical particles

$$i\hbar \frac{\partial}{\partial t} \Psi(1, 2, t) = H(1, 2)\Psi(1, 2, t), \tag{4.72}$$

that the symmetry of the wave function is a constant of the motion. If $\Psi(1, 2, t)$ is symmetric or antisymmetric under $1 \leftrightarrow 2$, then the right-hand side of Eq. (4.72) has the same symmetry since $H(1, 2)$ is symmetric. Hence the increment $d\Psi(1, 2, t)$ and the wave function $\Psi(1, 2, t + dt)$ at the slightly later time $t + dt$ have the same symmetry. Integrating Eq. (4.72) over a finite time interval, it follows that a symmetric wave function always stays symmetric, an antisymmetric one always antisymmetric. In fact, for electrons, only antisymmetric wave functions occur in nature as we shall discuss further in a moment.

First we consider a system of three identical particles. (The extension to more than three particles is trivial and introduces no new features.) An energy eigenvalue E of the completely symmetric Hamiltonian $H(1, 2, 3)$ is now six-fold degenerate corresponding to the six orderings of the particle labels:

$$(1, 2, 3) \quad (2, 1, 3) \quad (2, 3, 1) \quad (3, 2, 1) \quad (3, 1, 2) \quad (1, 3, 2). \tag{4.73}$$

If $\psi(1, 2, 3)$ is an eigenfunction of $H(1, 2, 3)$ with the eigenvalue E, the permutations (4.73) give five more eigenfunctions belonging to the same eigenvalue. From these six eigenfunctions we can again—as in the case of two particles—form one completely symmetric linear combination, namely

$$\psi(1, 2, 3) + \psi(2, 1, 3) + \psi(2, 3, 1) + \psi(3, 2, 1) + \psi(3, 1, 2) + \psi(1, 3, 2) \tag{4.74a}$$

and one completely antisymmetric linear combination, namely

$$\psi(1, 2, 3) - \psi(2, 1, 3) + \psi(2, 3, 1) - \psi(3, 2, 1) + \psi(3, 1, 2) - \psi(1, 3, 2). \tag{4.74b}$$

When any one pair of labels are interchanged, the symmetric wave function (4.74a) is unaltered, the antisymmetric wave function (4.74b) is multiplied by (-1). From the six degenerate eigenfunctions $\psi(1, 2, 3)$, $\psi(2, 1, 3)$, etc., we can form four more linearly independent combinations. These wave functions have much more complicated symmetry properties when particle labels are permuted; they are neither completely symmetric nor completely antisymmetric.

Similar results hold for more than three particles: we always obtain one completely symmetric wave function, one completely antisymmetric wave function and for $N > 2$ some more complicated wave functions. It makes life a lot simpler that nature does not employ these complicated wave functions: one knows from all experimental evidence that the wave functions of systems of identical particles are always symmetric or antisymmetric under the interchange of pairs of particle labels. Furthermore for a given kind of particle, the symmetry is always of one kind only. For example, for electrons, positrons, protons and neutrons the wave functions are always antisymmetric; for photons, pions and K-mesons they are always symmetric. Which type of symmetry holds for a particular particle depends on its spin angular momentum only. We shall deal with spin later. Here we merely quote some plausible results. Just as the possible values of the orbital angular momentum are $l = 0, 1, 2, \ldots$, so the possible values of the spin angular momentum are $s = 0, \frac{1}{2}, 1, \frac{3}{2}, \ldots$. It is an empirical fact that systems of identical particles with integer spin ($s = 0, 1, \ldots$) have symmetric wave functions, and with half-integer spin ($s = \frac{1}{2}, \frac{3}{2}, \ldots$) have antisymmetric wave functions. It was first shown by Pauli in 1940 that a relativistic quantum theory of identical particles can only be constructed if the above correlation between spin and symmetry holds. Pauli's proof was later much generalized and represents one of the great successes of quantum field theory.*

There exists another aspect to this spin–symmetry connection. Consider identical particles which interact very weakly with each other, so that as a first approximation we can neglect their interaction. For two particles the Hamiltonian can then be written

$$H(1, 2) = H'(1) + H'(2). \tag{4.75}$$

Suppose $H'(1)$ has the complete set of eigenfunctions $\psi_a(1)$ with eigenvalues E_a, i.e.

$$H'(1)\psi_a(1) = E_a\psi_a(1).$$

In terms of the single-particle states ψ_a, the symmetric and antisymmetric eigenfunctions of (4.75) are of the form

$$\Psi(1, 2) = \psi_a(1)\psi_b(2) \pm \psi_a(2)\psi_b(1). \tag{4.76}$$

In the antisymmetric case, $\Psi(1, 2)$ vanishes identically if $a = b$: two identical particles with half-integer spin (e.g. two electrons) cannot exist in the same single-particle state. This obviously generalizes to more than two identical particles with half-integer spin: in such a system at most one particle can occupy any one single-particle state. This is the Pauli exclusion principle. In this form it is restricted to non-interacting particles. The requirement of antisymmetric wave functions for identical particles with half-integer spin always holds and represents the general statement of the Pauli principle.

No such restriction applies to integer-spin particles. In the symmetric case, we can take $a = b$ in Eq. (4.76), and any number of such particles can occupy the same single-particle state. This difference leads to a connection between the spins of particles and the quantum statistics they obey. Half-integer spin

---

* For a simple discussion see Mandl and Shaw, pp. 72–73.

particles obey Fermi–Dirac statistics and are known as fermions. Integer spin particles obey Bose–Einstein statistics and are known as bosons.*

We have here talked of electrons, protons, pions, etc. as particles. What constitutes a particle in the present context? We can think of a proton as consisting of three (spin $\frac{1}{2}$) quarks; of a pion as consisting of two quarks. So is a hydrogen atom (consisting of a proton and an electron) or an alpha particle (consisting of two protons and two neutrons) a particle? The answer is a qualified 'yes': we can think of all these as composite particles and the above arguments about identical particles apply, provided the excitation energies of the internal degrees of freedom of the composite particles are large compared with the interaction energies of the system, i.e. provided the composite particles remain in their ground states. A composite particle possesses a spin, obtained by vector addition of the angular momenta of its constituents. If this spin has integer value, the composite particle behaves like a boson: if it has half-integer value, it behaves like a fermion. From the rules for the addition of angular momentum in quantum mechanics (to be discussed in Chapter 5) it follows that a composite particle containing an odd or an even number of fermions behaves like a fermion or a boson respectively, whatever number of bosons it contains. For example, the atoms of the common helium isotope ^{4}He (i.e. the nuclei are alpha particles) and of the isotope ^{3}He (whose nuclei consist of two protons and one neutron) obey Bose–Einstein and Fermi–Dirac statistics respectively. This difference in statistics explains the very different properties of liquid ^{4}He and of liquid ^{3}He. There is of course much more and quite overwhelming evidence for all the above conclusions. We mention the description of black-body radiation as a photon gas obeying Bose–Einstein statistics. That electrons are fermions obeying the Pauli exclusion principle is demonstrated by their behaviour in metals and in semiconductors, and by the explanation of the periodic table of elements in terms of the electronic structure of atoms; in turn the latter underlies our understanding of molecules, and so on.

4.4.1 The periodic table

The properties of atoms and therefore of all matter, i.e. the world as it exists, depends crucially on the Pauli exclusion principle. I would like to justify this claim and show how the Pauli principle explains the salient features of the periodic table in terms of the electronic structure of atoms.[†]

---

* For a discussion of Bose–Einstein and Fermi–Dirac statistics and their applications to black-body radiation, electrons in metals, etc., see Mandl, Chapters 9–11.

† For a fuller discussion of atomic structure see, for example, Fano and Fano, Chapter 18, or Gasiorowicz, Chapter 19, and at a much more advanced level the excellent account by Bethe and Jackiw, Chapter 4.

In the lowest approximation we neglect the Coulomb repulsion between the atomic electrons and treat each electron as moving independently in the Coulomb field of the point nuclear charge Ze. The wave functions and energy levels of each electron are those of the hydrogenic system, discussed in section 2.5.2. In the next approximation one must take some account of the electron–electron interaction. The nuclear charge Ze is surrounded by the negative electrons, so that the field which any one electron sees is the field due to the nuclear charge Ze screened by the surrounding electrons. This screened field which each electron experiences can be approximated by a central potential. (This is known as the central field approximation.) As a result one obtains single-particle states which can be labelled, as for hydrogen, by a principal quantum number $n (= 1, 2, \ldots)$, an orbital angular momentum quantum number $l (= 0, 1, \ldots, n - 1)$ and an orbital magnetic quantum number m_l $(= l, l - 1, \ldots, -l)$. In addition we must allow for the electron spin. In section 5.3, we shall find that for a particle with spin $\frac{1}{2}$ the z-component s_z of the spin angular momentum can only take on the two values $s_z = \hbar m_s$, with $m_s = \pm \frac{1}{2}$. These spin states with $m_s = \frac{1}{2}$ and $m_s = -\frac{1}{2}$ are referred to as spin up ($\uparrow$) and spin down ($\downarrow$) states respectively. We know from section 2.5.2 that for the hydrogenic system the energy levels depend on the principal quantum number n only but are degenerate with respect to l and m_l. They are also degenerate with respect to $m_s = \pm \frac{1}{2}$.*

For the screened potential the degeneracy with respect to l is removed. The levels now depend on n and l, and the level E_{nl} is $2(2l + 1)$-fold degenerate. One refers to states with the quantum numbers (n, l) as atomic (nl) orbitals and to the $2(2l + 1)$ such orbitals as the (nl) shell.

One can obtain the order of the levels as follows. An electron, when far from the nucleus, will see the nuclear charge Ze screened by $Z - 1$ electrons; effectively it will see a charge e. When close to the nucleus, it will experience the field of the full unscreened nuclear charge Ze. We know from Eq. (2.89) that the centrifugal barrier becomes larger as the orbital angular momentum l increases. The effect of this barrier is that, for a given value of n, electrons with larger l-values penetrate less deeply into the more attractive unscreened inner region of the atom. Hence for a given n the energies E_{nl} increase as l increases. This leads to the following order of the energy levels E_{nl} in increasing energy

$$1s, \ 2s, \ 2p, \ 3s, \ 3p, \ [4s, \ 3d], \ 4p, \ [5s, \ 4d], \ 5p, \ [6s, \ 4f, \ 5d], \ 6p, \ [7s, \ 5f, \ 6d]. \quad (4.77)$$

Levels which are bracketed together have very similar energies. The reason is that increasing n (e.g. from $n = 3$ to $n = 4$) increases the energy whereas decreasing l (e.g. from $l = 2$ to $l = 0$) decreases the energy. The two changes

---

* We are neglecting spin–orbit coupling. In this case the energy does not depend on the orientation of the electron spin, i.e. on the value of s_z.

work in opposing directions, and the order of the levels depends on details which cannot be obtained from these qualitative arguments.

The periodic table is obtained by filling the atomic orbitals in order of increasing energy, as given by Eq. (4.77). On account of the Pauli principle exactly one electron is placed in each occupied orbital. Table 4.1 shows how this works for the first few elements. The third column in the table gives the configurations of the atomic ground states, i.e. the list of orbitals occupied in the ground states. For example, the entry $(1s)^2(2s)$ for lithium means that two electrons occupy the (1s) shell (one with spin up and one with spin down). The (1s) shell is now full and the third electron must go into the (2s) state. A full shell is also called a closed shell.

The properties of atoms can largely be explained by their electron configurations. The rare gas atoms are particularly stable and chemically inert. They possess particularly stable configurations because the next unoccupied orbital lies considerably higher in energy so that the excitation energy to an excited

TABLE 4.1   The beginning part of the periodic table

Atomic number Z	Element	Configuration
1	H	(1s)
2	He	$(1s)^2$
3	Li	$(He)(2s) = (1s)^2(2s)$
4	Be	$(He)(2s)^2$
5	B	$(He)(2s)^2(2p)$
6	C	$(He)(2s)^2(2p)^2$
7	N	$(He)(2s)^2(2p)^3$
8	O	$(He)(2s)^2(2p)^4$
9	F	$(He)(2s)^2(2p)^5$
10	Ne	$(He)(2s)^2(2p)^6$
11	Na	$(Ne)(3s)$
12	Mg	$(Ne)(3s)^2$
13	Al	$(Ne)(3s)^2(3p)$
14	Si	$(Ne)(3s)^2(3p)^2$
15	P	$(Ne)(3s)^2(3p)^3$
16	S	$(Ne)(3s)^2(3p)^4$
17	Cl	$(Ne)(3s)^2(3p)^5$
18	Ar	$(Ne)(3s)^2(3p)^6$
19	K	$(Ar)(4s)$
20	Ca	$(Ar)(4s)^2$
21	Sc	$(Ar)(4s)^2(3d)$
22	Ti	$(Ar)(4s)^2(3d)^2$
⋮	⋮	⋮

state is large. For helium the (1s) shell is full. The lowest excited states correspond to the configurations (1s) (2s) and (1s) (2p) and lie about 20 eV above the ground state. Similarly the ionization energy has the exceptionally large value of 24.6 eV. The next rare gas, in order of increasing Z, is neon with the ground state configuration $(1s)^2(2s)^2(2p)^6$; all $n = 2$ shells are filled. At $Z = 18$, the rare gas argon has a ground state where all shells up to and including the (3p) shell are filled. This is surprising. One might have expected the energy gap to occur when all $n = 3$ shells are filled. In fact, the (3s) and (3p) shells are full but the (3d) shell is vacant. The explanation is that there is a considerable energy gap between the (3p) and (3d) orbitals due to the centrifugal barrier being much larger in a (3d) state than in a (3p) state. This again illustrates the l-dependence of the single-particle energy levels E_{nl} and the effect of the centrifugal barrier. But such arguments must be used cautiously and depend on a knowledge of the energy levels. For example, the centrifugal potential does not produce a large energy gap between the (2s) and (2p) shells; as a result the beryllium ground state $(1s)^2(2s)^2$ is fairly easily excited to the configuration $(1s)^2(2s)(2p)$ and beryllium is chemically active. Arguments similar to the ones above can be applied to the other inert gases.

Because of the great stability of these closed-shell configurations, most properties of atoms arise from the electrons in the outer incompletely filled shells. Atoms with similar electron configurations of the partially filled shells possess similar properties; for example, similar optical spectra and similar chemical behaviour. Thus the alkali atoms have the ground state configurations

$$\text{(rare gas configuration) } (ns)$$

with $n = 2, 3, 4, \ldots$ for Li, Na, K, $\ldots$. In the excited states the (ns) electron is promoted to a more highly excited orbital, and the line spectra correspond to transitions between these states accompanied by photon emission. The level schemes of different alkali atoms only differ in the values of the principal quantum number n of the electron in the outer (ns) shell. Hence their level schemes and optical spectra are very similar. The chemical properties of an element also are determined by the valence electrons, i.e. by the electrons outside the stable closed shells. The group of alkali atoms have the one valence electron in the outer (ns) shell and possess very similar chemical properties. Similar arguments apply to other groups of elements; for example, the halogens F, Cl, $\ldots$ with outer shell configurations $(np)^5$ for $n = 2, 3, \ldots$, or the alkali earth elements Mg, Ca, $\ldots$ which have the configurations

$$\text{(rare gas configuration) } (ns)^2.$$

The description of many-electron atoms in terms of a central potential (the screened Coulomb potential) and a wave function which is a product of Z different atomic orbitals is only approximate. Although this wave function satisfies the Pauli principle, it does not allow fully for the effects of an antisymmetric wave function. In addition, the electron–electron repulsion leads to correlation effects between electrons which are not taken into account in our approximation in which the electrons move in a central potential independently of each other. The true Hamiltonian contains these electron–electron repulsion terms $(\Sigma e^2/4\pi\varepsilon_0 r_{ij})$ and is not separable; consequently the wave function is not a product of Z single-particle wave functions but a sum of such products. Lastly, we have neglected the spin–orbit interaction. While this is a small correction for light atoms, it produces large effects in heavy atoms and must be taken into account.

PROBLEMS 4

4.1 Under a rotation through an infinitesimal angle α about the z-axis, the wave function $f(\mathbf{r}_1, \ldots, \mathbf{r}_N)$ of a system of N particles transforms into

$$f'(\mathbf{r}_1, \ldots, \mathbf{r}_N) = f(\mathbf{r}_1, \ldots, \mathbf{r}_N) - \alpha \frac{i}{\hbar} L_z f(\mathbf{r}_1, \ldots, \mathbf{r}_N). \tag{4.78}$$

Derive the operator L_z and interpret its meaning.

4.2 The angular momentum operators $\mathbf{l}_1$ and $\mathbf{l}_2$ of two particles satisfy the usual angular momentum commutation relations (3.44), i.e.

$$[l_{nx}, l_{ny}] = i\hbar l_{nz}, \quad \text{etc.,} \quad [l_{nx}, \mathbf{l}_n^2] = 0, \quad \text{etc.,} \quad n = 1, 2, \ldots \tag{4.79a}$$

where etc. stands for similar equations obtained by cyclic permutations of x, y and z. In addition, $\mathbf{l}_1$ and $\mathbf{l}_2$ commute with each other:

$$[l_{1j}, l_{2k}] = 0, \quad j, k = x, y, z. \tag{4.79b}$$

Show that the operator $\mathbf{L} = \mathbf{l}_1 + \mathbf{l}_2$ also satisfies the angular momentum commutation relations

$$[L_x, L_y] = i\hbar L_z, \quad \text{etc.,} \quad [L_x, \mathbf{L}^2] = 0, \quad \text{etc.,} \tag{4.80a}$$

as well as

$$[\mathbf{l}_n^2, L_j] = [\mathbf{l}_n^2, \mathbf{L}^2] = 0, \tag{4.80b}$$

but that $\mathbf{L}^2$ does not commute with the components of $\mathbf{l}_1$ or $\mathbf{l}_2$, e.g.

$$[\mathbf{L}^2, l_{1x}] \neq 0. \tag{4.80c}$$

4.3 Use symmetry arguments to solve problem 3.2.

4.4 The lowest-lying energy levels of the lithium atom have the configurations $(1s)^2(nl)$. Obtain the parities of these states, and hence determine which transitions from an excited state to the ground state of lithium are allowed and which forbidden by the parity selection rule for electric dipole transitions.

4.5 There exist mechanisms other than electric dipole transitions for the emission and absorption of radiation in atomic transitions. The matrix element for an electric quadrupole transition from the state $\psi_i(\mathbf{r}_1, \ldots, \mathbf{r}_Z)$ to the state $\psi_f(\mathbf{r}_1, \ldots, \mathbf{r}_Z)$ of an atom is given by

$$Q_{fi} = \int d^3\mathbf{r}_1 \ldots d^3\mathbf{r}_Z \psi_f^*(\mathbf{r}_1, \ldots, \mathbf{r}_Z) \left[\sum_{j=1}^{Z} x_j z_j \right] \psi_i(\mathbf{r}_1, \ldots, \mathbf{r}_Z).$$

Derive the parity selection rule for electric quadrupole transitions.

CHAPTER

Angular momentum I: theory

In this chapter we shall develop the theory of angular momentum in quantum mechanics in considerable generality and detail. Its importance is due to the fact that it depends only on the rotational symmetry properties of systems and conservation laws. It therefore enables one to deduce exact results about complicated systems which admit only approximate calculations. These exact results are often very important for our understanding and the approximate treatments of these systems. Within the scope of this book it is not possible to give a comprehensive treatment of angular momentum or complete proofs of all results. Fortunately, for many purposes an understanding and knowledge of the results suffice.

The complexity of the systems to be studied in this chapter, and later, makes the notation which has been used so far for wave functions and states rather cumbersome. Dirac's notation is much more convenient and widely used, and it will be explained in the first section of the chapter.

If mathematical formalism appears to dominate this chapter, the balance will be redressed in the next chapter where some realistic applications of angular momentum theory will illustrate its importance—and where the density of equations will be quite low!

5.1 DIRAC NOTATION

I now want to introduce the Dirac notation. Instead of writing $\psi_a, \psi_b, \ldots$ for the wave functions which specify different states $a, b, \ldots$, we shall write $|\psi_a\rangle, |\psi_b\rangle, \ldots$. Alternatively, we could write $|a\rangle, |b\rangle, \ldots$; here $a, b, \ldots$ simply label the states whose wave functions are $\psi_a, \psi_b, \ldots$. For $a, b, \ldots$, any labels which specify these states unambiguously will do. For example, for a particle in a central potential, with wave functions $\psi_{nlm}(\mathbf{r})$, Eq. (2.92), we could write $|n, l, m,\rangle$ instead of $|\psi_{nlm}\rangle$. The symbol $|\ \rangle$ was introduced by Dirac who called it a ket (or a state-ket since it specifies the state, i.e. the wave function, of a system). We shall use the Dirac notation as a compact way of writing rather elaborate mathematics. This is not doing justice to Dirac. For him it was a convenient notation for a very general and quite profound formulation of quantum mechanics.*

We can now express our earlier results in Dirac notation. With $|a\rangle = |\psi_a\rangle$ and $|b\rangle = |\psi_b\rangle$, the scalar product of the kets $|a\rangle$ and $|b\rangle$ is denoted by $\langle b|a\rangle$ and defined in the usual way[†] by

$$\langle b|a\rangle = \langle \psi_b|\psi_a\rangle = \int d^3r\, \psi_b^*(\mathbf{r})\psi_a(\mathbf{r}). \tag{5.1}$$

From this explicit definition, the basic properties of the scalar product follow:

(i)
$$\langle b|a\rangle^* = \langle a|b\rangle. \tag{5.2}$$

(ii) The scalar product of any ket $|d\rangle$ with

$$|c\rangle = \lambda|a\rangle + \mu|b\rangle,$$

where λ and μ are complex numbers, is given by

$$\langle d|c\rangle = \lambda\langle d|a\rangle + \mu\langle d|b\rangle. \tag{5.3}$$

One obtains $\langle c|d\rangle$ from this equation by taking its complex conjugate and using Eq. (5.2):

$$\langle c|d\rangle = \lambda^*\langle a|d\rangle + \mu^*\langle b|d\rangle. \tag{5.4}$$

(iii)
$$\langle a|a\rangle > 0 \tag{5.5}$$

for all wave functions $\psi_a(\mathbf{r})$.

A ket $|a\rangle$ is called normalized to unity (or normed) if

$$\langle a|a\rangle = 1, \tag{5.6}$$

---

* In Chapter 12, I shall give an introduction to the Dirac formalism. Although this material is not needed elsewhere in this book, a reader could study it with benefit at this stage as it deepens one's understanding. (Chapter 12 depends only on Chapters 1 to 3—see the flow diagram on the inside front cover.)

† It is obvious from Eq. (5.1) that the ordering of a and b is significant. When talking of the scalar product of $|a\rangle$ and $|b\rangle$, or of $|a\rangle$ with $|b\rangle$, I shall not necessarily mean $\langle b|a\rangle$ rather than $\langle a|b\rangle$, as it will be clear from the context which is meant. The symbol $\langle\ |$ is called a bra. But it is not necessary to introduce bras as independent entities. They will merely occur as the left-hand sides of scalar products. The names bra and ket originate, of course, from $\langle\ \rangle =$ bra(c)ket.

and two kets $|a\rangle$ and $|b\rangle$ are orthogonal if

$$\langle a|b\rangle = \langle b|a\rangle = 0. \tag{5.7}$$

If $\xi_1(\mathbf{r})$, $\xi_2(\mathbf{r})$, ... are a complete set of orthonormal wave functions, we have

$$\langle \xi_m|\xi_n\rangle = \delta_{mn} \tag{5.8}$$

and the expansion of any wave function $\psi_a(\mathbf{r})$ in terms of this complete set $\xi_1(\mathbf{r})$, $\xi_2(\mathbf{r})$, ... in Dirac notation takes the form

$$|a\rangle = \sum_n c_a(n)|\xi_n\rangle. \tag{5.9a}$$

Taking the scalar product of this equation with $|\xi_m\rangle$, we obtain the expansion coefficient

$$c_a(m) = \langle \xi_m|a\rangle, \tag{5.9b}$$

so that we can write Eq. (5.9a) as

$$|a\rangle = \sum_n |\xi_n\rangle\langle\xi_n|a\rangle. \tag{5.10}$$

It is usual to simplify the notation and write $|n\rangle$ for $|\xi_n\rangle$, etc. (just as we wrote $|a\rangle$ for $|\psi_a\rangle$); Eq. (5.10) then becomes

$$|a\rangle = \sum_n |n\rangle\langle n|a\rangle. \tag{5.11}$$

In section 1.2 we defined operators, and in particular Hermitian operators which represent observables. An operator A is a prescription for transforming wave functions: A operating on a wave function ψ transforms it into the wave function $\chi = A\psi$. Translated into Dirac notation, A operating on $|\psi\rangle$ transforms it into $|\chi\rangle = A|\psi\rangle$. The state into which A transforms $|\psi\rangle$ has here been labelled χ. Alternatively, we can label this state $A\psi$ and write

$$|\chi\rangle = |A\psi\rangle = A|\psi\rangle. \tag{5.12}$$

Taking the scalar product of Eq. (5.12) with an arbitrary ket $|\phi\rangle$ gives

$$\langle \phi|\chi\rangle = \langle \phi|A\psi\rangle = \langle \phi|A|\psi\rangle. \tag{5.13}$$

For example, for a single spinless particle the wave functions are functions of $\mathbf{r}$ ($\phi = \phi(\mathbf{r})$, etc.), the operator A is a function of $\mathbf{r}$ and $\mathbf{p} = -i\hbar\nabla$: $A = A(\mathbf{r}, -i\hbar\nabla)$, and Eq. (5.13) stands for

$$\int d^3r\,\phi^*(\mathbf{r})\chi(\mathbf{r}) = \int d^3r\,\phi^*(\mathbf{r})A(\mathbf{r}, -i\hbar\nabla)\psi(\mathbf{r}).$$

It follows from Eqs. (5.13) and (5.2) that

$$\langle\phi|A|\psi\rangle^* = \langle\phi|A\psi\rangle^* = \langle A\psi|\phi\rangle. \tag{5.14}$$

Note, we do not have $\langle A|\psi|\phi\rangle$ on the right-hand side of this equation which would be meaningless. $\langle\phi|A|\psi\rangle$ is called the matrix element of the operator A between the states $|\phi\rangle$ and $|\psi\rangle$.

The definition (1.13) of a Hermitian operator A is easily translated into Dirac notation: for any two kets $|\phi\rangle$ and $|\psi\rangle$

$$\langle\phi|A\psi\rangle = \langle A\phi|\psi\rangle. \tag{5.15a}$$

On account of Eqs. (5.13) and (5.14) this can be written

$$\langle\phi|A|\psi\rangle = \langle\psi|A|\phi\rangle^*. \tag{5.15b}$$

5.2   THE ANGULAR MOMENTUM OF A SYSTEM

In section 4.3, the angular momentum operators $\mathbf{J}$ were defined as three Hermitian operators J_x, J_y and J_z which satisfy the commutation relations

$$[J_x, J_y] = i\hbar J_z, \quad \text{etc.,} \tag{5.16}$$

where etc. stands for two similar equations obtained by cyclic permutations of x, y, z (i.e. $x \to y \to z \to x$) carried out once or twice. All angular momentum properties of a system follow from these commutation relations directly. The operators $\mathbf{J}$ are not to be thought of as differential operators, as they were in section 2.3 where the properties of orbital angular momentum were derived from explicit expressions like $l_z = -i\hbar\,\partial/\partial\phi$. Consequently, the following results will have much wider applicability.

With $\mathbf{J}^2$ defined by

$$\mathbf{J}^2 = J_x^2 + J_y^2 + J_z^2, \tag{5.17}$$

one can show that

$$[J_j, \mathbf{J}^2] = 0, \quad j = x, y, z. \tag{5.18}$$

In section 4.3 we deduced this result from the rotational invariance of $\mathbf{J}^2$, but it can also be obtained from the commutation relations (5.16), using the commutator identities (3.5). (See problem 3.1 where, writing $\mathbf{l}$ for $\mathbf{J}$, the same result is derived, i.e. Eq. (3.44d) is obtained from Eq. (3.44c) using the commutator identities (3.5) only.)

Since the components of $\mathbf{J}$ do not commute with each other, only one of them can be specified simultaneously with $\mathbf{J}^2$. Conventionally, this component is called J_z. To determine the simultaneous eigenstates of $\mathbf{J}^2$ and J_z we must study the eigenvalue equations

$$\mathbf{J}^2|J, M\rangle = \hbar^2 J(J + 1)|J, M\rangle \tag{5.19}$$

$$J_z|J, M\rangle = \hbar M|J, M\rangle \tag{5.20}$$

i.e. $|J, M\rangle$ is the mutual eigenstate of $\mathbf{J}^2$ and J_z with the eigenvalues $\hbar^2 J(J + 1)$ and $\hbar M$. At this stage J and M are unknown parameters, still to be determined. Denoting the eigenvalues of $\mathbf{J}^2$ by $\hbar^2 J(J + 1)$ will turn out to be convenient, as the reader no doubt already appreciates.*

5.2.1   The basic results

For most purposes it is more important to know and understand the basic results about the angular momentum of a system than to have studied a detailed derivation. I shall therefore start by stating and discussing the basic results:

(1) The eigenvalues of $\mathbf{J}^2$ are of the form $\hbar^2 J(J + 1)$, with J taking on the values

$$J = 0, \tfrac{1}{2}, 1, \tfrac{3}{2}, 2, \ldots. \tag{5.22a}$$

(2) Corresponding to a given eigenvalue of $\mathbf{J}^2$, i.e. to one of the values (5.22a) for J, J_z possesses the eigenvalues $\hbar M$ where M takes on the $2J + 1$ values

$$M = J, J - 1, \ldots, -J. \tag{5.22b}$$

Eigenstates belonging to different eigenvalues of an observable are orthogonal. We shall also assume them normed, so that

$$\langle J', M'|J, M\rangle = \delta_{JJ'}\delta_{MM'}. \tag{5.23}$$

J and M are called the angular momentum and magnetic quantum numbers, in analogy to the corresponding names in the case of orbital angular momentum. Whereas the orbital angular momentum quantum number l can only have integer values, $l = 0, 1, 2, \ldots$, we see from Eq. (5.22a) that J can also take on half-integer values $J = \tfrac{1}{2}, \tfrac{3}{2}, \ldots$. These half-integer values are related to spin, as we shall see. Thus, not all values of J occur for a particular system.

---

* $\mathbf{J}^2$ and J_z do not suffice to determine the state of a system completely. They will have to be augmented by other commuting observables $A, B, \ldots$—which we shall denote collectively by Q—to constitute a complete set of commuting observables. (For example, for a particle in a central potential a complete set of commuting observables consists of $\mathbf{l}^2$, l_z and the Hamiltonian H.) The eigenkets should then be written $|q, J, M\rangle$, with

$$Q|q, J, M\rangle = q|q, J, M\rangle. \tag{5.21}$$

As we shall be considering one eigenvalue q only, we shall omit the label q from the kets.

Which values actually occur depends on the system. In contrast, for a given value of J, J_z always takes on all the $2J + 1$ values $\hbar J$, $\hbar(J - 1), \ldots, -\hbar J$. Furthermore, the $2J + 1$ eigenstates $|J, M\rangle$, with $M = J, \ldots, -J$, have the same energy, i.e. are degenerate, unless there is a preferred direction in space which destroys the spherical symmetry; for example, this happens for an atom placed in an applied electric or magnetic field. The reason for this degeneracy is, of course, that for an isotropic system the choice of the direction for the z-axis—and this determines the value of M—cannot be significant.

5.2.2   The raising and lowering operators $J_\pm$.

The derivation of the values (5.22) for the quantum numbers J and M depends on a property of the angular momentum operators which is very useful in practice. I shall now derive this property. In the starred section 5.8, at the end of this chapter, this property will be used to obtain the values (5.22) of the angular momentum quantum numbers.

Defining the operators

$$J_\pm = J_x \pm iJ_y \tag{5.24}$$

one easily shows from the commutation relations (5.16) and (5.18) that

$$[J_z, J_\pm] = \pm \hbar J_\pm, \qquad [\mathbf{J}^2, J_\pm] = 0. \tag{5.25}$$

(Note that in equations like (5.24) or (5.25) the upper signs or the lower signs always go together.) It follows from Eqs. (5.24) and (5.25) that

$$\mathbf{J}^2 J_\pm |J, M\rangle = J_\pm \mathbf{J}^2 |J, M\rangle = \hbar^2 J(J + 1) J_\pm |J, M\rangle \tag{5.26}$$

and

$$J_z J_\pm |J, M\rangle = J_\pm (J_z \pm \hbar)|J, M\rangle = \hbar(M \pm 1) J_\pm |J, M\rangle. \tag{5.27}$$

Thus $J_\pm |J, M\rangle$ are eigenstates of $\mathbf{J}^2$ with the eigenvalue $\hbar^2 J(J + 1)$, and of J_z with the eigenvalues $\hbar(M \pm 1)$, i.e. apart from factors of proportionality $J_+ |J, M\rangle$ is the state $|J, M + 1\rangle$, and $J_- |J, M\rangle$ the state $|J, M - 1\rangle$, i.e.

$$J_\pm |J, M\rangle \propto |J, M \pm 1\rangle. \tag{5.28a}$$

$J_\pm$ are called ladder operators or raising and lowering operators. Operating with $J_\pm$ repeatedly on $|J, M\rangle$ generates eigenstates $|J, M \pm 1\rangle$, $|J, M \pm 2\rangle, \ldots$ with the same eigenvalue $\hbar^2 J(J + 1)$ for $\mathbf{J}^2$ and with eigenvalues $\hbar(M \pm 1)$, $\hbar(M \pm 2), \ldots$ for J_z.

We shall see that Eqs. (5.28a)—even without the factors of proportionality—are very useful when dealing with angular momentum problems; for example, when considering the addition of angular momentum. For completeness we quote these factors; they will be derived in section 5.8. With the conventional definitions of the phase factors* of the eigenkets $|J, M\rangle$, Eqs. (5.28a) become

$$J_{\pm}|J, M\rangle = \hbar[J(J + 1) - M(M \pm 1)]^{1/2}|J, M \pm 1\rangle$$
$$= \hbar[(J \mp M)(J \pm M + 1)]^{1/2}|J, M \pm 1\rangle. \qquad (5.28b)$$

We see from the last line that

$$J_{+}|J, J\rangle = 0, \quad J_{-}|J, -J\rangle = 0. \qquad (5.29)$$

These relations show that the procedures (5.28) for generating eigenstates $|J, M \pm 1\rangle$ from $|J, M\rangle$ by means of the operators $J_{\pm}$ break off at $M = \pm J$: for a given eigenvalue $\hbar^2 J(J + 1)$ of $\mathbf{J}^2$, the largest and smallest eigenvalues of J_z are $\hbar M = \pm \hbar J$, leading to $2J + 1$ states $|J, M\rangle$ with $M = J, J - 1, \ldots, -J$.

5.3   SPIN $\frac{1}{2}$

Quantization of orbital angular momentum led to the angular momentum quantum number l having integer values $l = 0, 1, 2, \ldots$. In the last section we found the additional possibilities of half-integer values for the angular momentum quantum number of a system: $J = \frac{1}{2}, \frac{3}{2}, \ldots$. It is a remarkable fact that nature utilizes these half-integer values. They are connected with the intrinsic angular momenta of particles in their own rest frames, i.e. with their spins. There is overwhelming evidence that electrons, protons, neutrons and many of the more unstable particles of nature possess spin $\frac{1}{2}$.

The theory of the last section for the case $J = \frac{1}{2}$ provides the formalism for handling the angular momentum of a spin $\frac{1}{2}$ particle. Conventionally, one writes $s = \frac{1}{2}$ instead of $J = \frac{1}{2}$ for the spin angular momentum quantum number and $\mathbf{s}$ for the spin angular momentum operators, instead of $\mathbf{J}$. $\mathbf{s}^2$ has the value $\hbar^2 s(s + 1) = \frac{3}{4}\hbar^2$ and s_z has just two eigenvalues $\hbar m_s$, with $m_s = \pm\frac{1}{2}$, which correspond to spin parallel (spin up: ↑) and spin antiparallel (spin down: ↓) to the z-axis. The corresponding eigenstates of s_z will be denoted by

$$|\alpha\rangle = |m_s = \tfrac{1}{2}\rangle \qquad |\beta\rangle = |m_s = -\tfrac{1}{2}\rangle \qquad (5.30)$$

i.e.

$$s_z|\alpha\rangle = \frac{\hbar}{2}|\alpha\rangle \qquad s_z|\beta\rangle = -\frac{\hbar}{2}|\beta\rangle. \qquad (5.31a)$$

Since the states $|\alpha\rangle$ and $|\beta\rangle$ belong to different eigenvalues of s_z, they are orthogonal. We shall also assume them normed, so that

$$\langle\alpha|\alpha\rangle = 1, \qquad \langle\beta|\beta\rangle = 1, \qquad \langle\alpha|\beta\rangle = 0. \qquad (5.32)$$

---

* Just as two wave functions ψ and $\psi e^{i\alpha}$, with α real, represent the same state, so do two kets $|\psi\rangle$ and $e^{i\alpha}|\psi\rangle$.

For spin $\frac{1}{2}$, the angular momentum commutation relations (5.16) are written

$$[s_x, s_y] = i\hbar s_z, \quad \text{etc.,} \tag{5.33}$$

and Eqs. (5.28b) for the raising and lowering operators $s_\pm$ become

$$s_\pm |m_s\rangle = (s_x \pm is_y)|m_s\rangle = \hbar[\tfrac{3}{4} - m_s(m_s \pm 1)]^{1/2}|m_s \pm 1\rangle. \tag{5.34}$$

From these equations one finds that

$$s_x|\alpha\rangle = \frac{\hbar}{2}|\beta\rangle \qquad s_x|\beta\rangle = \frac{\hbar}{2}|\alpha\rangle \tag{5.31b}$$

$$s_y|\alpha\rangle = i\frac{\hbar}{2}|\beta\rangle \qquad s_y|\beta\rangle = -i\frac{\hbar}{2}|\alpha\rangle. \tag{5.31c}$$

An arbitrary spin state $|\chi\rangle$ can be written as a linear superposition of $|\alpha\rangle$ and $|\beta\rangle$:

$$|\chi\rangle = |\alpha\rangle a_1 + |\beta\rangle a_2 \tag{5.35}$$

where a_1 and a_2 are complex numbers. They are obtained by taking the scalar product of $|\chi\rangle$ with $|\alpha\rangle$ or $|\beta\rangle$:

$$a_1 = \langle\alpha|\chi\rangle, \qquad a_2 = \langle\beta|\chi\rangle. \tag{5.36}$$

Substituting these expansion coefficients in Eq. (5.35) gives

$$|\chi\rangle = |\alpha\rangle\langle\alpha|\chi\rangle + |\beta\rangle\langle\beta|\chi\rangle \tag{5.37}$$

which has the form (5.11) of an expansion in eigenstates. If $|\chi\rangle$ is normed it satisfies the condition

$$1 = \langle\chi|\chi\rangle = \langle\chi|\alpha\rangle\langle\alpha|\chi\rangle + \langle\chi|\beta\rangle\langle\beta|\chi\rangle$$
$$= |\langle\alpha|\chi\rangle|^2 + |\langle\beta|\chi\rangle|^2 = |a_1|^2 + |a_2|^2. \tag{5.38}$$

Since Eqs. (5.31a–c) tell us the effect of operating with $\mathbf{s} = (s_x, s_y, s_z)$ on $|\alpha\rangle$ and $|\beta\rangle$, and since an arbitrary spin state can be expanded in terms of $|\alpha\rangle$ and $|\beta\rangle$, it follows that we know the effect of operating with $\mathbf{s}$ on any spin state $|\chi\rangle$. For example, we can calculate the expectation value of s_z in the normed state (5.37). From Eqs. (5.31a) and (5.37)

$$s_z|\chi\rangle = \frac{\hbar}{2}|\alpha\rangle\langle\alpha|\chi\rangle - \frac{\hbar}{2}|\beta\rangle\langle\beta|\chi\rangle \tag{5.39}$$

whence

$$\langle\chi|s_z|\chi\rangle = \frac{\hbar}{2}[|\langle\alpha|\chi\rangle|^2 - |\langle\beta|\chi\rangle|^2]. \tag{5.40}$$

This expression has the form we should have expected, since

$$|a_1|^2 = |\langle \alpha | \chi \rangle|^2, \qquad |a_2|^2 = |\langle \beta | \chi \rangle|^2 \tag{5.41}$$

are the probabilities that a particle in the spin state $|\chi\rangle$ has spin up and spin down respectively.

5.3.1 Matrix representation of spin

Spin $\tfrac{1}{2}$ angular momentum can be very conveniently handled by representing spin operators by 2×2 matrices and spin states by two-component column vectors. We represent the spin states $|\alpha\rangle$ and $|\beta\rangle$ by the column vectors

$$\alpha = \begin{bmatrix} 1 \\ 0 \end{bmatrix}, \qquad \beta = \begin{bmatrix} 0 \\ 1 \end{bmatrix}, \tag{5.42}$$

and the spin angular momentum operators by

$$\mathbf{s} = \tfrac{1}{2}\hbar\boldsymbol{\sigma} \tag{5.43a}$$

where $\boldsymbol{\sigma} = (\sigma_x, \sigma_y, \sigma_z)$ are the Pauli matrices

$$\sigma_x = \begin{bmatrix} 0, & 1 \\ 1, & 0 \end{bmatrix}, \quad \sigma_y = \begin{bmatrix} 0, & -i \\ i, & 0 \end{bmatrix}, \quad \sigma_z = \begin{bmatrix} 1, & 0 \\ 0, & -1 \end{bmatrix}. \tag{5.43b}$$

From Eqs. (5.42) and (5.43) one verifies at once all our earlier results. For example,

$$s_z \alpha = \tfrac{1}{2}\hbar \begin{bmatrix} 1, & 0 \\ 0, & -1 \end{bmatrix}\begin{bmatrix} 1 \\ 0 \end{bmatrix} = \tfrac{1}{2}\hbar\begin{bmatrix} 1 \\ 0 \end{bmatrix} = \tfrac{1}{2}\hbar\alpha$$

which is the matrix form of the first of Eqs. (5.31a). One similarly verifies that the spin matrices (5.43) satisfy the angular momentum commutation relations (5.33).* The general spin state $|\chi\rangle$, Eqs. (5.35) or (5.37), is now represented by

$$a_1\alpha + a_2\beta = a_1\begin{bmatrix} 1 \\ 0 \end{bmatrix} + a_2\begin{bmatrix} 0 \\ 1 \end{bmatrix} = \begin{bmatrix} a_1 \\ a_2 \end{bmatrix}. \tag{5.45}$$

---

* Of course, expressions like $s_x s_y$ must be calculated employing matrix multiplication. If A is the 2×2 matrix

$$\begin{bmatrix} A_{11}, & A_{12} \\ A_{21}, & A_{22} \end{bmatrix},$$

etc., then

$$(AB)_{mn} = \sum_{p=1}^{2} A_{mp}B_{pn}, \qquad m, n = 1, 2. \tag{5.44}$$

To complete the matrix formalism, we require the definition of the scalar product. If

$$|\eta\rangle = |\alpha\rangle\langle\alpha|\eta\rangle + |\beta\rangle\langle\beta|\eta\rangle = |\alpha\rangle b_1 + |\beta\rangle b_2 \qquad (5.46)$$

then the scalar product of $|\chi\rangle$, Eq. (5.37), with $|\eta\rangle$ is given by

$$\langle\eta|\chi\rangle = \langle\eta|\alpha\rangle\langle\alpha|\chi\rangle + \langle\eta|\beta\rangle\langle\beta|\chi\rangle$$

$$= b_1^* a_1 + b_2^* a_2 = [b_1^*, b_2^*]\begin{bmatrix} a_1 \\ a_2 \end{bmatrix}. \qquad (5.47)$$

The last expression is the matrix form for the scalar product $\langle\eta|\chi\rangle$. Just as the ket $|\chi\rangle$ is represented by a two-component column vector, with elements a_1 and a_2, so the bra part $\langle\eta|$ of the scalar product $\langle\eta|\chi\rangle$ is represented by a two-component row vector, with elements b_1^* and b_2^* (i.e. the elements for $\langle\eta|$ are the complex conjugates of those for $|\eta\rangle$), and the right-hand side of Eq. (5.47) must be interpreted according to the rules for matrix multiplication.

We illustrate the matrix formalism in two examples. Firstly, we calculate the expectation value of s_x in the spin state $|\chi\rangle$. From Eqs. (5.43) and (5.45), one has directly

$$\langle\chi|s_x|\chi\rangle = \tfrac{1}{2}\hbar[a_1^*, a_2^*]\begin{bmatrix} 0, & 1 \\ 1, & 0 \end{bmatrix}\begin{bmatrix} a_1 \\ a_2 \end{bmatrix}$$

$$= \tfrac{1}{2}\hbar[a_1^*, a_2^*]\begin{bmatrix} a_2 \\ a_1 \end{bmatrix} = \tfrac{1}{2}\hbar(a_1 a_2^* + a_2^* a_1). \qquad (5.48)$$

The reader should derive the result (5.40) for $\langle\chi|s_z|\chi\rangle$ using matrices.

Secondly, we consider a problem which will be of interest to us later: to find the eigenvalues and eigenstates of a component of the spin operator $\mathbf{s}$ in the direction of a unit vector $\hat{\mathbf{n}}$, i.e. we want to solve the eigenvalue problem

$$\hat{\mathbf{n}} \cdot \mathbf{s}|\chi\rangle = \tfrac{1}{2}\hbar\lambda|\chi\rangle. \qquad (5.49)$$

For simplicity we shall assume that $\hat{\mathbf{n}}$ lies in the (x, z)-plane, i.e. it has the Cartesian components $\hat{\mathbf{n}} = (\sin\theta, 0, \cos\theta)$, with $0 \leqslant \theta \leqslant \pi$, and

$$\hat{\mathbf{n}} \cdot \mathbf{s} = s_x \sin\theta + s_z \cos\theta. \qquad (5.50)$$

From Eqs. (5.43) the eigenvalue problem (5.49) now becomes

$$\begin{bmatrix} \cos\theta, & \sin\theta \\ \sin\theta, & -\cos\theta \end{bmatrix}\begin{bmatrix} a_1 \\ a_2 \end{bmatrix} = \lambda\begin{bmatrix} a_1 \\ a_2 \end{bmatrix}. \qquad (5.51)$$

Eq. (5.51) is easily solved. It is equivalent to

$$a_1 \cos \theta + a_2 \sin \theta = \lambda a_1 \tag{5.52a}$$

$$a_1 \sin \theta - a_2 \cos \theta = \lambda a_2. \tag{5.52b}$$

Each of these equations gives a value for a_1/a_2. For these to be the same, i.e. for Eqs. (5.52a) and (5.52b) to be consistent, λ must have one of the values $\lambda = \pm 1$, as the reader can easily check. Hence the eigenvalues of $\hat{\mathbf{n}} \cdot \mathbf{s}$ are $\pm \frac{1}{2}\hbar$, which are the same as those of s_z. Actually we could have obtained these values without calculation: there is nothing to choose between different directions, and so the component of $\mathbf{s}$ in *any* direction must have the eigenvalues $\pm \frac{1}{2}\hbar$. From Eq. (5.52a) we find

$$a_1 \sin (\tfrac{1}{2}\theta) = a_2 \cos (\tfrac{1}{2}\theta) \qquad \text{for } \lambda = +1 \tag{5.53a}$$

$$a_1 \cos (\tfrac{1}{2}\theta) = -a_2 \sin (\tfrac{1}{2}\theta), \qquad \text{for } \lambda = -1. \tag{5.53b}$$

The normalization condition $|a_1|^2 + |a_2|^2 = 1$ leads to the normed eigenvectors

$$\begin{bmatrix} \cos (\tfrac{1}{2}\theta) \\ \sin (\tfrac{1}{2}\theta) \end{bmatrix} \quad \text{and} \quad \begin{bmatrix} -\sin (\tfrac{1}{2}\theta) \\ \cos (\tfrac{1}{2}\theta) \end{bmatrix} \tag{5.54}$$

for $\lambda = +1$ and $\lambda = -1$ respectively. The arbitrary phase factor in each of these eigenvectors was chosen so that for $\theta = 0$ (i.e. $\hat{\mathbf{n}}$ in the direction of the z-axis) the vectors (5.54) reduce to the eigenvectors α and β, Eqs. (5.42). If $|\hat{\mathbf{n}}\pm\rangle$ denote the eigenkets of $\hat{\mathbf{n}} \cdot \mathbf{s}$ with eigenvalues $\pm\frac{1}{2}\hbar$, then it follows from Eqs. (5.54) that

$$|\hat{\mathbf{n}}+\rangle = \cos (\tfrac{1}{2}\theta)|\alpha\rangle + \sin (\tfrac{1}{2}\theta)|\beta\rangle \tag{5.55a}$$

$$|\hat{\mathbf{n}}-\rangle = -\sin (\tfrac{1}{2}\theta)|\alpha\rangle + \cos (\tfrac{1}{2}\theta)|\beta\rangle. \tag{5.55b}$$

From Eq. (5.55a) one easily obtains the probability $P(\hat{\mathbf{z}}, \hat{\mathbf{n}}+)$ that a measurement of the spin component $\hat{\mathbf{n}} \cdot \mathbf{s}$ of a particle in the state $|\alpha\rangle$ (i.e. with spin parallel to the unit vector $\hat{\mathbf{z}}$ in the direction of the positive z-axis) produces the result $+\frac{1}{2}\hbar$. This probability is given by* $|\langle \hat{\mathbf{n}} + |\alpha\rangle|^2$ so that from Eq. (5.55a)

$$P(\hat{\mathbf{z}}, \hat{\mathbf{n}}+) = \cos^2 (\tfrac{1}{2}\theta). \tag{5.56}$$

We shall use this result in the analysis of spin correlations in section 6.3.

---

* Remember that, in the expansion of $|\alpha\rangle$ in terms of the orthonormal spin eigenstates $|\hat{\mathbf{n}}\pm\rangle$, $\langle \hat{\mathbf{n}}+|\alpha\rangle$ is the expansion coefficient multiplying $|\hat{\mathbf{n}}+\rangle$.

5.3.2 Two-component wave functions

So far we have considered only the spin. The wave function of a spin $\frac{1}{2}$ particle has also a spatial dependence and the complete wave function has the form

$$\Psi = \psi_1(\mathbf{r})\alpha + \psi_2(\mathbf{r})\beta = \begin{bmatrix} \psi_1(\mathbf{r}) \\ \psi_2(\mathbf{r}) \end{bmatrix} \qquad (5.57a)$$

where α and β are the eigenvectors (5.42) of s_z. This is an obvious generalization of expression (5.45), with the spin up and spin down probability amplitudes a_1 and a_2 being functions of $\mathbf{r}$. In other words, a spin $\frac{1}{2}$ particle is described by a two-component wave function. The probability interpretation to which Eq. (5.57a) gives rise follows at once: $|\psi_i(\mathbf{r})|^2 d^3\mathbf{r}$ is the probability that the particle is in the volume element $d^3\mathbf{r}$ at $\mathbf{r}$, with spin up if $i = 1$, and with spin down if $i = 2$, and

$$P(\mathbf{r})d^3\mathbf{r} = \sum_{i=1}^{2} |\psi_i(\mathbf{r})|^2 d^3\mathbf{r} \qquad (5.58)$$

is the probability that the particle is in the volume element $d^3\mathbf{r}$ at $\mathbf{r}$ with either spin up or spin down. The normalization of the wave function Ψ which these probability statements presuppose is given by

$$\int P(\mathbf{r})d^3\mathbf{r} = \sum_{i=1}^{2} \int |\psi_i(\mathbf{r})|^2 d^3\mathbf{r} = 1. \qquad (5.59)$$

The ket $|\Psi\rangle$ for the two-component wave function (5.57a) can be written in various ways, such as

$$|\Psi\rangle = |\psi_1\alpha + \Psi_2\beta\rangle = |\psi_1, \psi_2\rangle. \qquad (5.57b)$$

The scalar product of $|\Psi\rangle$ with

$$|\Phi\rangle = |\phi_1\alpha + \phi_2\beta\rangle = |\phi_1, \phi_2\rangle \qquad (5.60)$$

is defined by

$$\begin{aligned} \langle\Phi|\Psi\rangle &= \langle\phi_1\alpha + \phi_2\beta|\psi_1\alpha + \psi_2\beta\rangle \\ &= \langle\phi_1|\psi_1\rangle\langle\alpha|\alpha\rangle + \langle\phi_1|\psi_2\rangle\langle\alpha|\beta\rangle \\ &\quad + \langle\phi_2|\psi_1\rangle\langle\beta|\alpha\rangle + \langle\phi_2|\psi_2\rangle\langle\beta|\beta\rangle \\ &= \sum_{i=1}^{2} \langle\phi_i|\psi_i\rangle = \sum_{i=1}^{2} \int d^3\mathbf{r}\,\phi_i^*(\mathbf{r})\psi_i(\mathbf{r}). \end{aligned} \qquad (5.61a)$$

This is an obvious combination of the definitions (5.47) and (5.1) of the scalar products for the two spin states and for the wave functions of two spinless particles respectively. For $\Phi = \Psi$, (5.61a) is the normalization integral (5.59). Alternatively, $\langle\Phi|\Psi\rangle$ can be written in matrix form as

$$\langle\Phi|\Psi\rangle = \langle\phi_1, \phi_2|\psi_1, \psi_2\rangle = \int d^3r[\phi_1^*(\mathbf{r}), \phi_2^*(\mathbf{r})]\begin{bmatrix}\psi_1(\mathbf{r})\\ \psi_2(\mathbf{r})\end{bmatrix}, \qquad (5.61b)$$

which, on multiplication, reduces to the result (5.61a).

The extension of the formalism to operators is obvious. For example,

$$\langle\Psi|s_z|\Psi\rangle = \tfrac{1}{2}\hbar\langle\psi_1, \psi_2|\sigma_z|\psi_1, \psi_2\rangle$$

$$= \tfrac{1}{2}\hbar\int d^3r[\psi_1^*(\mathbf{r}), \psi_2^*(\mathbf{r})]\begin{bmatrix}1 & 0\\ 0 & -1\end{bmatrix}\begin{bmatrix}\psi_1(\mathbf{r})\\ \psi_2(\mathbf{r})\end{bmatrix}$$

$$= \tfrac{1}{2}\hbar\left\{\int d^3r|\psi_1(\mathbf{r})|^2 - \int d^3r|\psi_2(\mathbf{r})|^2\right\} \qquad (5.62)$$

which should be compared with Eq. (5.40) and has a similar interpretation.

5.4   THE ANGULAR MOMENTUM ADDITION THEOREM

We now want to find the resultant angular momentum of a composite system consisting of two subsystems each with its own angular momentum. We could be considering the orbital angular momenta of two electrons, or their spins, or the spin and the orbital angular momentum of one electron. The extension of these results to more than two systems is straightforward in principle but can get very complicated in practice.

The angular momentum operators $\mathbf{J}_1$ and $\mathbf{J}_2$ of the two subsystems satisfy the standard angular momentum commutation relations (5.16) and (5.18):

$$[J_{nx}, J_{ny}] = i\hbar J_{nz}, \quad \text{etc.}, \qquad [J_{nj}, \mathbf{J}_n^2] = 0, \quad \text{etc.}, \qquad (5.63)$$

where $n = 1, 2$ labels the subsystems, $j = x, y, z$, and etc. stands for two similar equations obtained by cyclic permutations of x, y and z. The essential assumption is that the two subsystems are independent of each other, i.e. that any operator for the first subsystem is compatible with any operator for the second; in particular, that

$$[J_{1j}, J_{2k}] = 0, \qquad j, k = x, y, z. \qquad (5.64)$$

This is the case for the orbital angular momenta or the spins of two particles. It is also true for the orbital angular momentum and the spin of the same particle, since $\mathbf{l}$ operates on the space wave functions only, and $\mathbf{s}$ operates on the spin states $|\alpha\rangle$ and $|\beta\rangle$ only.

The resultant angular momentum operator $\mathbf{J}$ is defined by

$$\mathbf{J} = \mathbf{J}_1 + \mathbf{J}_2 \qquad (5.65)$$

and $\mathbf{J}^2$, as usual, by

$$\mathbf{J}^2 = J_x^2 + J_y^2 + J_z^2. \tag{5.66}$$

One can show directly from the commutation relations (5.63)–(5.64) that $\mathbf{J}$ itself satisfies the standard angular momentum commutation relations

$$[J_x, J_y] = i\hbar J_z, \quad \text{etc.,} \qquad [J_j, \mathbf{J}^2] = 0, \quad \text{etc.} \tag{5.67}$$

and that

$$[\mathbf{J}_n^2, J_z] = 0, \qquad [\mathbf{J}_n^2, \mathbf{J}^2] = 0. \tag{5.68}$$

Note, however, that $\mathbf{J}^2$ does not commute with the components of $\mathbf{J}_1$ or $\mathbf{J}_2$:

$$[\mathbf{J}^2, J_{nj}] \neq 0, \qquad j = x, y, z, \qquad n = 1, 2. \tag{5.68a}$$

The results (5.67), (5.68) and (5.68a) were in fact derived in problem 4.2. Relabelling $\mathbf{l}_1, \mathbf{l}_2$ and $\mathbf{L}$ in that problem $\mathbf{J}_1, \mathbf{J}_2$ and $\mathbf{J}$, Eqs. (4.79a–b) go over into Eqs. (5.63–64) and Eqs. (4.80a–c) go over into Eqs. (5.67), (5.68) and (5.68a); i.e. $\mathbf{J}_1$ and $\mathbf{J}_2$ satisfy the same commutation relations as $\mathbf{l}_1$ and $\mathbf{l}_2$: consequently $\mathbf{J}$ satisfies the same commutation relations as $\mathbf{L}$.

The commutation relations (5.67–(5.68) for $\mathbf{J}$ have two important consequences.

(1) Since $\mathbf{J}$ satisfies the standard angular momentum commutation relations, the basic results of section 5.2 apply: $\mathbf{J}^2$ and J_z possess mutual eigenstates with angular momentum quantum numbers J and M; for each value of J:

$$M = J, J - 1, \ldots, -J. \tag{5.69}$$

(2) $\mathbf{J}_1^2$, $\mathbf{J}_2^2$, $\mathbf{J}^2$ and J_z possess a complete set of simultaneous eigenstates. Instead of describing the system by states which are products of states for the two subsystems*

$$|q_1, J_1, M_1\rangle|q_2, J_2, M_2\rangle, \tag{5.70a}$$

the system can be specified by the simultaneous eigenstates

$$|q_1, q_2, J_1, J_2, J, M\rangle. \tag{5.70b}$$

q_1 and q_2 are the eigenvalues of other observables Q_1 and Q_2, for systems 1 and 2 respectively, which together with $\mathbf{J}_1^2$, J_{1z} and $\mathbf{J}_2^2$, J_{2z} form complete sets of commuting observables for these systems. Since we are only interested in the angular momentum properties we shall suppress the operators Q_1 and Q_2 and their eigenvalues, assuming that we are always dealing with the same eigenvalues q_1 and q_2. (For example, when considering the orbital angular momentum of the excited configuration (1s)(2s) of the helium atom, q_1 and q_2

---

* When writing products of states, like in (5.70a), the left-hand ket will always refer to the first system and the right-hand ket to the second.

would stand for the principal quantum numbers $q_1 = n_1 = 1$, $q_2 = n_2 = 2$; and $J_1 = l_1 = 0$, $J_2 = l_2 = 0$, with $M_1 = M_2 = 0$ in this case.)

The question now arises: if subsystems 1 and 2 are in states with definite quantum numbers J_1 and J_2, what values can the total angular momentum quantum number J assume? The answer is provided by the following *angular momentum addition theorem*:

From the $(2J_1 + 1)(2J_2 + 1)$ states

$$|J_1, M_1\rangle|J_2, M_2\rangle, \quad M_n = J_n, J_n - 1, \ldots, -J_n, \quad n = 1, 2, \quad (5.71)$$

one can form linear combinations $|J_1, J_2, J, M\rangle$ where J takes on the values

$$J = J_1 + J_2, J_1 + J_2 - 1, \ldots, |J_1 - J_2| \quad (5.72a)$$

and, for each of these values of J, M takes on the $2J + 1$ values

$$M = J, J - 1, \ldots, -J. \quad (5.72b)$$

The range of values (5.72b) for M we already obtained above, Eq. (5.69). The range of values for J, Eq. (5.72a), produces the correct number of states. We start with $(2J_1 + 1)(2J_2 + 1)$ states $|J_1, M_1\rangle |J_2, M_2\rangle$. After forming linear combinations from these to obtain the total angular momentum eigenstates $|J_1, J_2, J, M\rangle$, we must end up with the same number of states. This we do since from Eqs. (5.72) the number of states $|J_1, J_2, J, M\rangle$ is given by

$$\sum_{|J_1 - J_2|}^{J_1 + J_2} (2J + 1) = (2J_1 + 1)(2J_2 + 1). \quad (5.73)$$

The values $J = J_1 + J_2$ and $J = |J_1 - J_2|$ in (5.72a) correspond to the classical extreme situations of two angular momentum vectors $\mathbf{J}_1$ and $\mathbf{J}_2$ being parallel and antiparallel to each other. Classically, the resultant angular momentum $\mathbf{J}$ can have any magnitude between these extreme values, depending on the relative orientation of $\mathbf{J}_1$ and $\mathbf{J}_2$. The quantization of angular momentum leads to the restrictions imposed by (5.72) on the values of $\mathbf{J}^2$ and J_z.

The proof of the angular momentum addition theorem follows from

$$J_z|J_1, M_1\rangle|J_2, M_2\rangle$$
$$= (J_{1z} + J_{2z})|J_1, M_1\rangle|J_2, M_2\rangle = \hbar(M_1 + M_2)|J_1, M_1\rangle|J_2, M_2\rangle \quad (5.74)$$

where the last step follows since J_{1z} operates only on $|J_1, M_1\rangle$ and J_{2z} only on $|J_2, M_2\rangle$. Thus $|J_1, M_1\rangle|J_2, M_2\rangle$ is an eigenstate of J_z with the eigenvalue $\hbar M = \hbar(M_1 + M_2)$. We know that the magnetic quantum number M occurs in groups of $(2J + 1)$ values with the maximum value $M_{\max} = J$. Hence if a given value of M occurs, there must be at least one value of J greater than or equal to this value of M and corresponding to this J there exist $2J + 1$ states with $M = J, J - 1, \ldots, -J$. We shall see below how from these results Eq. (5.72a) is derived, essentially by counting states with different values of M.

I shall not prove the angular momentum addition theorem for the general case. Instead I shall in the next section prove it for two particular cases which are important in practice. The general result can be derived in exactly the same way.

5.5   EXAMPLES OF ANGULAR MOMENTUM ADDITION

5.5.1   Two spin $\frac{1}{2}$ particles

For our first example, we consider the addition of the spins of two electrons, $s_1 = s_2 = \frac{1}{2}$. According to the angular momentum addition theorem, the resultant spin S takes on the values $S = 1$ (comprising three spin states with spin magnetic quantum numbers $M_S = 1, 0, -1$) and $S = 0$ (with $M_S = 0$).

This result comes about as follows. Each electron can be in the spin-up or spin-down state, denoted by α and β in Eq. (5.42). For the two-electron system there are four spin states, 'up–up', 'up–down', 'down–up' and 'down–down', with spin functions $\alpha_1\alpha_2$, $\alpha_1\beta_2$, $\beta_1\alpha_2$ and $\beta_1\beta_2$ respectively. We define the resultant spin operator

$$\mathbf{S} = \mathbf{s}_1 + \mathbf{s}_2. \qquad (5.75)$$

Since $s_{iz}\alpha_i = \frac{1}{2}\hbar\alpha_i$, $s_{iz}\beta_i = -\frac{1}{2}\hbar\beta_i$, and $\mathbf{s}_i$ operates on the spin functions of the ith particle only, it follows that the spin functions

$$\alpha_1\alpha_2 \quad \alpha_1\beta_2 \quad \beta_1\alpha_2 \quad \beta_1\beta_2 \qquad (5.76)$$

are eigenfunctions of S_z with the magnetic quantum numbers

$$M_S = m_{s1} + m_{s2} = \quad 1 \quad 0 \quad 0 \quad -1 \qquad (5.77)$$

respectively. (This result is just Eq. (5.74) applied to the present case.)

Since $\alpha_1\alpha_2$ is an $M_S = 1$ state and there is no state with $M_S > 1$, it follows that the resultant spin $S = 1$ must occur and that corresponding to this value $S = 1$ there must be three spin states with $M_S = 1, 0, -1$. Denoting the spin functions with quantum numbers S and M_S by $\chi_{SM_S}(1, 2)$, it is clear that

$$\chi_{11}(1, 2) = \alpha_1\alpha_2, \quad \chi_{1-1}(1, 2) = \beta_1\beta_2 \qquad (5.78a)$$

We see from Eqs. (5.76–5.77) that there are two $M_S = 0$ states, $\alpha_1\beta_2$ and $\beta_1\alpha_2$. From these we can form two linearly independent combinations in many ways. In particular, there must be one linear combination which is the $S = 1$, $M_S = 0$ spin state $\chi_{10}(1, 2)$. We know from Eq. (5.28a) that $\chi_{10}(1, 2)$ is obtained from $\chi_{11}(1, 2)$ by means of the lowering operator S_-:

$$S_-\chi_{11}(1, 2) = (s_{1-} + s_{2-})\alpha_1\alpha_2 \propto \chi_{10}(1, 2). \qquad (5.79)$$

The operator S_- is symmetric in the labels 1 and 2 of the two electrons. Hence operating on the spin state $\alpha_1\alpha_2$, which is also symmetric under the interchange of 1 and 2, must produce a symmetric spin state, i.e.

$$\chi_{10}(1,2) = \frac{1}{\sqrt{2}}(\alpha_1\beta_2 + \beta_1\alpha_2), \tag{5.78b}$$

where the factor $1/\sqrt{2}$ has been added to normalize the state to unity. Note we were able to obtain the state $\chi_{10}(1,2)$ merely from the symmetry of the lowering operator S_- and of $\alpha_1\alpha_2$, without use of Eq. (5.34) for $s_-\alpha$.

This leaves just one other linear combination of $\alpha_1\beta_2$ and $\beta_1\alpha_2$ to account for. Since there is just one state left, it must be a $S = 0$, $M_S = 0$ state $\chi_{00}(1,2)$: if it were not $S = 0$, we would have to be able to form some states with $M_S > 0$ and $M_S < 0$. Symmetry arguments also at once lead to the spin function $\chi_{00}(1,2)$. It must be a linear combination of $\alpha_1\beta_2$ and $\beta_1\alpha_2$ which is orthogonal to $\chi_{10}(1,2)$, since it belongs to a different eigenvalue of $\mathbf{S}^2$. Since $\chi_{10}(1,2)$ is symmetric in 1 and 2, it follows that the antisymmetric combination of $\alpha_1\beta_2$ and $\beta_1\alpha_2$ is orthogonal to $\chi_{10}(1,2)$, i.e. the normed spin state is given by

$$\chi_{00}(1,2) = \frac{1}{\sqrt{2}}(\alpha_1\beta_2 - \beta_1\alpha_2). \tag{5.78c}$$

This result could also have been obtained by choosing the constants a and b so that S_+ or S_- operating on $a\alpha_1\beta_2 + b\beta_1\alpha_2$ give zero.

We summarize these important results. Combining two spins $s_1 = \frac{1}{2}$ and $s_2 = \frac{1}{2}$ gives $S = 1$ or $S = 0$. The $S = 1$ states, known as *triplet* states, have the symmetric spin functions

$$\chi_{11}(1,2) = \alpha_1\alpha_2, \quad \chi_{10}(1,2) = \frac{1}{\sqrt{2}}(\alpha_1\beta_2 + \beta_1\alpha_2), \quad \chi_{1-1}(1,2) = \beta_1\beta_2 \tag{5.80a}$$

and the $S = 0$ state, called a *singlet* state, has the antisymmetric spin function

$$\chi_{00}(1,2) = \frac{1}{\sqrt{2}}(\alpha_1\beta_2 - \beta_1\alpha_2). \tag{5.80b}$$

5.5.2   Two orbital angular momenta $l = 1$

For our second example, we consider the addition of the orbital angular momenta of two particles in p-states: $l_1 = l_2 = 1$. A p-state wave function is of the form $f(r)Y_1^m(\theta, \phi)$, where $Y_1^m(\theta, \phi)$ are the first-order spherical harmonics. Since the radial dependence of this wave function is not relevant to the considerations of angular momentum, we shall suppress the radial factor and write for the angular momentum wave function of one particle*

$$|1, m\rangle = Y_1^m(\Omega) \tag{5.81}$$

---

* Specifying a state by a wave function is not quite the same as specifying it by a ket. The latter is more general, but for many purposes this distinction can be ignored.

where Ω stands for the polar angles (θ, ϕ). The product functions

$$Y_1^{m_1}(\Omega_1)Y_1^{m_2}(\Omega_2), \qquad m_1, m_2 = 1, 0, -1 \qquad (5.82)$$

are eigenfunctions of $L_z = l_{1z} + l_{2z}$ with the magnetic quantum numbers $M_L = m_1 + m_2$. Corresponding to the $3 \times 3 = 9$ states (5.82) we expect states with resultant orbital angular momentum quantum numbers $L = 2, 1$ and 0. [The total number of such states is $5 + 3 + 1 = 9$, as required; see Eq. (5.73).]

The maximum value of M_L is $M_L = 2$, so the eigenvalue $L = 2$ must occur. Denoting the two-particle states by $|l_1, l_2, L, M_L\rangle$, the $L = 2, M_L = 2$ eigenstate is obviously

$$|1, 1, 2, 2\rangle = Y_1^1(\Omega_1)Y_1^1(\Omega_2). \qquad (5.83a)$$

This state is symmetric if Ω_1 and Ω_2 are interchanged, i.e. if the particle labels 1 and 2 are interchanged. Hence all the states $|1, 1, 2, M\rangle$ are symmetric, since they are obtained from $|1, 1, 2, 2\rangle$ by repeatedly applying the symmetric lowering operator $L_- = l_{1-} + l_{2-}$. I leave these and some subsequent details as an exercise for the reader (see problem 5.4).

The $L = 1$ state with $M = 1$ must be the linear combination of $Y_1^1(\Omega_1)Y_1^0(\Omega_2)$ and $Y_1^0(\Omega_1)Y_1^1(\Omega_2)$ which is orthogonal to the symmetric state $|1, 1, 2, 1\rangle$, i.e. it must be the antisymmetric combination

$$|1, 1, 1, 1\rangle = \frac{1}{\sqrt{2}}\{Y_1^1(\Omega_1)Y_1^0(\Omega_2) - Y_1^0(\Omega_1)Y_1^1(\Omega_2)\} \qquad (5.83b)$$

where the factor $1/\sqrt{2}$ ensures normalization of this state. Similarly, the states $|1, 1, 1, 0\rangle$ and $|1, 1, 1, -1\rangle$ are antisymmetric and are easily written down.

Finally, the $L = 0$ state $|1, 1, 0, 0\rangle$ is obtained by choosing numbers a, b and c such that either of the equations

$$L_{\pm}\{aY_1^1(\Omega_1)Y_1^{-1}(\Omega_2) + bY_1^0(\Omega_1)Y_1^0(\Omega_2) + cY_1^{-1}(\Omega_1)Y_1^1(\Omega_2)\} = 0 \quad (5.84)$$

holds. From Eq. (5.28b)

$$l_- Y_1^m(\Omega) = \hbar\sqrt{2}Y_1^{m-1}, \qquad \text{for } m = 0, 1, \qquad (5.85)$$

and from the last two equations one finds $a + b = 0$ and $b + c = 0$. Hence the normed state $|1, 1, 0, 0\rangle$ is given by

$$|1, 1, 0, 0\rangle = \frac{1}{\sqrt{3}}\{Y_1^1(\Omega_1)Y_1^{-1}(\Omega_2) - Y_1^0(\Omega_1)Y_1^0(\Omega_2) + Y_1^{-1}(\Omega_1)Y_1^1(\Omega_2)\}. \quad (5.83c)$$

An interesting feature of Eqs. (5.83) is that the states with $L = 2$ and $L = 0$ are symmetric with respect to the interchange of particle labels, $1 \leftrightarrow 2$, whereas the states with $L = 1$ are antisymmetric. In section 5.7 we shall utilize these properties to construct antisymmetric wave functions for the two-electron systems, as required by the Pauli exclusion principle.

In writing angular momentum quantum numbers in these two examples, I have used the standard spectroscopic convention of lower case letters $l, s, m, \ldots$ for one-particle quantities, and capitals $L, S, M, \ldots$ for systems of particles. One also refers to states with $L = 0, 1, 2, \ldots$ as S, P, D, $\ldots$ states, in analogy to the single-particle designations.*

5.6   THE GENERAL CASE

The two examples in the last section illustrate how to obtain the angular momentum eigenfunctions of a composite system from first principles, using the raising and lowering operators aided by orthogonality and symmetry arguments. In the general case we know from Eq. (5.74) that the states $|J_1, J_2, J, M\rangle$ are linear combinations of product states $|J_1, M_1\rangle|J_2, M_2\rangle$ with $M_1 + M_2 = M$, i.e. they are of the form

$$|J_1, J_2, J, M\rangle = \sum_{M_1, M_2} C(J_1, M_1, J_2, M_2; J, M)|J_1, M_1\rangle|J_2, M_2\rangle \quad (5.86a)$$

where the summations are over values of $M_1(= J_1, \ldots, -J_1)$ and $M_2(= J_2, \ldots, -J_2)$ such that

$$M_1 + M_2 = M, \quad (5.86b)$$

i.e. the coefficients $C(J_1, M_1, J_2, M_2; J, M)$ in Eq. (5.86a) vanish unless the condition (5.86b) is satisfied. For many purposes it suffices to know that the states $|J_1, J_2, J, M\rangle$ are of the form (5.86a–b) without knowing the values of the coefficients $C(J_1, M_1, J_2, M_2; J, M)$. These coefficients are known as Clebsch–Gordan or Wigner or vector coupling coefficients.[†]

The standard definition of the Clebsch–Gordan coefficients assumes certain phase conventions for the normed angular momentum eigenstates $|J_1, M_1\rangle$ and $|J_2, M_2\rangle$. The spherical harmonics in section 2.3 and the spin functions in section 5.3 were defined with these conventional phases. With these definitions, the Clebsch–Gordan coefficients are real and the resultant angular momentum eigenstates $|J_1, J_2, J, M\rangle$ are also normed. Explicit expressions exist for these coefficients and they are extensively tabulated. When using such tables, one must use the conventional phases for the angular momentum eigenfunctions.[‡]

---

* The reader should note that in the conventional notation the letter S occurs with two different meanings: either as the spin quantum number or as the orbital angular momentum quantum number $L = 0$.

[†] Many other notations are used for these coefficients. A frequently used notation is $\langle J_1, M_1, J_2, M_2|J_1, J_2, J, M\rangle$.

[‡] This also applies to our *ab initio* derivations. Eq. (5.85) and equations like it hold for the spherical harmonics as defined in section 2.3.1 but not with arbitrary phase factors. Correspondingly, Eq. (5.83c) which depended on Eq. (5.85) in its derivation, holds for the conventional phases but cannot hold for independently chosen arbitrary phases for Y_1^1, Y_1^0 and Y_1^{-1}; for example, Eq. (5.83c) cannot remain correct if we only replace Y_1^1 by $(-Y_1^1)$ in it.

5.7   ADDITION OF ANGULAR MOMENTUM FOR ELECTRONS

New considerations arise when adding the angular momenta of electrons or other identical spin $\frac{1}{2}$ particles. The resulting states must be antisymmetric with respect to the labelling of the particles and this may restrict the resultant angular momentum states.

To start with, we consider two electrons. Writing the normed single-particle wave functions

$$\psi_{nlm}(\mathbf{r}) = R_{nl}(\mathbf{r})Y_l^m(\Omega), \tag{5.87}$$

we can form resultant orbital angular momentum states for two electrons with orbitals $(n_1 l_1)$ and $(n_2 l_2)$

$$\Psi_{LM_L}(\mathbf{r}_1, \mathbf{r}_2) = \sum_{\substack{m_1, m_2 \\ m_1 + m_2 = M_L}} C(l_1, m_1, l_2, m_2; L, M_L)\psi_{n_1 l_1 m_1}(\mathbf{r}_1)\psi_{n_2 l_2 m_2}(\mathbf{r}_2). \tag{5.88}$$

For different shells, i.e. $(n_1 l_1) \neq (n_2 l_2)$, the state (5.88) and $\Psi_{LM_L}(\mathbf{r}_2, \mathbf{r}_1)$ are always orthogonal, and we can form the symmetric and antisymmetric normed combinations

$$\frac{1}{\sqrt{2}}\left[\Psi_{LM_L}(\mathbf{r}_1, \mathbf{r}_2) \pm \Psi_{LM_L}(\mathbf{r}_2, \mathbf{r}_1)\right]. \tag{5.89}$$

These space wave functions must be combined with the spin functions $\chi_{SM_s}(1, 2)$ for two electrons, Eqs. (5.80). The singlet $(S = 0)$ and triplet $(S = 1)$ spin functions are respectively antisymmetric and symmetric under the permutation $1 \leftrightarrow 2$. To form overall antisymmetric states, the symmetric space function (5.89) must be combined with the singlet spin function, and the antisymmetric space function (5.89) with the triplet spin states. Thus, for each value of $L(= l_1 + l_2, \ldots, |l_1 - l_2|)$ there exist both singlet and triplet states. For example, the first four excited energy levels of helium have the configurations (1s)(2s) and (1s)(2p); the former gives singlet and triplet S states (i.e. $L = 0$), the latter gives singlet and triplet P states (i.e. $L = 1$).

Electrons in the same (nl) shell are called equivalent. For two equivalent electrons, it can be shown quite generally that the wave function (5.88) is already symmetric or antisymmetric under the interchange $\mathbf{r}_1 \leftrightarrow \mathbf{r}_2$. If it is symmetric it must be combined with the antisymmetric singlet spin state, if antisymmetric with the triplet spin states. For example, in the helium ground state configuration (1s)^2 there is just one orbital $\psi_{100}(\mathbf{r}) = R_{10}(r)/\sqrt{(4\pi)}$ and we can form only one antisymmetric state

$$\frac{1}{4\pi} R_{10}(r_1)R_{10}(r_2) \frac{1}{\sqrt{2}}(\alpha_1\beta_2 - \beta_1\alpha_2). \tag{5.90}$$

For the (2p)^2 configuration of the valence electrons of the carbon atom in its ground state we found, in section 5.5, that the space wave functions of the $L = 2$ and $L = 0$ states [e.g. Eqs. (5.83a) and (5.83c)] are symmetric with respect to $\mathbf{r}_1 \leftrightarrow \mathbf{r}_2$, whereas the $L = 1$ wave functions [e.g. Eq. (5.83b)] are antisymmetric. Thus the S and D states must be singlet states, the P states must be triplet states. For the triplet P states, the magnetic quantum numbers take on the values $M_L = 1, 0, -1$ and $M_S = 1, 0, -1$, giving nine states with $L = S = 1$ and different pairs of values for (M_L, M_S). Similarly there are five different singlet D states $(L = 2, S = 0, M_L = 2, \ldots, -2, M_S = 0)$ and one singlet S state $(L = S = 0, M_L = M_S = 0)$. Altogether the $(2p)^2$ configuration comprises 15 states.

One easily sees that this is the correct total number of states. There are six different (2p) orbitals, specified by $m_l = 1, 0, -1$ and $m_S = \pm\frac{1}{2}$. To construct antisymmetric two-electron states from these, we can place the first electron into any one of these six states; the second electron can by placed in one of the remaining five states only. By labelling the electrons 'first' and 'second' each possibility has been counted twice, so that altogether there are $6 \times 5/2! = 15$ states, in agreement with our count above.

The above results for $(1s)^2$ and $(2p)^2$ can be generalized to $(nl)^2$ configurations. The resultant orbital angular momentum now takes on the values $L = 0, 1, \ldots, 2l$. One can show that for the states to be antisymmetric under particle interchange, one must have $S = 0$ for even values of L: S, D, G, ... states are singlet states; and one must have $S = 1$ for odd values of L: P, F, H, ... states are triplet states (see problem 5.9).

★5.8   DERIVATION OF THE EIGENVALUES OF $\mathbf{J}^2$ AND J_z

In section 5.2, Eqs. (5.22), we stated the eigenvalues of the angular momentum operators $\mathbf{J}^2$ and J_z. We shall now derive these. It will be helpful to write the eigenvalue equations (5.19) and (5.20) in a slightly modified notation as

$$\mathbf{J}^2|\lambda, M\rangle = \hbar^2\lambda|\lambda, M\rangle, \quad J_z|\lambda, M\rangle = \hbar M|\lambda, M\rangle, \tag{5.91}$$

i.e. the eigenvalues of $\mathbf{J}^2$ are denoted by $\hbar^2\lambda$ and the eigenkets are labelled by λ and M.

For a given value of λ, the magnetic quantum number M possesses a maximum and a minimum value. To prove this, we show first of all that for any state $|\psi\rangle$ the expectation value $\langle\psi|J_x^2|\psi\rangle$ is non-negative. This is just what one would expect for the expectation value of the square of an observable. With $|\chi\rangle = J_x|\psi\rangle$, we have

$$\langle\psi|J_x^2|\psi\rangle = \langle\psi|J_x|\chi\rangle = \langle\chi|J_x|\psi\rangle^*$$
$$= \langle\chi|\chi\rangle^* = \langle\chi|\chi\rangle \geq 0 \tag{5.92}$$

where the last step on the first line of this equation comes from the Hermiticity of J_x. The equality sign occurs in Eq. (5.92) if $|\chi\rangle = J_x|\psi\rangle = 0$, i.e. if $|\psi\rangle$ is an eigenstate of J_x with the eigenvalue 0. Similar arguments apply to $\langle\psi|J_y^2|\psi\rangle$

and $\langle \psi | J_z^2 | \psi \rangle$. It follows from

$$\langle \lambda, M | \mathbf{J}^2 | \lambda, M \rangle = \langle \lambda, M | J_x^2 + J_y^2 + J_z^2 | \lambda, M \rangle = \hbar^2 \lambda \qquad (5.93)$$

and

$$\langle \lambda, M | J_x^2 + J_y^2 | \lambda, M \rangle = \langle \lambda, M | \mathbf{J}^2 - J_z^2 | \lambda, M \rangle = \hbar^2 (\lambda - M^2) \qquad (5.94)$$

that

$$\lambda \geqslant 0 \quad \text{and} \quad M^2 \leqslant \lambda. \qquad (5.95)$$

We see from the last inequality that, for a given value of λ, M must possess a maximum value $M_{\max}$ and a minimum value $M_{\min}$.

We next relate $M_{\max}$ and $M_{\min}$ to λ. Let us write Eq. (5.28a) more completely as

$$J_{\pm} | \lambda, M \rangle = c_{\pm}(\lambda, M) | \lambda, M \pm 1 \rangle \qquad (5.96)$$

where the constants $c_{\pm}(\lambda, M)$ still remain to be determined. For M to possess a maximum value $M_{\max}$, which we shall denote by J, the procedure (5.96) for generating eigenkets $|\lambda, M + 1\rangle$ from $|\lambda, M\rangle$ by means of J_+ must break off for $M = J$. This occurs if in Eq. (5.96) $c_+(\lambda, J) = 0$, so that

$$J_+ | \lambda, J \rangle = 0, \qquad (5.97a)$$

which also implies

$$J_- J_+ | \lambda, J \rangle = 0. \qquad (5.98)$$

From $J_{\pm} = J_x \pm iJ_y$ and the commutation relations (5.16) one finds that

$$J_- J_+ = \mathbf{J}^2 - J_z^2 - \hbar J_z, \qquad (5.99)$$

whence Eq. (5.98) becomes

$$\hbar^2 [\lambda - J(J + 1)] | \lambda, J \rangle = 0,$$

i.e.

$$\lambda = J(J + 1) \qquad (5.100)$$

and the eigenvalues of $\mathbf{J}^2$ are of the form $\hbar^2 J(J + 1)$. Hereafter we revert from $|\lambda, M\rangle$ to the conventional notation $|J, M\rangle$ for the mutual eigenkets of $\mathbf{J}^2$ and J_z with eigenvalues $\hbar^2 J(J + 1)$ and $\hbar M$.

It follows from symmetry arguments that $M_{\min} = -M_{\max} = -J$. This result can also be derived directly from

$$J_- | J, M_{\min} \rangle = 0. \qquad (5.97b)$$

The values (5.22) for J and M are now easily obtained. Repeatedly applying J_+ to $|J, -J\rangle$ generates $|J, -J + 1\rangle, |J, -J + 2\rangle, \ldots, |J, J\rangle$. Hence $2J$ must be zero or a positive integer, i.e. J assumes the values (5.22a) and M takes on the corresponding $2J + 1$ values (5.22b).

Lastly, we derive the constants in Eq. (5.96) which we now denote by $c_\pm(J, M)$. From Eq. (5.99) we have

$$\langle J, M | J_- J_+ | J, M \rangle = \hbar^2 [J(J + 1) - M(M + 1)].\qquad (5.101)$$

We can also write this matrix element, using Eq. (5.96), as

$$\langle J, M | J_- J_+ | J, M \rangle = c_+(J, M)\langle J, M | J_- | J, M + 1 \rangle.\qquad (5.102)$$

We next show that for any two kets $|\phi\rangle$ and $|\psi\rangle$

$$\langle \phi | J_- | \psi \rangle = \langle \psi | J_+ | \phi \rangle^*.\qquad (5.103)$$

J_+ and J_-, unlike J_x and J_y, are not Hermitian operators. To prove Eq. (5.103), we therefore write

$$\langle \phi | J_- | \psi \rangle = \langle \phi | J_x | \psi \rangle - i\langle \phi | J_y | \psi \rangle$$
$$= [\langle \psi | J_x | \phi \rangle + i\langle \psi | J_y | \phi \rangle]^* = \langle \psi | J_+ | \phi \rangle^*.$$

On account of Eqs. (5.103) and (5.96) we have

$$\langle J, M | J_- | J, M + 1 \rangle = \langle J, M + 1 | J_+ | J, M \rangle^* = [c_+(J, M)]^*.\qquad (5.104)$$

Substituting Eq. (5.104) in Eq. (5.102) and comparing the resulting equation with Eq. (5.101), one obtains

$$|c_+(J, M)|^2 = \hbar^2 [J(J + 1) - M(M + 1)].\qquad (5.105)$$

With the usual convention, one chooses $c_+(J, M)$ real and positive, so that

$$c_+(J, M) = \hbar[J(J + 1) - M(M + 1)]^{1/2}.\qquad (5.106a)$$

In the same way one shows that

$$c_-(J, M) = \hbar[J(J + 1) - M(M - 1)]^{1/2}.\qquad (5.106b)$$

Substituting expressions (5.106) in Eqs. (5.96) one obtains Eqs. (5.28b).

PROBLEMS 5

5.1 Verify that the Pauli matrices (5.43b) satisfy the following relations:

(i) $[\sigma_x, \sigma_y] = 2i\sigma_z$, etc.,   (ii) $\sigma_x^2 = I$, etc.,

(iii) $\sigma_x \sigma_y + \sigma_y \sigma_x = 0$, etc.,   (iv) $\sigma_x \sigma_y = i\sigma_z$, etc.,

where etc. stands for similar equations with x, y and z cyclicly permuted and I denotes the 2×2 unit matrix.

5.2 Obtain the eigenvalues and the corresponding normed eigenstates for each of the Pauli matrices (5.43b).

5.3 A spin $\frac{1}{2}$ particle is in the spin eigenstate with spin parallel to the positive z-axis. What is the average value of its spin component in a direction making an angle θ with the z-axis?

5.4 For two particles in p-states, construct the normalized simultaneous eigenfunctions of L^2 and L_z, where L is the resultant orbital angular momentum operator of the two particles.

5.5 For a spin $\frac{1}{2}$ particle in a p-state, with wave functions $\psi_m(\mathbf{r}) = f(r)Y_1^m(\theta, \phi)$, $m = 1, 0, -1$, find the simultaneous eigenstates of $\mathbf{j}^2$ and j_z, where $\mathbf{j}$ is the total angular momentum operator of the particle.

5.6 The ^{3}He nucleus consists of two protons and a neutron. Let $|S, M\rangle$ be the normalized spin state of this system with total spin quantum number S and z-component of spin $\hbar M$. By first adding the proton spins and then combining their resultant spin with that of the neutron:

   (i) list the resulting states by their pairs of quantum numbers (S, M);
   (ii) obtain the corresponding spin eigenstates $|S, M\rangle$.

5.7 Positronium consists of an electron and a positron bound together like the electron and proton in the hydrogen atom. The spin interaction energy of the electron and positron can be written

$$H = A\mathbf{s}_1 \cdot \mathbf{s}_2$$

where $\mathbf{s}_1$ and $\mathbf{s}_2$ are the spin operators of the electron and the positron and A is a constant. Obtain the interaction energies of positronium in the singlet state $|0, 0\rangle$ and in the triplet state $|1, M\rangle$, $M = 1, 0, -1$.

   If a uniform magnetic field B is applied parallel to the z-axis, the Hamiltonian becomes

$$H' = H + \lambda B(s_{1z} - s_{2z})$$

where λ is a constant. Show that, in non-zero magnetic field, the triplet states $|1, 1\rangle$ and $|1, -1\rangle$ are still energy eigenstates while $|1, 0\rangle$ and $|0, 0\rangle$ are not. Obtain the energies of the states $|1, 1\rangle$ and $|1, -1\rangle$ and of the other two energy eigenstates.

5.8 Suppose the neutron–proton force in the deuteron can be described by the potential

$$V = V_1(r) + V_2(r)\mathbf{s}_1 \cdot \mathbf{s}_2,$$

where $\mathbf{s}_1$ and $\mathbf{s}_2$ are the spin operators of the proton and the neutron and the potentials V_1 and V_2 depend on the neutron–proton separation r only. What are the good quantum numbers of the deuteron?

   If the additional term (called a tensor force)

$$V' = V_3(r)\left\{\frac{3(\mathbf{s}_1 \cdot \mathbf{r})(\mathbf{s}_2 \cdot \mathbf{r})}{r^2} - \mathbf{s}_1 \cdot \mathbf{s}_2\right\}$$

is added to the above potential V, what now are the good quantum numbers?

5.9 The Clebsch–Gordan coefficients possess the symmetry property

$$C(l_1, m_1, l_2, m_2; L, M) = (-1)^{L - l_1 - l_2}C(l_2, m_2, l_1, m_1; L, M). \qquad (5.107)$$

Use this property to show that for two electrons in equivalent orbitals the resultant orbital angular momentum quantum number L must have even values for singlet states and odd values for triplet states.

Angular momentum II: applications

In the last chapter, the general theory of angular momentum was developed. In this chapter I shall illustrate the versatility and power of these methods by some realistic applications.

In the first two sections, I shall consider two important topics in atomic physics: the classification of atomic states according to their angular momentum quantum numbers, and electric dipole selection rules. But these methods and ideas have much wider applicability and are of great importance in molecular, solid state, nuclear and elementary particle physics.

Quantum mechanics affords a description of a system which by classical standards is incomplete. Is this a shortcoming of quantum physics or is it inherent in nature? A criterion to settle this question is provided by Bell's inequality. This is most simply formulated in terms of the spin correlations of two spin $\frac{1}{2}$ particles, and we shall consider this exciting topic in the last section of this chapter.

6.1 ATOMIC STRUCTURE

In section 4.4.1 we specified the states of an atomic electron by atomic orbitals and described the states of a Z-electron atom in terms of configurations $(n_1 l_1)(n_2 l_2) \ldots (n_Z l_Z)$. We now want to assign angular momentum quantum numbers to atomic states.

6.1.1   Spectroscopic notation

First, I shall explain the standard spectroscopic notation. In it, one denotes states with quantum numbers L and S by ^{2S+1}L, called a *multiplet*. $2S + 1$ is called the *multiplicity*, and one writes S, P, D, ... for $L = 0, 1, 2, \ldots$. A ^{2S+1}L multiplet comprises $(2L + 1)(2S + 1)$ states corresponding to the values $M_L = L, \ldots, -L$, $M_S = S, \ldots, -S$ of the orbital and spin magnetic quantum numbers M_L and M_S. For example, the carbon atom has the ground state configuration $(1s)^2(2s)^2(2p)^2$. Since the electrons in the (1s) and (2s) shells have zero orbital angular momentum and are paired into singlet $S = 0$ states, the angular momentum of the atom stems entirely from the two (2p) electrons. We saw in section 5.7 that the $(2p)^2$ configuration leads to multiplets ^{1}S, ^{1}D and ^{3}P. To show from which configuration these multiplets stem, one would write $(2p)^2$ ^{1}S [or more fully $(1s)^2(2s)^2(2p)^2$ ^{1}S], etc. Combining L and S into a total angular momentum J, where

$$J = L + S, L + S - 1, \ldots, |L - S|, \tag{6.1}$$

one denotes the states to which the multiplet ^{2S+1}L gives rise by $^{2S+1}L_J$ which is called a spectroscopic term (or just a term). For example, carbon has the ground state $^3P_0$, i.e. it is an $L = 1$, $S = 1$ state with total angular momentum $J = 0$. For a single electron, one writes l, s, j instead of L, S, J, and combining its orbital angular momentum l with its spin $s = \frac{1}{2}$ gives a resultant angular momentum

$$j = l \pm \tfrac{1}{2} \quad \text{for } l \neq 0, \qquad j = \tfrac{1}{2} \quad \text{for } l = 0. \tag{6.2}$$

6.1.2   Atomic multiplets

The assignment of configurations to atomic states is based on the central field approximation, as explained in section 4.4.1. Each electron moves independently of the others and all orbitals in a given shell (nl) have the same energy. Thus all states originating from a given configuration have the same energy, i.e. are degenerate. For example, all the 15 states of the carbon ground state configuration have the same energy in this approximation. The real atomic Hamiltonian H, Eq. (4.23), differs from the central field approximation and the electron–electron repulsion term in it partly removes this degeneracy. The atomic Hamiltonian (4.23) is invariant under rotations and spin-independent. Hence it commutes not only with the total angular momentum operators $\mathbf{J}$ of the atom but also with its resultant orbital angular momentum operators $\mathbf{L}$, as was discussed in section 4.3. The energy eigenstates can therefore be chosen as simultaneous eigenstates of H, $\mathbf{L}^2$ and L_z, i.e. L and M_L are good quantum numbers. Since H is spin-independent, it also commutes with the total spin operators $\mathbf{S} = \mathbf{s}_1 + \mathbf{s}_2$, and S and M_S are also good quantum numbers.* The energy levels can therefore be classified according to the quantum numbers L and S. A given multiplet ^{2S+1}L continues to be degenerate with respect to M_L and M_S: since there is no preferred direction in space (i.e. we are not considering the case where an applied electric or magnetic field is present) the energy cannot depend on the values of the components of $\mathbf{L}$ or $\mathbf{S}$ in an arbitrary direction. In this way, the $(2p)^2$ ground state configuration of carbon leads to three levels: ^3P, ^1D and ^1S.

Similarly, the $(1s)(2s)\ ^1S$ and $(1s)(2s)\ ^3S$ levels of helium are separated in energy as a consequence of the electron–electron interaction, as are the $(1s)(2p)\ ^1P$ and $(1s)(2p)\ ^3P$ levels. Experimentally, the ^3S level lies below the ^1S level, and the ^3P level lies below the ^1P level. This agrees with Hund's rule which says that for a given electron configuration the level of highest multiplicity has the lowest energy. The ^3P ground state level of carbon is another example of Hund's rule, which is an empirical rule; though largely obeyed, there are exceptions to it.[†]

The above results enable us to assign angular momentum quantum numbers to atomic states. We note first of all that a closed shell always has $L = S = 0$. In a closed shell there are as many electrons as there are orbitals in the shell, with one electron in each orbital so that the Pauli exclusion principle is satisfied. Hence there is just one completely antisymmetric wave function one can construct from the orbitals and for it $M_L = M_S = 0$: hence $L = S = 0$ for a closed shell. Consequently one need only take incomplete shells into account when constructing the multiplets of a given electron configuration.

We next discuss the ground state configurations of the elements boron to neon in the periodic table for which the $(2p)$ shell is progressively filled. Boron with just one $(2p)$ electron has the ground state ^2P. Carbon has already been dealt with and shown to have the ground state ^3P. Nitrogen with three $(2p)$ electrons has the ground state ^4S and its wave function in terms of atomic orbitals is easily written down (see problem 6.1). Oxygen, with the outer shell configuration $(2p)^4$, we can think of as a closed $(2p)$ shell with two holes, i.e. two electrons missing from it. When adding angular momenta, a hole behaves like an electron except that the signs of m_l and m_s are reversed. This does not affect the resultant L and S values. Hence the configurations $(2p)^{6-k}$ and $(2p)^k$ give rise to the same multiplets. The three lowest levels of oxygen are ^3P, ^1D and ^1S. Similarly one concludes that the ground state of fluorine is ^2P, like boron. Neon with the 2p shell closed of course has the ground state $^1S_0$.

---

* However, one cannot use this type of reasoning to conclude, for example, that s_{1z} and s_{2z} are also constants of the motion. The observables for identical particles must be symmetric under particle interchange, i.e. S_z is a permissible observable, but s_{1z} or s_{2z} are not.

† The traditional intuitively convincing explanation of Hund's rule has been discredited by accurate computations. For a more elaborate explanation the reader is referred to R.J. Boyd, *Nature*, **310** (1984) 480.

6.1.3 Spin–orbit interaction

Instead of specifying $(2L + 1)(2S + 1)$ states for a ^{2S+1}L multiplet by the quantum numbers L, S, M_L and M_S, we can use L, S, J and M, where the total angular momentum and total magnetic quantum numbers J and M assume the values

$$M = J, J - 1, \ldots, -J, \qquad J = L + S, L + S - 1, \ldots, |L - S|. \qquad (6.3)$$

For the atomic Hamiltonian (4.23) all these states have the same energy. This degeneracy is partly removed by the spin–orbit interaction. (See section 7.4.3.) This is a relativistic correction to (4.23) which is very small for all but the heaviest atoms. The spin–orbit interaction can then be treated as a perturbation which splits a ^{2S+1}L multiplet into several levels characterized by L, S and J, and denoted by $^{2S+1}L_J$, with the values of J given by Eq. (6.3). For $L \geqslant S$ there are $2S + 1$ levels, for $S \geqslant L$ there are $2L + 1$ of them. Each level $^{2S+1}L_J$ is still $(2J + 1)$-fold degenerate, corresponding to the values of M in Eq. (6.3). This degeneracy is removed if an external field is applied, as in the Zeeman effect. (See section 7.5.)

We illustrate the effect of the spin–orbit interaction for the $(2p)^2$ configuration of carbon. The energy levels, without and with spin–orbit interaction, are shown schematically, i.e. *not* to scale, in Fig. 6.1. Without spin–orbit interaction, we obtained the three energy levels ^3P, $^1D_2$ and $^1S_0$. From spectroscopy one finds that the $^1D_2$ and $^1S_0$ levels lie respectively 1.3 eV and 2.7 eV above the ^3P ground state. The spin–orbit interaction splits the ^3P ground state into three levels $^3P_0$, $^3P_1$ and $^3P_2$. Experimentally, the $^3P_1$ and $^3P_2$ levels are found to lie 2.0 meV and 5.4 meV above the $^3P_0$ ground state. The smallness of these splittings (of the order of milli-electron-volts) compared with the separation between different ^{2S+1}L multiplets (of the order of electron-volts) stems from the weakness of the spin–orbit interaction. This splitting of a multiplet into

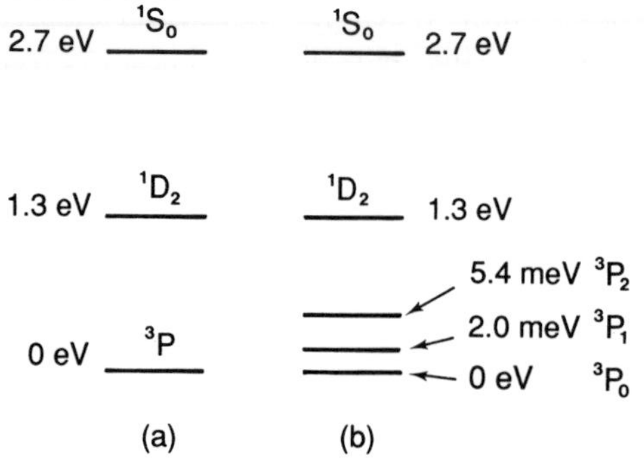

Fig. 6.1 The ^3P, $^1D_2$ and $^1S_0$ energy levels of carbon, *not* to scale; (a) and (b) are without and with spin–orbit interaction respectively.

several closely-spaced levels with different values of J is responsible for the fine structure of atomic spectra.

The angular momentum coupling scheme we have developed involves first adding the orbital angular momenta l_i of the electrons to give L, their spins s_i to give S, and then to add L to S to obtain J. This coupling scheme is called LS coupling or Russell–Saunders coupling. In the alternative scheme one would first add the orbital and spin angular momenta l_i and s_i of each electron to give j_i, and then combine the angular momenta j_i of the individual electrons to give J. This scheme is called jj coupling. If the spin–orbit interaction is weak, LS coupling is a good approximation, as we have seen. In the opposite extreme, the spin–orbit interaction of the orbital angular momentum l_i of each electron with its spin s_i dominates and one must first add l_i and s_i for each electron to give a resultant j_i and then add the j_i of the different electrons, i.e. one has a case of jj coupling. The strength of the spin–orbit interaction depends on the atomic number Z. For light atoms, it is weak and LS coupling is appropriate. As Z increases, the spin–orbit interaction gets increasingly more important, and one moves from LS coupling to a situation where jj coupling is appropriate. However, even for large Z, one usually deals with an intermediate situation, where accurate calculations are much more difficult.

The failure of LS coupling is illustrated by the fine structure splittings of the ^3P level of the group of elements with the outer-shell configurations $(np)^2$, to which carbon belongs. If one treats the spin–orbit interaction as weak, one can use perturbation theory to calculate the ratio

$$r = \frac{E(^3P_1) - E(^3P_2)}{E(^3P_0) - E(^3P_1)} \tag{6.4}$$

of the fine-structure level splittings where $E(^3P_J)$ is the energy of the ^3P_J level. In LS coupling this ratio has the exact value $r = 2$. (See page 177.) Table 6.1 shows the observed values of r for the elements with the ground state configuration $(np)^2$. These values demonstrate the progressive break-down of LS coupling as the atomic number Z increases.

TABLE 6.1 The ratio r, Eq. (6.4), of fine-structure level splittings for elements with the outer-shell configurations $(np)^2$. For LS coupling $r = 2$.

Element	C	Si	Ge	Sn	Pb
n	2	3	4	5	6
Z	6	14	32	50	82
r	1.65	1.89	1.53	1.03	0.36

★6.2   ELECTRIC DIPOLE SELECTION RULES

When discussing the parity selection rules for electric dipole transitions, in section 4.1, I stated that there are also angular momentum selection rules. We are now in a position to discuss these.

6.2.1   Derivation from conservation laws

Selection rules stem from conservation laws for systems as a whole, in our case atom plus radiation. The photons which are emitted or absorbed in radiative transitions in atoms possess definite parity and angular momentum. A photon with one unit of angular momentum (in units of $\hbar$) and negative parity is called an electric dipole photon. In an electric dipole transition, one such photon is emitted or absorbed. Because angular momentum and parity are conserved in electromagnetic interactions, electric dipole transitions in atoms at once lead to selection rules for the atomic transitions. In atoms the probability per unit time for radiative transitions involving electric dipole photons is very much larger than for transitions involving other types of photons.* It is only when electric dipole transitions are forbidden by conservation laws, i.e. by the selection rules, that other types of photons have to be considered.

Suppose we are considering a transition from an atomic state $|\lambda_i, J_i, M_i\rangle$ to a state $|\lambda_f, J_f, M_f\rangle$, where J_i, J_f and M_i, M_f are the total angular momentum and total magnetic quantum numbers of the initial (i) and final (f) states, and λ_i and λ_f are other parameters to specify these states completely. (They could, for example, be the configurations of the two states.) From the rules for addition of angular momentum in quantum mechanics, it follows that if an emitted or absorbed photon possesses one unit of angular momentum, the change $\Delta J = J_f - J_i$ in the total angular momentum quantum number is restricted to one of the values

$$\Delta J = 0, \pm 1. \qquad (6.5a)$$

We also know from the angular momentum addition rules that for a transition from an atomic state with $J_i = 0$ to one with $J_f = 0$ to occur with the emission or absorption of one photon, that photon must possess zero angular momentum. There exist no photons with zero angular momentum; photons always have an angular momentum of at least one unit. (This is a consequence of the transverse nature of the electromagnetic waves.) Hence not only for an electric

---

* This statement is true for photons in the visible and ultraviolet region of the spectrum; it is not necessarily true for x-rays. Similarly, radiative transitions in nuclei involving γ-rays are not dominated to the same degree by electric dipole transitions.

dipole transition but for any transition involving only one photon do we have the selection rule

$$J_i = 0 \rightarrow J_f = 0 \text{ is strictly forbidden.} \tag{6.5b}$$

The selection rule for the total magnetic quantum number follows similarly. Since the z-component of the angular momentum of an electric dipole photon can only assume the values $\hbar$, 0 and $-\hbar$, the change in the magnetic quantum number, $\Delta M = M_f - M_i$, is restricted to

$$\Delta M = 0, \pm 1. \tag{6.6}$$

This selection rule only becomes significant when an external field is present, so as to remove the degeneracies of energy levels with respect to different values of the magnetic quantum numbers.

The selection rules (6.5) and (6.6) were here deduced from the conservation of angular momentum and the angular momentum properties of electric dipole photons. They also follow directly from the electric dipole matrix element, as will now be shown.

6.2.2 Hydrogenic systems

We shall start by considering a hydrogenic system. The initial and final states $|i\rangle$ and $|f\rangle$ factorize into space and spin parts

$$|i\rangle = |n_i l_i m_i\rangle|m_{si}\rangle, \qquad |f\rangle = |n_f l_f m_f\rangle|m_{sf}\rangle, \tag{6.7}$$

where the spatial wave functions are of the form

$$|nlm\rangle = R_{nl}(r)Y_l^m(\theta, \phi). \tag{6.8}$$

The probability per unit time for an electric dipole transition from the state $|i\rangle$ to the state $|f\rangle$ can be shown* to be proportional to $|M_{fi}|^2$, where

$$M_{fi} = \langle f|\varepsilon \cdot (-e\mathbf{r})|i\rangle \tag{6.9}$$

is the electric dipole matrix element for this transition. The unit vector ε is the polarization vector of the radiation, i.e. of the photon, involved in the transition. The electric dipole operator $(-e\mathbf{r})$ is spin-independent: the electron spin does not flip in an electric dipole transition. Hence we must have $m_{sf} = m_{si}$, and in this case Eq. (6.9) reduces to

$$M_{fi} = -e\langle n_f l_f m_f|\varepsilon \cdot \mathbf{r}|n_i l_i m_i\rangle. \tag{6.10}$$

The selection rule for the magnetic quantum number depends on the polarization of the photon. Let $\varepsilon_x, \varepsilon_y$ and ε_z be unit vectors in the directions

---

* This result will be derived in Chapter 9, section 9.5.

of the x-, y- and z-axes. For linear polarization in the z direction $\varepsilon = \varepsilon_z$, and $\varepsilon \cdot \mathbf{r}$ becomes

$$\varepsilon \cdot \mathbf{r} = \varepsilon_z \cdot \mathbf{r} = z = r \cos \theta = r \left(\frac{4\pi}{3} \right)^{1/2} Y_1^0(\theta, \phi). \tag{6.11a}$$

(See Eqs. (2.56) for the explicit definitions of $Y_1^\mu(\theta, \phi)$, $\mu = 0, \pm 1$.) Instead of linear polarizations in the x or y directions, it is more appropriate to consider photons circularly polarized in the xy plane. Such photons possess ± 1 unit of angular momentum about the z-axis and lead naturally to the selection rule for the magnetic quantum number. The polarization vectors for circular polarization in the xy plane are $\varepsilon = (\varepsilon_x \pm i\varepsilon_y)/\sqrt{2}$ and

$$\varepsilon \cdot \mathbf{r} = \frac{1}{\sqrt{2}} (\varepsilon_x \pm i\varepsilon_y) \cdot \mathbf{r} = \frac{1}{\sqrt{2}} (x \pm iy) = \mp r \left(\frac{4\pi}{3} \right)^{1/2} Y_1^{\pm 1}(\theta, \phi). \tag{6.11b}$$

Substituting Eqs. (6.11) and the wave functions (6.8) in the matrix element in (6.10), it becomes

$$\int \mathrm{d}r r^3 R_{n_f l_f}^*(r) R_{n_i l_i}(r) \int \mathrm{d}\phi \sin \theta \; \mathrm{d}\theta Y_{l_f}^{m_f *}(\theta, \phi) Y_1^\mu(\theta, \phi) Y_{l_i}^{m_i}(\theta, \phi) \tag{6.12}$$

where $\mu = 0$ corresponds to photons linearly polarized in the z direction, and $\mu = \pm 1$ to photons circularly polarized in the xy plane.

The integral with respect to ϕ in Eq. (6.12) is

$$\int_0^{2\pi} \mathrm{d}\phi \exp[i(m_i + \mu - m_f)\phi]$$

and this vanishes unless $m_f = m_i + \mu$, with $\mu = 0, \pm 1$. Hence $\Delta m_l = m_f - m_i$ must satisfy the selection rule

$$\Delta m_l = 0, \pm 1 \tag{6.13}$$

for the orbital magnetic quantum number m_l.

The selection rule for the orbital angular momentum quantum number also follows from Eq. (6.12). From the properties of the spherical harmonics one can show* that

$$Y_1^\mu(\theta, \phi) Y_l^m(\theta, \phi) = a Y_{l+1}^{m+\mu}(\theta, \phi) + b Y_{l-1}^{m+\mu}(\theta, \phi) \tag{6.14}$$

where a and b are constants which depend on l, m and μ. Substituting Eq. (6.14) in Eq. (6.12), it follows from the orthogonality relation (2.57) of the spherical

---

* The general formula of which Eq. (6.14) is a special case is, for example, derived in Sakurai, p. 216.

harmonics that the angular integral vanishes unless $l_f = l_i \pm 1$, i.e. $\Delta l = l_f - l_i$ must satisfy the selection rule

$$\Delta l = \pm 1. \tag{6.15}$$

(The selection rule for the magnetic quantum number of course also follows from Eq. (6.14). But as it can easily be derived *ab initio*, I preferred to do so.)

Since the spherical harmonic $Y_l^m(\theta, \phi)$ has parity $(-1)^l$, Eq. (2.63), it follows from Eqs. (6.14) and (6.12) that the initial and final states $|i\rangle$ and $|f\rangle$ in an electric dipole transition must have opposite parities and that the angular momentum selection rule (6.15) already encompasses the parity selection rule (4.33). This is peculiar to one-electron systems and does not hold in general, as we shall see.

The angular momentum and parity selection rules (6.13) and (6.15) are fully borne out by the line spectra of hydrogenic systems (atomic hydrogen, He^+, etc.) and of the alkali atoms, whose spectra are qualitatively very similar. Of course, in a hydrogenic system the energy levels depend only on the principal quantum number n and are degenerate with respect to the orbital quantum number l. This degeneracy is removed for the valence electron of an alkali atom moving in the Coulomb field of the nucleus screened by the electrons in the closed shells. As a consequence the energies of an alkali atom with the valence electron in (np) and (ns) orbitals differ greatly; for example, in sodium the first excited state, with the valence electron in a $(3p)$ orbital, lies 2.1 eV above the $(3s)$ ground state, and the transition $(3p) \rightarrow (3s)$ is responsible for the characteristic yellow sodium light. In hydrogen these levels are degenerate, if fine structure is neglected as we are doing at present. In Figs. 6.2(a) and (b) we show the

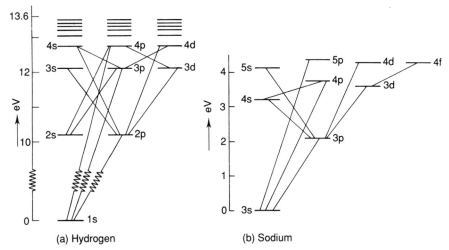

(a) Hydrogen   (b) Sodium

Fig. 6.2 The lowest-lying energy levels and some of their electric dipole transitions for atomic hydrogen and sodium. Fine structure has been neglected.

lowest lying energy levels and their electric dipole transitions for hydrogen and sodium. The levels are labelled by the orbital (nl) of the valence electron. For sodium they are usually labelled $n\,^2L$, e.g. the ground state is $3\,^2S$. It is of course the total orbital angular momentum L, and not the orbital angular momentum l of the valence electron, which is a good quantum number, although it is a very good approximation to describe the core electrons in terms of a fixed central potential and use a one-electron approximation.

6.2.3   Many-electron atoms

We now turn to atoms with more than one electron. If LS coupling is valid, as is often the case, further selection rules arise for electric dipole transitions in addition to those for J, M and parity, Eqs. (6.5), (6.6) and (4.30). If we write the initial and final states

$$|i\rangle = |\lambda_i, L_i, M_{Li}, S_i, M_{Si}\rangle, \quad |f\rangle = |\lambda_f, L_f, M_{Lf}, S_f, M_{Sf}\rangle, \qquad (6.16)$$

where L_i, S_i, L_f, S_f are the orbital and spin angular momentum quantum numbers of the initial and final states, and $M_{Li}, M_{Si}, M_{Lf}, M_{Sf}$ are the corresponding magnetic quantum numbers, then the dipole transition matrix element is given by

$$\langle \lambda_f, L_f, M_{Lf}, S_f, M_{Sf}| \sum_{j=1}^{z} \mathbf{r}_j |\lambda_i, L_i, M_{Li}, S_i, M_{Si}\rangle. \qquad (6.17)$$

Since the dipole operator is spin independent, the selection rules

$$\Delta M_S = M_{Sf} - M_{Si} = 0, \qquad \Delta S = S_f - S_i = 0 \qquad (6.18)$$

follow at once and the matrix element (6.17) reduces to

$$\delta(S_f, S_i)\delta(M_{Sf}, M_{Si})\langle \lambda_f', L_f, M_{Lf}| \sum_{j=1}^{z} \mathbf{r}_j |\lambda_i', L_i, M_{Li}\rangle, \qquad (6.19)$$

where, for brevity, we have written λ_i' for λ_i, S_i, M_{Si} and λ_f' for λ_f, S_f, M_{Sf}, and $\delta(a, b)$ for δ_{ab}.

The selection rule $\Delta S = 0$ means that electric dipole transitions only occur between states of the same multiplicity. For example, in helium the low-lying excited states have the configurations $(1s)(nl)$ with $n \geqslant 2$. Each of these configurations gives rise to a singlet and a triplet multiplet. The spectrum of helium divides into two separate systems: transitions between singlet states and transitions between triplet states. This is illustrated in Fig. 6.3. In addition to the electric dipole transitions, shown as continuous lines in the figure, the transition

$$(1s)(2p)\,^3P_1 \rightarrow (1s)^2\,^1S_0, \qquad (6.20)$$

shown as a dashed line, is observed as a very weak spectral line. This transition is from a triplet to a singlet state and so cannot occur as an electric dipole

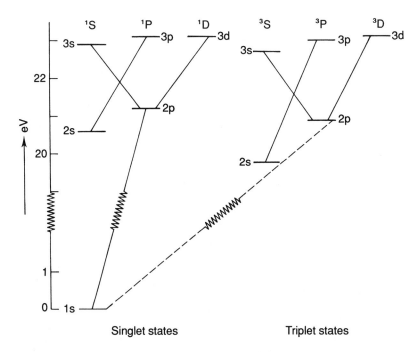

Fig. 6.3   The lowest-lying energy levels and some of their electric dipole transitions for helium. The levels correspond to the configurations (1s)(nl) and are labelled nl.

transition if LS coupling holds exactly. This line—known as an intercombination line or an intersystem line—must either be due to the emission of a different type of photon, i.e. not an electric dipole photon, or to a break-down of LS coupling. Both mechanisms can be shown to be very weak, resulting in a low intensity line. There is of course a spin–orbit interaction. This is very weak in light atoms such as helium but as a consequence L and S are only approximate good quantum numbers and the excited state in the transition (6.20) contains a small admixture of the corresponding singlet state (1s)(2p) ^{1}P which permits an electric dipole transition to the ground state. In contrast, the spin–orbit interaction is large in heavy atoms and LS coupling is not a good approximation. Mercury ($Z = 80$) has the ground-state outer-shell configuration (6s)^2 and the transition (6s)(6p) ^{3}P$_1$ → (6s)^2 ^{1}S$_0$, which is analogous to (6.20), is responsible for an intense line in the ultraviolet spectrum: the ^{3}P state has a considerable admixture of the corresponding ^{1}P state.

The matrix element (6.19) leads to the selection rules for L and M_L:

$$\Delta M_L = M_{Lf} - M_{Li} = 0, \; \pm 1, \qquad (6.21a)$$

$$\Delta L = L_f - L_i = 0, \; \pm 1. \qquad (6.21b)$$

To derive the selection rule (6.21a) for M_L, we again replace $\mathbf{r}_j$ by $r_j Y_1^\mu(\theta_j, \phi_j)$ in the matrix element in (6.19), giving

$$\langle \lambda_f', L_f, M_{Lf} | \sum_{j=1}^{z} r_j Y_1^\mu(\theta_j, \phi_j) | \lambda_i', L_i, M_{Li} \rangle. \tag{6.22}$$

Since

$$L_z = \sum_{k=1}^{z} -i\hbar \frac{\partial}{\partial \phi_k}$$

and $|\lambda_i', L_i, M_{Li}\rangle$ is an eigenstate of L_z with the eigenvalue $\hbar M_{Li}$, it follows that

$$L_z Y_1^\mu(\theta_j, \phi_j) = \hbar\mu Y_1^\mu(\theta_j, \phi_j)$$

and

$$L_z\{Y_1^\mu(\theta_j, \phi_j) | \lambda_i', L_i, M_{Li}\rangle\}$$
$$= \{L_z Y_1^\mu(\theta_j, \phi_j)\} | \lambda_i', L_i, M_{Li} \rangle + Y_1^\mu(\theta_j, \phi_j) L_z | \lambda_i', L_i, M_{Li} \rangle$$
$$= \hbar(\mu + M_{Li}) Y_1^\mu(\theta_j, \phi_j) | \lambda_i', L_i, M_{Li} \rangle.$$

Hence $\sum r_j Y_1^\mu(\theta_j, \phi_j) | \lambda_i', L_i, M_{Li} \rangle$ is an eigenstate of L_z with the eigenvalue $\hbar(\mu + M_{Li})$. It follows that the matrix element (6.22) is the scalar product of $|\lambda_f', L_f, M_{Lf}\rangle$, which is an eigenstate of L_z with the eigenvalue $\hbar M_{Lf}$, and of an eigenstate of L_z with the eigenvalue $\hbar(\mu + M_{Li})$. Hence this matrix element is zero unless

$$M_{Lf} = \mu + M_{Li},$$

which at once leads to the selection rule (6.21a) since μ can take on the values $0, \pm 1$ only. For circularly polarized photons $\mu = \pm 1$ and $\Delta M_L = \pm 1$; for linearly polarized photons $\mu = 0$ and $\Delta M_L = 0$.

A proper derivation of the selection rule (6.21b) for L is beyond the scope of this book and we shall assume it.* Note that this selection rule for L allows $\Delta L = 0$, in contrast to the selection rule $\Delta l = \pm 1$, Eq. (6.15), for a one-electron system. The difference is of course due to the fact that in the latter case l determines the parity of a state and $\Delta l = 0$ would mean no parity change. On the other hand, L is not related to the parity of a many-electron atom. For example, all states arising from the $(2p)^2$ ground state configuration of carbon have positive parity: the wave functions are linear combinations of products $Y_1^{m_1}(\theta_1, \phi_1) Y_1^{m_2}(\theta_2, \phi_2)$ (see Eqs. (5.83a–c)) and under inversion each spherical harmonic is multiplied by (-1). Thus the multiplets ^1S, ^3P and ^1D which originate from the $(2p)^2$ configuration all have the same parity. More generally, if an atomic state is specified by a single configuration $(n_1 l_1)(n_2 l_2)\ldots(n_z l_z)$ its parity is given by

$$\Pi = (-1)^{\Sigma l_j}. \tag{6.23}$$

---

* For a derivation see, for example, Bethe and Jackiw, Chapter 11.

We summarize the electric dipole selection rules. From the conservation laws and therefore quite generally:

$$\Delta J = 0, \pm 1 \tag{6.24a}$$

$$J_i = 0 \to J_f = 0 \text{ is strictly forbidden} \tag{6.24b}$$

$$\Delta M = 0, \pm 1 \tag{6.24c}$$

$$\Pi_f = -\Pi_i. \tag{6.24d}$$

In *LS* coupling we have the additional selection rules:

$$\Delta S = 0: \quad \text{no change in multiplicity} \tag{6.24e}$$

$$\Delta M_S = 0 \tag{6.24f}$$

$$\Delta L = 0, \pm 1 \tag{6.24g}$$

$$\Delta M_L = 0, \pm 1. \tag{6.24h}$$

6.2.4 Forbidden transitions

I conclude this section with a few comments on atomic transitions other than electric dipole transitions. A transition is called forbidden if the electric dipole matrix element is zero. In this case the transition cannot occur by emission or absorption of an electric dipole photon. When an excited state of an atom cannot decay by an electric dipole emission, i.e. if the transitions to all lower lying states are forbidden, the atom must lose energy by some other processes. For example, a different type of photon, not possessing one unit of angular momentum and negative parity, may be emitted but such processes are much slower than electric dipole transitions. Depending on the conditions, other competing processes may dominate; for example, atoms may be de-excited in collisions with other atoms or with the walls of a discharge tube.

A typical example where de-excitation cannot occur by electric dipole transitions is afforded by the excited states of an atom which stem from its ground-state configuration since such states all have the same parity. For example, the ^{3}P ground state, the ^{1}D first excited state and the ^{1}S second excited state of oxygen have the common $(2p)^4$ configuration. The separation between neighbouring levels is about 2 eV in both cases. The most favoured radiative decay of the ^{1}S state is to the ^{1}D first excited state, emitting a photon of even parity and with two units of angular momentum; but this process, known as an electric quadrupole transition, is very slow. The most likely ^{1}D $\to$ ^{3}P radiative decay is by emission of a photon of even parity and with one unit of angular momentum, a so-called magnetic dipole transition. One can show that, if *exact LS* coupling holds, magnetic dipole transitions cannot occur (see problem 6.4). Hence this transition depends on the fact that the spin–orbit interaction in oxygen leads to a slight break-down of *LS* coupling. Again, the

^{1}D state has a very long lifetime. In a discharge tube the ^{1}S and ^{1}D states will usually lose their energy through collisions, but at the extremely low pressures which exist in the upper atmosphere the spectral lines of these transitions are observed, for example in the light of the night sky and of the aurora borealis.

★6.3 BELL'S INEQUALITY

According to quantum mechanics, the state of a system is fully specified by its wave function, yet the properties (i.e. the observables) of the system do not have definite values but only probability distributions in general. Two interpretations suggest themselves. One is that quantum mechanics is an inherently probabilistic theory, reflecting the way nature works. The alternative viewpoint is that there exist parameters which fully determine the values of observables but quantum mechanics is an incomplete theory which does not involve these hidden parameters, as they are called, and the wave functions reflect the probability distributions with respect to these parameters: the stochastic nature of quantum mechanics is due to our ignorance. This situation would be quite similar to that in classical statistical mechanics where a fully deterministic causal microscopic theory leads to statistical laws only.

In a famous paper of 1935, Einstein, Podolsky and Rosen argued that quantum mechanics is an incomplete theory and that the conventional quantum-mechanical interpretation of measurements leads to conclusions which appear quite unreasonable. I shall explain their argument in the form due to Bohm.

Suppose in some process pairs of spin $\frac{1}{2}$ particles, say two neutrons, are produced in the singlet spin state at a source S (see Fig. 6.4). The neutrons fly away in opposite directions and their individual spins are analysed by two Stern–Gerlach magnets at A and B. The orientations of the magnets can be altered so that the spin components in any direction can be measured. Suppose both magnets are oriented to measure the spin components in the direction defined by a unit vector $\hat{\mathbf{n}}$. For the singlet spin state, the spins of the two neutrons always point in opposite directions. Hence a measurement of the spin component in the $\hat{\mathbf{n}}$ direction of the neutron at A, with the result $+\frac{1}{2}\hbar$, must

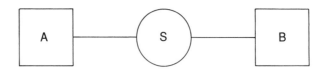

Fig. 6.4 Schematic diagram of the spin correlation experiment. A neutron pair is produced in the singlet state by the source at S. The two neutrons travel to the Stern–Gerlach magnets at A and at B. The magnets can be oriented to measure the spin components of the neutrons in any directions.

lead to the result $-\frac{1}{2}\hbar$ in a similar measurement on the other neutron of the pair at B. (For simplicity we shall call these spin states $+$ and $-$ spin states in the direction $\hat{n}$.) This conclusion holds for any direction $\hat{n}$: if the measurement at A gives $+$, the one at B gives $-$, and vice versa. When we have measured the spin component in the direction $\hat{n}$ of the neutron at A, we *know* the spin component in the same direction of the neutron at B. It is fully determined. Did this determination only arise as the result of the measurement at A? If so, how did the measuring process at A get communicated to the neutron at B? If the measurements at A and B are arranged to be simultaneous, this would require some instantaneous action at a distance! Similar conclusions follow for the magnets at A and B oriented along different directions. If the magnet at A is set along the x direction and that at B along the z direction, a $+$ result in the x direction at A implies equal probabilities (of $\frac{1}{2}$ each) of a $+$ or $-$ result in the z direction at B. The result at B is no longer predetermined. How do the neutron and the magnet at B 'know' which way the magnet at A was pointing? (We could even arrange for the orientation of the magnet at A to be chosen so late—by some random process—that the information of its setting could not reach B in time for the measurement there, unless transmitted at superluminal speed.) It seems more 'reasonable' to assume that for both neutrons of a pair the spins are fully determined (i.e. any spin component has a definite value) from the moment the pair is formed and that the measurement at A does not influence the measurement at B, and vice versa.

Einstein, to the end of his life, was dissatisfied with quantum mechanics as a fundamental theory. To him the real world consisted of systems (particles, fields) possessing objective properties which exist independently of any measurements by observers. It is a second essential feature of such real properties that the result of a measurement of a property at a point B cannot depend on an event at a point A, sufficiently far away from B, so that information about the event, travelling with the speed of light, could not reach B till after the measurement has taken place. Theories meeting these two requirements are called realistic local theories. At the operational level, quantum mechanics is universally successful. Can it be modified so as to become a realistic local theory and yet retain the mathematical formalism on which its success depends?

It was John S. Bell's great contribution, in 1964, to this debate to produce a criterion—now known as Bell's inequality—which any realistic local theory must obey. I shall derive Bell's inequality, and we shall see that the predictions which quantum mechanics makes for certain correlation experiments violate Bell's inequality. Hence quantum mechanics cannot be modified into a realistic local theory. The fact that there exist experiments for which quantum mechanics and a realistic local theory predict different results means that we can choose between them on the basis of experiment rather than of metaphysics. These experiments have now been done. They too violate Bell's inequality, and the size of the violation is that required by quantum mechanics. In other words, quantum mechanics is in accord with nature. Neither quantum mechanics nor nature are realistic and local in character; both display very strange behaviour.

I shall derive Bell's inequality for the case, considered above, of the spin correlations of two spin $\frac{1}{2}$ particles in a singlet state. I shall give the particularly simple proof due to Wigner.* However, the result is not peculiar to spin $\frac{1}{2}$ particles, and in the paper quoted in the footnote Bell obtains the inequality under very general conditions.

To derive Bell's inequality for the above spin correlation experiment, we require that the Stern-Gerlach magnets at A and B (Fig. 6.4) can be oriented so as to measure the spin components in three directions, specified by the unit vectors $\hat{n}_1$, $\hat{n}_2$ and $\hat{n}_3$. It is the hallmark of a realistic theory that any individual neutron in the correlation experiment has definite values (each either $+$ or $-$) for the spin components in all three directions $\hat{n}_1$, $\hat{n}_2$ and $\hat{n}_3$. It is not assumed that we can measure all these components simultaneously, only that if any one of them is measured, by orienting the Stern–Gerlach magnet appropriately, the outcome of the measurement is predictable with certainty. For example, a neutron could be characterized by $(+ \; - \; +)$, meaning that for this neutron a measurement of the spin component in the $\hat{n}_1$ direction is certain to produce the result $+$, in the $\hat{n}_2$ direction the result $-$, and in the $\hat{n}_3$ direction the result $+$. The neutron pairs can thus be divided into groups specified by $(\sigma_1\sigma_2\sigma_3; \tau_1\tau_2\tau_3)$, where σ_i and τ_i (which can each take on the values $+$ or $-$) denote the spin components in the $\hat{n}_i$ direction of the neutrons travelling to the magnet at A and at B respectively. Let $f(\sigma_1\sigma_2\sigma_3; \tau_1\tau_2\tau_3)$ be the fraction of neutron pairs, produced at S, belonging to the group $(\sigma_1\sigma_2\sigma_3; \tau_1\tau_2\tau_3)$. The values of these fractions f will depend on the process from which the neutron pairs originate. In the present case, the neutrons are produced in the singlet state and so must have opposite spins along the same direction. For example, if $\sigma_3 = +, \tau_3$ must be $-$, and $f(+ \; - \; +; \; - \; + \; +) = 0$. Thus

$$f(\sigma_1\sigma_2\sigma_3; \tau_1\tau_2\tau_3) = 0, \quad \text{unless } \sigma_i = -\tau_i, \quad i = 1, 2, 3. \tag{6.25}$$

Note that the locality requirement is satisfied in this description: the result of a spin measurement at A depends only on the values of σ_1, σ_2 and σ_3 and on the orientation of the magnet at A, but it is independent of the orientation of the magnet at B; similarly the result of a measurement of a spin at B is independent of the magnet setting at A.

---

* E.P. Wigner, 'On hidden variables and quantum mechanical probabilities', *Am. J. Phys.*, **38** (1970), 1005. For a full discussion of this and other problems in the interpretation of quantum mechanics, the reader is referred to Bell's collected papers on quantum philosophy. I especially recommend the paper entitled 'Bertlemann's socks and the nature of reality' which is great fun to read and very illuminating, yet fairly elementary. Good non-technical articles on Bell's inequality are: B. d'Espagnat, *Scientific American*, **241** (1979), 128; N.D. Mermin, *Physics Today*, April 1985, 38; A. Shimony, *Scientific American*, **258** (1988), 36.

We can now easily obtain the $(+ +)$ spin correlations, i.e. the probabilities $\langle \hat{\mathbf{n}}_i +; \hat{\mathbf{n}}_j + \rangle$ that, for a neutron pair, measurements of the spin components at A along $\hat{\mathbf{n}}_i$ and at B along $\hat{\mathbf{n}}_j$ both give $+$ results. We have

$$\langle \hat{\mathbf{n}}_1 +; \hat{\mathbf{n}}_2 + \rangle = \sum_{\sigma_2 \sigma_3} \sum_{\tau_1 \tau_3} f(+ \sigma_2 \sigma_3; \tau_1 + \tau_3). \tag{6.26}$$

On account of condition (6.25), the only non-zero terms in the summation in (6.26) are those with $\sigma_2 = -, \tau_1 = -$ and $\sigma_3 = -\tau_3 = \pm$. Hence Eq. (6.26) becomes

$$\langle \hat{\mathbf{n}}_1 +; \hat{\mathbf{n}}_2 + \rangle = f(+ - +; - + -) + f(+ - -; - + +). \tag{6.27a}$$

In the same way one obtains

$$\langle \hat{\mathbf{n}}_3 +; \hat{\mathbf{n}}_2 + \rangle = f(+ - +; - + -) + f(- - +; + + -) \tag{6.27b}$$

$$\langle \hat{\mathbf{n}}_1 +; \hat{\mathbf{n}}_3 + \rangle = f(+ + -; - - +) + f(+ - -; - + +). \tag{6.27c}$$

The two terms on the right-hand side of Eq. (6.27a) occur in one or other of Eqs. (6.27b and c). Since the fractions $f(\sigma_1 \sigma_2 \sigma_3; \tau_1 \tau_2 \tau_3)$ are necessarily non-negative, it follows from Eqs. (6.27) that

$$\langle \hat{\mathbf{n}}_1 +; \hat{\mathbf{n}}_2 + \rangle \leqslant \langle \hat{\mathbf{n}}_3 +; \hat{\mathbf{n}}_2 + \rangle + \langle \hat{\mathbf{n}}_1 +; \hat{\mathbf{n}}_3 + \rangle. \tag{6.28}$$

This is Bell's inequality.

The quantum-mechanical values of the correlations $\langle \hat{\mathbf{n}}_i +; \hat{\mathbf{n}}_j + \rangle$ are easily calculated from first principles (see problem 6.5) or, using an earlier result, as follows. In the singlet state, the probability is $\frac{1}{2}$ that a measurement of the $\hat{\mathbf{n}}_i$ spin component of the neutron arriving at A gives the result $+$. If this result is obtained, a measurement of the $\hat{\mathbf{n}}_i$ spin component of the neutron at B necessarily gives the result $-$. In Eq. (5.56) we obtained the probability that, for a neutron in the $+$ state in the $\hat{\mathbf{z}}$ direction, a measurement of the spin component in the $\hat{\mathbf{n}}$ direction gives the result $+$. Taking $\hat{\mathbf{z}} = -\hat{\mathbf{n}}_i$ and $\hat{\mathbf{n}} = \hat{\mathbf{n}}_j$, we can write Eq. (5.56)

$$P(-\hat{\mathbf{n}}_i, \hat{\mathbf{n}}_j +) = \cos^2(\tfrac{1}{2}\theta) = \sin^2(\tfrac{1}{2}\theta_{ij}) \tag{6.29}$$

where θ is the angle between $-\hat{\mathbf{n}}_i$ and $\hat{\mathbf{n}}_j$, and $\theta_{ij} = \pi - \theta$ that between $\hat{\mathbf{n}}_i$ and $\hat{\mathbf{n}}_j$; see Fig. 6.5. It follows that

$$\langle \hat{\mathbf{n}}_i +; \hat{\mathbf{n}}_j + \rangle = \tfrac{1}{2} P(-\hat{\mathbf{n}}_i, \hat{\mathbf{n}}_j +) = \tfrac{1}{2} \sin^2(\tfrac{1}{2}\theta_{ij}). \tag{6.30}$$

For the predictions of quantum mechanics to be compatible with those of a realistic local theory, the probabilities (6.30) must satisfy the Bell inequality (6.28), i.e. the inequality

$$\sin^2(\tfrac{1}{2}\theta_{12}) \leqslant \sin^2(\tfrac{1}{2}\theta_{23}) + \sin^2(\tfrac{1}{2}\theta_{13}) \tag{6.31}$$

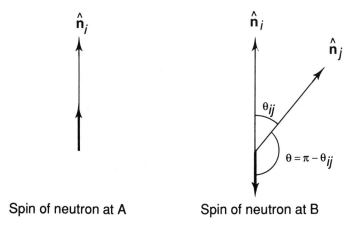

Spin of neutron at A   Spin of neutron at B

Fig. 6.5   Spin correlation of two neutrons in the singlet state.

must hold. It is easy to see that there are geometries for which the inequality (6.31) is violated; for example, if $\hat{n}_1$, $\hat{n}_2$ and $\hat{n}_3$ are coplanar and $\hat{n}_3$ bisects the angle between $\hat{n}_1$ and $\hat{n}_2$:

$$\theta_{13} = \theta_{23} = \tfrac{1}{2}\theta_{12}.$$

For these angles, Eq. (6.31) reduces to

$$\sin^2 \theta_{13} \leqslant 2 \sin^2(\tfrac{1}{2}\theta_{13})$$

which simplifies to

$$\cos^2(\tfrac{1}{2}\theta_{13}) \leqslant \tfrac{1}{2}.$$

But $\cos(\tfrac{1}{2}\theta_{13})$ is greater than $1/\sqrt{2}$ for $0 < \tfrac{1}{2}\theta_{13} < \tfrac{1}{4}\pi$, and the inequality (6.31) is violated for this range of angles: quantum mechanics violates Bell's inequality and some of its predictions are not compatible with a realistic local theory.

   At least twelve correlation experiments have now been performed to test Bell's inequality. Only one of these used protons, all the others observed the correlations of the polarizations of two photons. (Roughly, right- and left-circular polarizations take the place of the two spin states.) Except for two very early experiments, all agree with quantum mechanics within the experimental error and violate Bell's inequality, usually by several standard deviations. All but one of these experiments employed static polarization analysers. This means that the measurement at A could be affected by the orientation of the analyser at B (and vice versa) through some, as yet unknown, long-distance interaction. However, this influence need not travel faster than light and so would not violate the special theory of relativity. The experiment of Aspect's group* tests

---

* A. Aspect, J. Dalibard and G. Roger, *Phys. Rev. Lett.*, **49** (1982), 1804.

just this point and was therefore of crucial importance. In this experiment the orientations of the polarization analysers are selected only after the photon pair has left the source and too late for this information, travelling with the velocity of light, from one analyser to reach the other in time for the polarization measurement there. The results of this experiment are in good agreement with quantum mechanics but violate Bell's inequality by five standard deviations. One must probably conclude that nature is as weird as quantum mechanics makes her out to be.

PROBLEMS 6

6.1 Obtain the spectroscopic term of the ground state of the nitrogen atom, assuming Hund's rule applies. Write down the part of the ground state wave function representing the p-state electrons in terms of their wave functions $\psi_{2,1,m}(\mathbf{r})$, $m = 1, 0, -1$.

6.2 To which multiplets do three equivalent p-electrons, i.e. the configuration $(np)^3$, give rise?

6.3 Which of the following transitions in carbon are electric dipole transitions?
   (a) $(2s)^2(2p)(3d)\ ^3D \rightarrow (2s)^2(2p)^2\ ^3P$
   (b) $(2s)^2(2p)(3s)\ ^3P \rightarrow (2s)^2(2p)^2\ ^1S$
   (c) $(2s)^2(2p)(3d)\ ^1D \rightarrow (2s)^2(2p)(3s)\ ^1P$
   (d) $(2s)(2p)^3\ ^3D \rightarrow (2s)^2(2p)^2\ ^3P$
   (e) $(2s)^2(2p)(3p)\ ^3P \rightarrow (2s)^2(2p)^2\ ^3P$
   (f) $(2s)(2p)^3\ ^5S \rightarrow (2s)^2(2p)^2\ ^3P$
   (g) $(2s)^2(2p)(3d)\ ^1D \rightarrow (2s)^2(2p)^2\ ^1S$
   (h) $(2s)^2(2p)(3s)\ ^1P \rightarrow (2s)^2(2p)^2\ ^1S$.

6.4 A magnetic dipole photon possesses positive parity and angular momentum $\hbar$. Obtain the selection rules for a transition between initial and final states $|i\rangle$ and $|f\rangle$ of an atom in which a magnetic dipole photon is emitted.

The matrix element responsible for the emission of the magnetic dipole radiation in this transition is proportional to

$$\langle f | \mathbf{L} + 2\mathbf{S} | i \rangle, \tag{6.32}$$

where $\mathbf{L}$ and $\mathbf{S}$ are the resultant orbital and resultant spin angular momentum operators of the atom. Show that if LS coupling holds exactly for the atom there can be no transition $|i\rangle \rightarrow |f\rangle$ with emission of magnetic dipole radiation.

6.5 Two spin $\frac{1}{2}$ particles, labelled 1 and 2, are in the singlet spin state χ_0. $\hat{\mathbf{a}}$ and $\hat{\mathbf{b}}$ are two unit vectors. By expanding χ_0 in terms of the eigenstates of $\boldsymbol{\sigma}_1 \cdot \hat{\mathbf{a}}$ and $\boldsymbol{\sigma}_2 \cdot \hat{\mathbf{b}}$, calculate the probability that the spins of particles 1 and 2 are parallel to $\hat{\mathbf{a}}$ and $\hat{\mathbf{b}}$ respectively. Interpret your result for the cases $\hat{\mathbf{a}} = \pm \hat{\mathbf{b}}$.

6.6 Two spin $\frac{1}{2}$ particles, labelled 1 and 2, are in the singlet spin state χ_0. $\hat{\mathbf{a}}$ and $\hat{\mathbf{b}}$ are two unit vectors. Calculate the expectation value of the product $(\boldsymbol{\sigma}_1 \cdot \hat{\mathbf{a}})(\boldsymbol{\sigma}_2 \cdot \hat{\mathbf{b}})$ in this state.

CHAPTER

Bound-state perturbation theory

As in classical physics, very few quantum-mechanical problems possess exact solutions. For most systems, the Schrödinger equation can only be solved by approximation methods. Many different methods exist, reflecting the diversity of applications: to stationary states, to transitions, to collision problems. In the following chapters we shall study some of these methods.

In this chapter, I shall develop perturbation theory for bound states. We shall want to find the stationary states (i.e. the energy eigenvalues and eigenfunctions) for a system which differs only slightly from a system about which 'everything' is known, or at any rate whatever we need to know. I shall mainly consider the lowest order of perturbation theory. The result for this case is very simple to apply and is one of the most useful formulas of quantum mechanics. In contrast, higher orders of perturbation theory lead to very elaborate calculations.

The results of this chapter, together with angular momentum theory of earlier chapters, enable us to treat quite complicated realistic systems. This is illustrated in sections 7.4 and 7.5 where the spin–orbit interaction and the Zeeman effect in atoms are considered. A reader who finds these applications too formidable or tedious may give up at any time; they will not be required later in this book.

7.1 THE NON-DEGENERATE CASE

We consider a Hamiltonian H which can be split into two parts

$$H = H_0 + \lambda V. \tag{7.1}$$

H_0 is an approximate unperturbed Hamiltonian and λV is a small perturbation. H, H_0 and V are observables and so are Hermitian operators; λ is a real parameter. Calling the perturbation λV, rather than V, will facilitate keeping track of different orders of perturbation theory, i.e. different powers of λ. It will also allow us to vary the strength of the perturbation by varying λ continuously. For example, if λV represents the effect of an applied magnetic field on an atom, we can vary the strength of the field; if λV represents the spin–orbit interaction in an atom we do not have such control and switching the spin–orbit interaction on gradually becomes a theoretical concept.

To begin with, we shall assume that H_0 and H possess completely discrete spectra of energy eigenvalues. In this section, we shall also assume that the eigenvalues of H_0 are non-degenerate. We shall denote the eigenvalues of H_0 by $E_n^{(0)}$, $n = 1, 2, \ldots$, and the corresponding eigenstates by $|\phi_n\rangle$, i.e.

$$H_0|\phi_n\rangle = E_n^{(0)}|\phi_n\rangle. \tag{7.2}$$

The states $|\phi_n\rangle$ are orthogonal; we shall also take them as normalized. The effect of the small perturbation λV is to modify slightly each state $|\phi_n\rangle$ and its energy: $|\phi_n\rangle$ will go over into $|\psi_n\rangle$, and $E_n^{(0)}$ will shift slightly to E_n, i.e. $|\psi_n\rangle$ is an eigenfunction of H with the eigenvalue E_n:

$$H|\psi_n\rangle = E_n|\psi_n\rangle. \tag{7.3}$$

In order to solve this equation, we shall expand E_n and $|\psi_n\rangle$ in power series in λ:

$$E_n = E_n^{(0)} + \lambda E_n^{(1)} + \lambda^2 E_n^{(2)} + \cdots \tag{7.4a}$$

$$|\psi_n\rangle = |\phi_n\rangle + \lambda|\phi_n^{(1)}\rangle + \lambda^2|\phi_n^{(2)}\rangle + \cdots. \tag{7.4b}$$

These equations define the corrections $E_n^{(1)}$, $E_n^{(2)}$, $\ldots$ and $|\phi_n^{(1)}\rangle$, $|\phi_n^{(2)}\rangle$, $\ldots$ to the unperturbed energies and states. The terms in λ represent the first-order corrections, the terms in λ^2 the second-order corrections, and so on. In Eq. (7.4b) we can choose the corrections $|\phi_n^{(1)}\rangle$, $|\phi_n^{(2)}\rangle$, $\ldots$ to be orthogonal to the zero-order state $|\phi_n\rangle$:

$$\langle\phi_n|\phi_n^{(1)}\rangle = \langle\phi_n|\phi_n^{(2)}\rangle = \cdots = 0. \tag{7.5}$$

The effect of this choice, which will be convenient in what follows, is that the state $|\psi_n\rangle$, given by (7.4b) is not normed to unity. (It can of course be normed, if required, by dividing by $\sqrt{\langle\psi_n|\psi_n\rangle}$.)

Substituting the expansions (7.4) in Eq. (7.3) leads to

$$(H_0 + \lambda V)(|\phi_n\rangle + \lambda|\phi_n^{(1)}\rangle + \lambda^2|\phi_n^{(2)}\rangle + \cdots)$$

$$- (E_n^{(0)} + \lambda E_n^{(1)} + \lambda^2 E_n^{(2)} + \cdots)(|\phi_n\rangle + \lambda|\phi_n^{(1)}\rangle + \lambda^2|\phi_n^{(2)}\rangle + \cdots) = 0. \tag{7.6}$$

For this equation to hold for all values of λ, the coefficients of each power of λ must vanish separately. The terms independent of λ are just the zero-order equation (7.2) for the unperturbed system. The terms linear and quadratic in λ respectively give the equations

$$(H_0 - E_n^{(0)})|\phi_n^{(1)}\rangle = -(V - E_n^{(1)})|\phi_n\rangle, \tag{7.7a}$$

$$(H_0 - E_n^{(0)})|\phi_n^{(2)}\rangle = -(V - E_n^{(1)})|\phi_n^{(1)}\rangle + E_n^{(2)}|\phi_n\rangle. \tag{7.7b}$$

First-order perturbation theory follows from Eq. (7.7a). Taking the scalar product of this equation with $|\phi_n\rangle$ gives

$$\langle\phi_n|H_0 - E_n^{(0)}|\phi_n^{(1)}\rangle = -\langle\phi_n|V|\phi_n\rangle + E_n^{(1)}\langle\phi_n|\phi_n\rangle. \tag{7.8}$$

Now it follows from the Hermiticity of H_0 [Eqs. (5.15)] that for any state $|\chi\rangle$ and any eigenstate $|\phi_p\rangle$ of H_0 with energy $E_p^{(0)}$

$$\langle\phi_p|H_0|\chi\rangle = E_p^{(0)}\langle\phi_p|\chi\rangle. \tag{7.9}$$

In particular

$$\langle\phi_n|H_0|\phi_n^{(1)}\rangle = E_n^{(0)}\langle\phi_n|\phi_n^{(1)}\rangle.$$

Hence the left-hand side of Eq. (7.8) vanishes and, since $|\phi_n\rangle$ is normed, this equation leads to

$$\lambda E_n^{(1)} = \langle\phi_n|\lambda V|\phi_n\rangle. \tag{7.10}$$

This is the equation I referred to as one of the most useful in quantum mechanics: the first-order shift $\lambda E_n^{(1)}$ of the energy level $E_n^{(0)}$ is given by the expectation value of the perturbation λV with respect to the unperturbed state $|\phi_n\rangle$.

To obtain the first-order correction $\lambda|\phi_n^{(1)}\rangle$ to the state $|\phi_n\rangle$, we take the scalar product of Eq. (7.7a) with $|\phi_p\rangle$ for $p \neq n$:

$$\langle\phi_p|H_0 - E_n^{(0)}|\phi_n^{(1)}\rangle = -\langle\phi_p|V - E_n^{(1)}|\phi_n\rangle. \tag{7.11}$$

Expanding $|\phi_n^{(1)}\rangle$ in terms of the unperturbed states $|\phi_p\rangle$,

$$|\phi_n^{(1)}\rangle = \sum_{p \neq n} a_{np}|\phi_p\rangle, \tag{7.12}$$

we can write the left-hand side of Eq. (7.11)

$$\langle\phi_p|H_0 - E_n^{(0)}|\phi_n^{(1)}\rangle = (E_p^{(0)} - E_n^{(0)})a_{np}, \tag{7.13}$$

where we have used Eq. (7.9). Since $|\phi_p\rangle$ and $|\phi_n\rangle$ are orthogonal, the right-hand side of Eq. (7.11) reduces to

$$-\langle\phi_p|V - E_n^{(1)}|\phi_n\rangle = -\langle\phi_p|V|\phi_n\rangle \equiv -V_{pn} \tag{7.14}$$

which defines the matrix element V_{pn}. Combining Eqs. (7.11), (7.13) and (7.14), we obtain

$$a_{np} = \frac{V_{pn}}{E_n^{(0)} - E_p^{(0)}}; \tag{7.15}$$

hence the first-order correction to the state $|\phi_n\rangle$ is, from Eq. (7.12), given by

$$\lambda|\phi_n^{(1)}\rangle = \sum_{p \neq n} \frac{\lambda V_{pn}}{E_n^{(0)} - E_p^{(0)}} |\phi_p\rangle. \tag{7.16}$$

For the first-order wave function $|\psi_n\rangle = |\phi_n\rangle + \lambda|\phi_n^{(1)}\rangle$, we have

$$\langle\psi_n|\psi_n\rangle = (\langle\phi_n| + \lambda\langle\phi_n^{(1)}|)(|\phi_n\rangle + \lambda|\phi_n^{(1)}\rangle) = 1 + O(\lambda^2), \tag{7.17}$$

since $\langle\phi_n|\phi_n^{(1)}\rangle = 0$. Hence the first-order wave function is already normed to unity. (In higher orders, $|\psi_n\rangle$, defined by Eq. (7.4b), is not normed to unity.)

Eqs. (7.10) and (7.16) state the results for first-order perturbation theory. Unlike the energy shift (7.10), which involves the state $|\phi_n\rangle$ only, the correction (7.16) to the wave function depends on the complete set of unperturbed states and in general leads to a very formidable calculation.

The second-order results follow in a similar way from Eq. (7.7b). We shall only derive the energy shift $\lambda^2 E_n^{(2)}$. Taking the scalar product of (7.7b) with $|\phi_n\rangle$, we obtain

$$\langle\phi_n|H_0 - E_n^{(0)}|\phi_n^{(2)}\rangle = -\langle\phi_n|V|\phi_n^{(1)}\rangle + E_n^{(1)}\langle\phi_n|\phi_n^{(1)}\rangle + E_n^{(2)}. \tag{7.18}$$

The term on the left-hand side and the second term on the right-hand side of this equation vanish, so that

$$E_n^{(2)} = \langle\phi_n|V|\phi_n^{(1)}\rangle = \sum_{p \neq n} a_{np} V_{np}, \tag{7.19}$$

where we substituted the expansion (7.12) for $|\phi_n^{(1)}\rangle$. With a_{np} given by Eq. (7.15) and using the Hermiticity condition $V_{np} = V_{pn}^*$ (i.e. Eq. (5.15b) applied to $\langle\phi_n|V|\phi_p\rangle$), we obtain from Eq. (7.19) as our final result for the second-order energy shift

$$\lambda^2 E_n^{(2)} = \sum_{p \neq n} \frac{\lambda^2 |V_{pn}|^2}{E_n^{(0)} - E_p^{(0)}}. \tag{7.20}$$

For perturbation theory to be useful, it must produce small corrections, so that calculations in the lowest order of perturbation theory suffice.* For the first-order changes to the wave function $|\phi_n\rangle$ to be small, we obtain from Eq. (7.16) the necessary condition

$$|\lambda V_{pn}| \ll |E_n^{(0)} - E_p^{(0)}| \qquad \text{for all } p \neq n. \tag{7.21}$$

If this condition holds, the admixture to the unperturbed state $|\phi_n\rangle$ of other states $|\phi_p\rangle$, $p \neq n$, is small. Similarly, we require a small level shift $\lambda E_n^{(1)}$. In

---

* It may happen that the first-order energy shift $E_n^{(1)}$, Eq. (7.10), vanishes exactly; for example, on symmetry grounds. In this case one has to go to the second-order correction (7.20).

general, the diagonal matrix element V_{nn} will be of the same order of magnitude as the non-diagonal elements V_{pn}. Condition (7.21) then implies

$$|\lambda V_{nn}| \ll |E_n^{(0)} - E_p^{(0)}|, \qquad \text{for all } p \neq n,$$

and therefore

$$|\lambda E_n^{(1)}| = |\lambda V_{nn}| \ll \text{Min} \, |E_n^{(0)} - E_p^{(0)}|, \tag{7.22}$$

i.e. the first-order level shift $\lambda E_n^{(1)}$ must be small compared to the level spacing $\text{Min} \, |E_n^{(0)} - E_p^{(0)}|$ between $E_n^{(0)}$ and the level lying nearest to it in energy.

The conditions (7.21) and (7.22) break down if the level $E_n^{(0)}$ is degenerate. In this case, there are some states $|\phi_p\rangle$, $p \neq n$, with the same energy $E_p^{(0)} = E_n^{(0)}$, some denominators in (7.16) and (7.20) vanish, and these results become meaningless. We shall see in section 7.3 how to handle the degenerate case.*

It is now clear that we can relax the condition, which we initially stipulated, that all the unperturbed energy levels are non-degenerate. We need only assume that the level $E_n^{(0)}$, whose energy shift we are calculating, is non-degenerate. Difficulties with Eqs. (7.16) and (7.20) arise only from the degeneracy of $E_n^{(0)}$, and these equations remain valid even if other levels $E_p^{(0)} \neq E_n^{(0)}$ are degenerate.

Finally, we note that the restriction to completely discrete energy spectra can be relaxed, as long as $E_n^{(0)}$ is a discrete level. If H_0 possesses a partly discrete and a partly continuous spectrum of energy eigenvalues (as in atomic hydrogen, for example), the summations in Eqs. (7.16) and (7.20) are replaced by summations over the discrete states plus integrals over the continuum of states. The important first-order result $E_n^{(1)} = V_{nn}$ holds as before.

7.2 THE HELIUM GROUND STATE

A simple application of non-degenerate perturbation theory is afforded by the ground state of the helium atom or a helium-like ion. The Hamiltonian of this two-electron system is

$$H = H_0 + V, \tag{7.23a}$$

with

$$H_0 = -\frac{\hbar^2}{2m}(\nabla_1^2 + \nabla_2^2) - \frac{Ze^2}{4\pi\varepsilon_0}\left(\frac{1}{r_1} + \frac{1}{r_2}\right), \tag{7.23b}$$

$$V = \frac{e^2}{4\pi\varepsilon_0 r_{12}}. \tag{7.23c}$$

---

* In fact, perturbation theory needs modifying if one or more levels lie close to $E_n^{(0)}$, so that some energy denominators in Eqs. (7.16) and (7.20) become small. For an analysis of this interesting situation see, for example, Davydov, pp. 175–178.

For $Z = 2$, we are dealing with the helium atom, for $Z > 2$ with helium-like ions. The unperturbed Hamiltonian H_0 represents two non-interacting hydrogenic systems. In the ground state each electron is in the (1s) state whose wave function $\psi_{100}(r)$ and energy E_1 were given in Eqs. (2.92), (2.100) and (2.95):

$$\psi_{100}(r) = \left(\frac{Z^3}{\pi a_0^3}\right)^{1/2} e^{-Zr/a_0}, \qquad E_1 = -Z^2 \text{ Ry}. \qquad (7.24)$$

The ground state of the two-electron Hamiltonian H_0 is the product wave function $\psi_{100}(r_1)\psi_{100}(r_2)$ with the energy

$$2E_1 = -2Z^2 \text{ Ry}. \qquad (7.25)$$

The complete antisymmetric ground state is the $(1s)^2\ {}^1S_0$ state

$$\Phi(1, 2) = \frac{Z^3}{\pi a_0^3} e^{-Z(r_1 + r_2)/a_0} \frac{1}{\sqrt{2}} (\alpha_1\beta_2 - \beta_1\alpha_2). \qquad (7.26)$$

The first-order correction to the ground state energy (7.25) is, from Eq. (7.10), given by

$$E^{(1)} = \langle \Phi(1, 2)| \frac{e^2}{4\pi\varepsilon_0 r_{12}} |\Phi(1, 2)\rangle$$

$$= \frac{e^2}{4\pi\varepsilon_0} \left(\frac{Z^3}{\pi a_0^3}\right)^2 \int d^3 r_1 d^3 r_2 \frac{1}{r_{12}} e^{-2Z(r_1 + r_2)/a_0}. \qquad (7.27)$$

The integral in (7.27) can be evaluated analytically* and gives the result

$$E^{(1)} = \tfrac{5}{4}Z \text{ Ry}. \qquad (7.28)$$

Combining Eqs. (7.25) and (7.28), we obtain for the ground state energy in first-order perturbation theory

$$2E_1 + E^{(1)} = (-2Z^2 + \tfrac{5}{4}Z) \text{ Ry}. \qquad (7.29)$$

For helium this gives the ground state energy -5.5 Ry, which lies only 6 per cent above the experimental value of -5.81 Ry. This is a substantial improvement on the zero-order energy of -8 Ry, Eq. (7.25). It is a surprisingly good result, considering that the electron–electron repulsion (7.23c) is quite comparable in size to the Coulomb interaction between the electrons and the nucleus. In the next chapter, we shall develop the variational method for calculating ground state energies, and we shall see that this easily gives very much better results.

---

* See, for example, Davydov, p. 344.

7.3   THE DEGENERATE CASE

We are now considering the case where the eigenvalue $E_n^{(0)}$ of the unperturbed Hamiltonian H_0 is s-fold degenerate, so that there are s linearly independent eigenfunctions $|u_{n\alpha}\rangle$, $\alpha = 1, \ldots, s$, belonging to this eigenvalue. The $|u_{n\alpha}\rangle$ can be chosen orthonormal

$$\langle u_{n\alpha} | u_{n\beta} \rangle = \delta_{\alpha\beta}, \qquad \alpha, \beta = 1, \ldots, s, \tag{7.30}$$

and any linear combination of the $|u_{n\alpha}\rangle$, $\alpha = 1, \ldots, s$, is also an eigenfunction of H_0 with the energy $E_n^{(0)}$. Thus the states which are to serve as zero-order states in the perturbation expansion are not uniquely determined. It is our first task to find s normed states

$$|\phi_{ni}\rangle = \sum_{\alpha=1}^{s} c_{i\alpha} |u_{n\alpha}\rangle, \qquad i = 1, \ldots, s, \tag{7.31}$$

where the $c_{i\alpha}$ are constant coefficients, which are the correct linear combinations for perturbation theory, i.e. they lead to sensible perturbation expansions

$$E_{ni} = E_n^{(0)} + \lambda E_{ni}^{(1)} + \lambda^2 E_{ni}^{(2)} + \cdots, \qquad i = 1, \ldots, s, \tag{7.32a}$$

$$|\psi_{ni}\rangle = |\phi_{ni}\rangle + \lambda |\phi_{ni}^{(1)}\rangle + \lambda^2 |\phi_{ni}^{(2)}\rangle + \cdots, \qquad i = 1, \ldots, s, \tag{7.32b}$$

of the eigenvalues E_{ni} and eigenstates $|\psi_{ni}\rangle$ of H:

$$H|\psi_{ni}\rangle = (H_0 + \lambda V)|\psi_{ni}\rangle = E_{ni}|\psi_{ni}\rangle, \qquad i = 1, \ldots, s. \tag{7.33}$$

The states $|\psi_{ni}\rangle$, $i = 1, \ldots, s$, will in general have different energies E_{ni}, i.e. the effect of the perturbation V is to split the s-fold degenerate level $E_n^{(0)}$ into several levels, at most s of them. In some cases, V only shifts the level $E_n^{(0)}$ without splitting it.

We saw in section 7.1 that in the case of degeneracy the expansions (7.16) and (7.20) for the first-order corrections to the states and the second-order corrections to the energies will in general contain terms with zero denominators. Hence these expansions will only make sense if these terms also have vanishing numerators. We can use this criterion to determine the correct zero-order linear combinations (7.31); we demand that

$$\langle \phi_{ni} | V | \phi_{nj} \rangle = \langle \phi_{ni} | V | \phi_{ni} \rangle \delta_{ij}, \qquad i, j = 1, \ldots, s, \tag{7.34}$$

i.e. that the $s \times s$ matrix $\langle \phi_{ni} | V | \phi_{nj} \rangle$, $i, j = 1, \ldots, s$, is a diagonal matrix. When this condition is satisfied, one easily obtains the first-order level shifts. Eq. (7.7a) is now replaced by

$$(H_0 - E_n^{(0)})|\phi_{ni}^{(1)}\rangle = -(V - E_{ni}^{(1)})|\phi_{ni}\rangle. \tag{7.35}$$

Taking the scalar product with $|\phi_{ni}\rangle$, the left-hand side of this equation vanishes, as before, and we obtain

$$E_{ni}^{(1)} = \langle \phi_{ni} | V | \phi_{ni} \rangle, \tag{7.36}$$

which agrees with our earlier result (7.10) for the first-order level shift.

These are two ways of finding the correct zero-order linear combinations (7.31): using 'brute force' or using symmetry arguments. The latter approach is more fun (it requires thought), is simpler and frequently suffices; I shall deal with it first.

Suppose we can find a Hermitian operator A which commutes with H_0 and with V:

$$[H_0, A] = 0, \quad [V, A] = 0. \tag{7.37}$$

We know from section 3.1 that this implies that H_0 and A possess a complete set of mutual eigenstates, and we can choose the eigenstates $|u_{n\alpha}\rangle$, $\alpha = 1, \ldots, s$, belonging to the eigenvalue $E_n^{(0)}$ of H_0, so that they are also eigenfunctions of A:

$$A|u_{n\alpha}\rangle = a_{n\alpha}|u_{n\alpha}\rangle, \quad \alpha = 1, \ldots, s, \tag{7.38}$$

where $a_{n\alpha}$ are the corresponding eigenvalues. It follows from Eqs. (7.37) and (7.38) that

$$0 = \langle u_{n\alpha}|[V, A]|u_{n\beta}\rangle = \langle u_{n\alpha}|VA - AV|u_{n\beta}\rangle$$

$$= (a_{n\beta} - a_{n\alpha})\langle u_{n\alpha}|V|u_{n\beta}\rangle, \quad \alpha, \beta = 1, \ldots, s,$$

so that

$$V_{\alpha\beta} \equiv \langle u_{n\alpha}|V|u_{n\beta}\rangle = 0, \quad \text{if } a_{n\beta} \neq a_{n\alpha}, \quad \alpha, \beta = 1, \ldots, s. \tag{7.39}$$

Hence, if all the s eigenvalues $a_{n1}, \ldots, a_{ns}$ differ from each other, then the $s \times s$ matrix $V_{\alpha\beta}$ is a diagonal matrix and the $|u_{n1}\rangle, \ldots, |u_{ns}\rangle$ are already the correct zero-order states for perturbation theory. (Even if only some of the eigenvalues are different, this is a useful procedure for reducing the size of the problem to be solved by 'brute force'.)

We illustrate this result by a simple example. A particle moving in a central potential $V_0(r)$ is perturbed by a potential

$$V = \text{const.}\, l_z \tag{7.40}$$

where l_z is the z-component of the orbital angular momentum. Since

$$H_0 = -\frac{\hbar^2}{2m}\nabla^2 + V_0(r)$$

is spherically symmetric it commutes with l_z, as does V of course. Hence the mutual eigenfunctions $|u_{nlm}(\mathbf{r})\rangle$ of H_0, $\mathbf{l}^2$ and l_z are the correct linear combinations for perturbation theory; we obviously have

$$\langle u_{nlm'}|V|u_{nlm}\rangle = \text{const.}\, \hbar m \delta_{mm'}$$

and the potential (7.40) lifts the degeneracy of any level $E_{nl}^{(0)}$ of H_0 completely, i.e. it splits the level $E_{nl}^{(0)}$ into $(2l + 1)$ levels E_{nlm}, $m = l, \ldots, -l$.

This example illustrates a general result: a degeneracy due to a symmetry of H_0 is reduced or removed altogether by a perturbation of lower symmetry. In the above case, H_0 is spherically symmetric whereas V, and hence H, possess only axial symmetry about the z-axis. On the other hand, a perturbation V with the same symmetry as H_0 will in general not affect the degeneracy. If, in the above example, the perturbation (7.40) is replaced by a central potential $V(r)$, a level $E_{nl}^{(0)}$ of H_0 is shifted by $V(r)$ but not split, since the full Hamiltonian H is also rotationally invariant.* In this case, the level shift is given by

$$E_{nl}^{(1)} = \langle \phi_{nl}| V(r)|\phi_{nl}\rangle \tag{7.41a}$$

where $|\phi_{nl}\rangle$ is *any* normed linear combination

$$|\phi_{nl}\rangle = \sum_{m=-l}^{l} c_m |u_{nlm}(\mathbf{r})\rangle \tag{7.41b}$$

(see problem 7.3).

Finally, we develop the 'brute force' method for finding the correct zero-order linear combinations of states for perturbation theory. We again start from Eq. (7.35):

$$(H_0 - E_n^{(0)})|\phi_{ni}^{(1)}\rangle = -(V - E_{ni}^{(1)})|\phi_{ni}\rangle, \tag{7.35}$$

with the zero-order states given by Eq. (7.31):

$$|\phi_{ni}\rangle = \sum_{\alpha=1}^{s} c_{i\alpha}|u_{n\alpha}\rangle, \qquad i = 1,\ldots,s, \tag{7.31}$$

and we must determine the coefficients $c_{i\alpha}$. Taking the scalar product of Eq. (7.35) with $|u_{n\beta}\rangle$, the left-hand side of this equation vanishes, as before, and we obtain

$$\langle u_{n\beta}|V|\phi_{ni}\rangle = E_{ni}^{(1)}\langle u_{n\beta}|\phi_{ni}\rangle.$$

If we substitute (7.31) in this equation and use the orthonormality (7.30) of the $|u_{n\alpha}\rangle$, this equation becomes

$$\sum_{\alpha=1}^{s} \langle u_{n\beta}|V|u_{n\alpha}\rangle c_{i\alpha} = E_{ni}^{(1)} c_{i\beta}$$

which can also be written

$$\sum_{\alpha=1}^{s} (V_{\beta\alpha} - E_{ni}^{(1)}\delta_{\beta\alpha})c_{i\alpha} = 0, \qquad \beta = 1,\ldots,s, \tag{7.42}$$

where $V_{\beta\alpha} \equiv \langle u_{n\beta}|V|u_{n\alpha}\rangle$ as previously.

---

* Of course, if $V_0(r)$ is the point Coulomb potential, then the energy levels of H_0 depend on the principal quantum number n only and are degenerate with respect to $l = 0, 1, \ldots, (n-1)$. In this case, $V(r)$ will split each unperturbed hydrogenic level $E_n^{(0)}$ into n levels E_{nl}, $l = 0, \ldots, (n-1)$, each of which continues to be $(2l+1)$-fold degenerate with respect to $m = l, l-1, \ldots, -l$.

Eqs. (7.42) are a set of s homogeneous equations for s unknowns $c_{i\alpha}$, $\alpha = 1, \ldots, s$. It possesses solutions (other than all $c_{i\alpha} = 0$) for certain values of $E_{ni}^{(1)}$ only. These values are determined by the vanishing of the $s \times s$ coefficient determinant of Eqs. (7.42), i.e. by

$$\det(V_{\beta\alpha} - E_{ni}^{(1)}\delta_{\beta\alpha}) = 0. \tag{7.43}$$

Eq. (7.43) is an equation of the sth degree in $E_{ni}^{(1)}$. It possesses s roots $E_{ni}^{(1)}$, $i = 1, \ldots, s$, not necessarily all different, and $\lambda E_{n1}^{(1)}$, $\lambda E_{n2}^{(1)}, \ldots$ are the first-order level shifts for the states $|\phi_{ni}\rangle$, given by Eqs. (7.31) with the coefficients $c_{i\alpha}$ the solutions of Eqs. (7.42). When the $|u_{n\alpha}\rangle$ are already the correct zero-order states, which diagonalize the $s \times s$ matrix $V_{\beta\alpha}$, Eq. (7.43) reduces to our earlier result (7.36).

★7.4 SPIN–ORBIT INTERACTION

In section 6.1.3 I gave a qualitative discussion of the spin–orbit interaction in atoms. Perturbation theory enables us to treat it quantitatively.

The spin–orbit interaction is a magnetic interaction between the magnetic moments associated with the spins and the orbital motions of the electrons. The spin magnetic moment of the electron, like the spin itself, is a relativistic phenomenon, i.e. it is described correctly by Dirac's relativistic equation of the electron. The proper derivation of the spin–orbit interaction follows as the non-relativistic limit of the Dirac equation for an electron in a central potential.* Being a relativistic effect, it is a very small effect except in heavy atoms. One can estimate the spin–orbit interaction from the above picture. The interaction energy of two magnetic moments M_1 and M_2, a distance R apart, is of the order

$$U_{\text{SO}} \sim \frac{M_1 M_2}{4\pi\varepsilon_0 c^2 R^3}.$$

For atomic hydrogen both magnetic moments are of the order of the Bohr magneton $\mu_{\text{B}} = e\hbar/(2m)$ and R is of the order of the Bohr radius $a_0 = 4\pi\varepsilon_0\hbar^2/(me^2)$, so that

$$U_{\text{SO}} \sim \frac{\mu_{\text{B}}^2}{4\pi\varepsilon_0 c^2 a_0^3} = \tfrac{1}{2}\alpha^2 \,\text{Ry}$$

where $\alpha = e^2/(4\pi\varepsilon_0\hbar c) \approx 1/137$ is the fine-structure constant and $\text{Ry} = \tfrac{1}{2}e^2/(4\pi\varepsilon_0 a_0) = 13.6\ \text{eV}$ is the Rydberg unit of energy. Thus the spin–orbit

---

* For a derivation, see Bethe and Jackiw, pp. 376–379. On pp. 152–153 they also give what they call a pseudo-derivation.

interaction U_{SO} is of the order of 10^{-4} Ry which indeed is small compared with typical electronic energies (of the order of electron-volts). For many-electron atoms, the spin–orbit interaction can be shown to be roughly proportional to the square of the atomic number Z

$$U_{SO} \sim Z^2 \alpha^2 \text{ Ry.} \quad * \qquad (7.44)$$

For heavy atoms, U_{SO} becomes comparable to electronic energies, but for light and medium size atoms the spin-orbit interaction is small and can be treated in perturbation theory.

In sections 7.4.1 and 7.4.2 we shall study the spin–orbit interaction and the fine structure for the hydrogen atom. In section 7.4.3. we shall consider briefly the much more complicated case of atoms with several electrons.

7.4.1 Spin–orbit interaction in hydrogen

It follows from the Dirac equation that for an electron moving in a central potential $V_c(r)$ the Schrödinger Hamiltonian

$$H_0 = -\frac{\hbar^2}{2m} \nabla^2 + V_c(r) \qquad (7.45)$$

is augmented by the spin–orbit interaction term

$$V_{SO} = \frac{1}{2m^2 c^2} \frac{1}{r} \frac{d V_c(r)}{dr} \mathbf{l} \cdot \mathbf{s} \qquad (7.46)$$

where $\mathbf{l}$ and $\mathbf{s}$ are the orbital and spin angular momentum operators. The eigenvalues E_{nl} of H_0 are $2(2l + 1)$-fold degenerate. The corresponding eigenfunctions, which are also eigenfunctions of l_z and s_z, do not diagonalize V_{SO}. On the other hand, if one introduces the total angular momentum operator

$$\mathbf{j} = \mathbf{l} + \mathbf{s}$$

one can write the spin–orbit interaction in the form

$$V_{SO} = \frac{1}{4m^2 c^2} \frac{1}{r} \frac{d V_c(r)}{dr} (\mathbf{j}^2 - \mathbf{l}^2 - \mathbf{s}^2). \qquad (7.47)$$

From this expression one easily sees that the correct zero-order wave functions, which diagonalize V_{SO}, are the angular momentum states $|nlsjm\rangle$ which are

---

* See Landau and Lifschitz, p. 266.

eigenstates of j^2 and j_z with the angular momentum quantum numbers $j(=l\pm\frac{1}{2}$, remember $s=\frac{1}{2})$ and $m(=j, j-1, \ldots, -j)$. The level shift produced by the spin–orbit interaction follows from our basic result (7.36):

$$E_{SO}^{(1)}(nlj) = \langle nlsjm | V_{SO} | nlsjm \rangle$$

$$= \frac{\hbar^2}{4m^2c^2} [j(j+1) - l(l+1) - \tfrac{3}{4}] \left\langle \frac{1}{r} \frac{dV_c(r)}{dr} \right\rangle_{nl} \qquad (7.48)$$

where $\langle \cdots \rangle_{nl}$ denotes the expectation value in an (nl) orbital.*

For s-states $(l = 0, j = s = \frac{1}{2})$ there is no level shift. Energy levels E_{nl} with $l \neq 0$ are split into two levels with $j = l \pm \frac{1}{2}$. Each of these levels continues to be $(2j + 1)$-fold degenerate: the degeneracy with respect to the total magnetic quantum number $m(=j, j-1, \ldots, -j)$ is not lifted since the spin–orbit interaction, like H_0, is rotationally invariant.

For hydrogen $V_c(r) = -e^2/4\pi\varepsilon_0 r$.† One can show that for a hydrogenic (nl) orbital

$$\left\langle \frac{1}{r^3} \right\rangle_{nl} = \frac{1}{a_0^3 n^3 l(l+1)(l+\frac{1}{2})},$$

so that Eq. (7.48) becomes

$$E_{SO}^{(1)}(nlj) = \frac{1}{4m^2c^2} \frac{\hbar^2 e^2}{4\pi\varepsilon_0 a_0^3 n^3} \frac{[j(j+1) - l(l+1) - \tfrac{3}{4}]}{l(l+1)(l+\frac{1}{2})}, \qquad j = l \pm \tfrac{1}{2}, \quad l \neq 0.$$

This expression can be simplified. The expression in square brackets assumes the value l for $j = l + \frac{1}{2}$ and the value $-(l+1)$ for $j = l - \frac{1}{2}$. Introducing the hydrogenic energy eigenvalues $E_n = -e^2/(4\pi\varepsilon_0 a_0 \cdot 2n^2)$ and the fine-structure constant $\alpha = e^2/(4\pi\varepsilon_0 \hbar c) = \hbar/(mca_0)$, we finally obtain

$$E_{SO}^{(1)}(nlj) = \frac{|E_n|}{n} \alpha^2 \left(\frac{1}{l+\frac{1}{2}} - \frac{1}{j+\frac{1}{2}} \right), \qquad j = l \pm \tfrac{1}{2}, \quad l \neq 0. \qquad (7.49a)$$

7.4.2   The fine structure in hydrogen

The spin–orbit level shift (7.49a) is not the only relativistic correction to the hydrogenic levels. The relativistic dependence of the electron's mass on its

---

* This follows since $|nlsjm\rangle$ is a superposition of spatial wave functions $R_{nl}(r) Y_l^{m_l}(\theta, \phi)$ for all of which $1/r[dV_c(r)/dr]$, being a function of r only, has the same expectation value, independent of m_l.

† For hydrogen, the energy levels depend on the principal quantum number n only and are degenerate with respect to $l = 0, 1, \ldots, n-1$. The states $|nlsjm\rangle$ also diagonalize V_{SO} with respect to different values of l, as follows from Eq. (7.47), and are therefore the correct states for perturbation theory.

velocity leads to a correction to the energy levels which, to order α^2, is given by (see problem 7.4)

$$E_{\text{mass}}^{(1)}(nl) = -\frac{|E_n|}{n}\alpha^2\left(\frac{1}{l+\frac{1}{2}} - \frac{3}{4n}\right). \tag{7.49b}$$

The fine structure, i.e. the relativistic corrections to order α^2, of the hydrogenic energy levels E_n is obtained by adding the corrections (7.49a) and (7.49b)

$$\Delta E_{\text{FS}}(nj) = E_{\text{SO}}^{(1)}(nlj) + E_{\text{mass}}^{(1)}(nl)$$

$$= \frac{|E_n|}{n}\alpha^2\left(\frac{3}{4n} - \frac{1}{j+\frac{1}{2}}\right). \tag{7.50}$$

Eq. (7.50) is valid for all values of l and j.* The fine-structure shift (7.50) depends on n and j only and is independent of l, i.e. states with given values of n and j and $l = j \pm \frac{1}{2}$ are degenerate. Furthermore this result, which we derived in perturbation theory to order α^2 only, holds to all orders in α^2: it is an exact result of the Dirac equation for the hydrogen atom. Thus the $n = 2$ level of the Schrödinger Hamiltonian H_0, Eq. (7.45), which has the energy $E_2 = -\text{Ry}/4$ and corresponds to the degenerate states 2s and 2p, is split into two levels; one level corresponds to the $2p_{3/2}$ state, and the other level corresponds to the degenerate $2s_{1/2}$ and $2p_{1/2}$ states. Fig. 7.1 illustrates qualitatively the fine structure of the energy levels of hydrogen for $n = 2$ and $n = 3$.

A small but significant discrepancy between the hydrogenic energy levels as described here and experiment was found by Lamb and Retherford in 1947. Using microwave techniques, they showed that the $2s_{1/2}$ and $2p_{1/2}$ states of hydrogen are not exactly degenerate but that the $2s_{1/2}$ level lies about 1000 MHz (i.e. about 4×10^{-6} eV) above the $2p_{1/2}$ level. This shift of the bound-state energy levels and the resulting splitting are known as the Lamb shift. Historically this discrepancy between theory and experiment gave the main impetus to the development of modern quantum electrodynamics, in which the electromagnetic field as well as the motion of the electrons are quantized. Quantum electrodynamics restores the agreement between theory and experiment. Because of its great importance as a test of quantum electrodynamics, measurements and calculations of the Lamb shift have been greatly refined. The current most complete calculations and the most recent experimental data determine the $2s_{1/2}$—$2p_{1/2}$ energy difference to better than two parts in 10^5 and are in agreement with each other. The precision of both experiment and theory, and

---

* Eq. (7.49a) holds for $l \neq 0$ only. [For $l = 0$, the spin–orbit interaction vanishes and $E_{\text{SO}}^{(1)}(n, l = 0, j = \frac{1}{2}) = 0$.] This suggests that Eq. (7.50) only holds for states other than s states. However for s states the non-relativistic limit of the Dirac equation contains an additional term, as a result of which the fine-structure formula (7.50) is valid for all values of l including $l = 0$. (See Davydov, sections 66 and 70.)

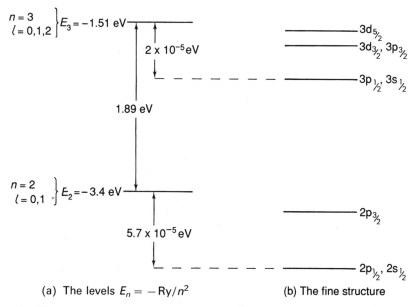

Fig. 7.1   The $n = 2$ and $n = 3$ levels of hydrogen; *not* to scale. The Lamb shifts are not shown.

their agreement, can only be described as stunning. Lamb shift measurements and calculations also exist for other levels in hydrogen, for deuterium and for the He^+ ion, and in all cases theory and experiment are in good agreement.

7.4.3   The fine structure in many-electron atoms

In section 6.1.3 we discussed the fine structure of energy levels in many-electron atoms qualitatively. The quantitative analysis is difficult and I shall only make some general remarks and quote some useful and plausible results.

The spin–orbit interaction is now given by the natural generalization of Eq. (7.46) for hydrogen, that is by

$$V_{SO} = \frac{1}{2m^2c^2} \sum_{i=1}^{Z} \frac{1}{r_i} \frac{dV_i(r_i)}{dr_i} \mathbf{l}_i \cdot \mathbf{s}_i \tag{7.51}$$

where $V_i(r_i)$ is the potential which the electron in the ith orbital experiences due to the other electrons and the nucleus. This spin–orbit term (7.51) must be added to the atomic Hamiltonian (4.23):

$$H_0(\mathbf{r}_1, \dots, \mathbf{r}_Z) = -\frac{\hbar^2}{2m} \sum_{i=1}^{Z} \nabla_i^2 + \sum_{i=1}^{Z} \frac{-Ze^2}{4\pi\varepsilon_0 r_i} + \sum_{\substack{i,j \\ i<j}}^{Z} \frac{e^2}{4\pi\varepsilon_0 |\mathbf{r}_i - \mathbf{r}_j|}. \tag{7.52}$$

For the Hamiltonian (7.52), the orbital and spin angular momentum quantum numbers L and S are good quantum numbers. Hence we can write the energy levels of H_0 as $E_0(\lambda LS)$, where λ denotes any other parameters necessary to specify the levels.* [For example, λ might stand for the atomic configuration $(n_1 l_1)(n_2 l_2) \ldots (n_z l_z)$.] The level $E_0(\lambda LS)$ is $(2L + 1)(2S + 1)$-fold degenerate, corresponding to the ranges $M_L = L, L - 1, \ldots, -L$ and $M_S = S, S - 1, \ldots, -S$ of the magnetic quantum numbers. The corresponding states will be written $|\lambda LSM_L M_S\rangle$.

If the spin–orbit interaction is small compared to the electrostatic interaction, i.e. if it is small compared to the level separation between different multiplets, then it can be treated as a perturbation of H_0. For light and medium-sized atoms this is usually the case and we shall consider this situation only. To apply perturbation theory, we require the correct linear combinations of the degenerate zero-order wave functions $|\lambda LSM_L M_S\rangle$. For hydrogen we found these correct linear combinations to be the eigenfunctions $|nlsjm\rangle$ of $\mathbf{j}^2$ and j_z. In the present case, it can be shown that the correct zero-order wave functions are similarly the eigenfunctions $|\lambda LSJM\rangle$ of $\mathbf{J}^2$ and J_z, where $\mathbf{J} = \mathbf{L} + \mathbf{S}$ and J and M are the quantum numbers associated with $\mathbf{J}^2$ and J_z. The spin–orbit interaction (7.51) causes a shifting and splitting of the level $E_0(\lambda LS)$ given by

$$\Delta E_{\text{FS}}(\lambda LSJ) = \langle \lambda LSJM | V_{\text{SO}} | \lambda LSJM \rangle, \qquad (7.53a)$$

where

$$M = J, J - 1, \ldots, -J, \qquad J = L + S, L + S - 1, \ldots, |L - S|. \qquad (7.53b)$$

A proper evaluation of (7.53a) is fairly complicated and I shall only quote the result[†]

$$\Delta E_{\text{FS}}(\lambda LSJ) = A(\lambda LS)\langle \lambda LSJM | \mathbf{L} \cdot \mathbf{S} | \lambda LSJM \rangle \qquad (7.54)$$

with M and J taking on the values (7.53b). Here $A(\lambda LS)$ is a constant for a given multiplet which is independent of J and M. It is essentially a suitably weighted average of the spin–orbit interaction of the individual electrons allowing for the angular momentum couplings in the state $|\lambda LSJM\rangle$. We now proceed as for hydrogen. With

$$\mathbf{L} \cdot \mathbf{S} = \tfrac{1}{2}(\mathbf{J}^2 - \mathbf{L}^2 - \mathbf{S}^2), \qquad (7.55)$$

Eq. (7.54) at once gives the fine structure of the unperturbed level $E_0(\lambda LS)$:

$$\Delta E_{\text{FS}}(\lambda LSJ) = \tfrac{1}{2}A(\lambda LS)[J(J + 1) - L(L + 1) - S(S + 1)], \qquad (7.56)$$

with M and J taking on the values (7.53b).

---

* The use of λ here must not be confused with its use, in sections 7.1 and 7.3, as the expansion parameter in the perturbation theory.

[†] For a derivation see Bethe and Jackiw, pp. 152–168.

Eq. (7.56) gives the splitting of the level $E_0(\lambda LS)$ into a fine structure multiplet of levels. For $S \leqslant L$, the multiplet consists of $2S + 1$ levels with $J = L + S, L + S - 1, \ldots, L - S$; if $L \leqslant S$, it consists of $2L + 1$ levels with $J = S + L, S + L - 1, \ldots, S - L$. Each level is $(2J + 1)$-fold degenerate, i.e. the level shift (7.56) has the same value for $M = J, J - 1, \ldots, -J$, since V_{SO}, like H_0, is rotationally invariant.

The quantitative evaluation of the level shift (7.56), i.e. of the constant A is difficult, since it involves a knowledge of the atomic wave functions. Even without knowledge of A from calculations, Eq. (7.56) is very useful in the interpretation of spectroscopic data. It only holds if LS coupling is a good approximation. It can therefore be used as a test for this, as we did in section 6.1.3 for elements with the outer-shell configuration $(np)^2$. The value $r = 2$, which was quoted in section 6.1.3 as the LS coupling limit of the ratio r, Eq. (6.4), follows from Eq. (7.56).

From Eq. (7.56) one obtains for the spacing of neighbouring levels of the same multiplet

$$\Delta_J = \Delta E_{FS}(\lambda LSJ) - \Delta E_{FS}(\lambda LS\ J - 1) = A(\lambda LS)J. \tag{7.57}$$

The result $\Delta_J \propto J$ is Landé's interval rule. From Eq. (7.57) one has $\Delta_J/\Delta_{J-1} = J/(J - 1)$ which allows the determination of J from experimental values of the fine structure splittings; one can then use (7.57) to find the value of A.

★7.5 THE ZEEMAN EFFECT

We next consider the energy levels of an atom in an applied uniform magnetic field **B**. There is now a preferred direction in space—that of **B**—and the system is invariant under rotations about the direction of **B** but not about other axes. Consequently the component in the direction of **B** of the total angular momentum **J** of the atom is conserved but $\mathbf{J}^2$ is not. We expect the $(2J + 1)$-fold degeneracy of the energy levels with respect to the magnetic quantum number M to be lifted.

The Hamiltonian of the system now becomes

$$H = H_0 + V_{SO} + V_{mag} \tag{7.58}$$

where H_0 and V_{SO} are given by Eqs. (7.52) and (7.51) and V_{mag} is the interaction energy of the magnetic moment μ of the atom with the applied field **B**:

$$V_{mag} = -\mathbf{\mu} \cdot \mathbf{B}. \tag{7.59}$$

μ is given by

$$\mathbf{\mu} = \sum_{i=1}^{Z} (\mathbf{\mu}_i^{orb} + \mathbf{\mu}_i^{sp}) \tag{7.60}$$

where

$$\boldsymbol{\mu}_i^{\text{orb}} = -\frac{e}{2m}\mathbf{l}_i, \qquad \boldsymbol{\mu}_i^{\text{sp}} = -\frac{e}{m}\mathbf{s}_i \qquad (7.61)$$

are the orbital and spin magnetic moment operators for the ith electron, and $\mathbf{l}_i$ and $\mathbf{s}_i$ are the corresponding angular momentum operators. The orbital magnetic moment $\boldsymbol{\mu}_i^{\text{orb}}$ for an electron moving in an orbit with angular momentum $\mathbf{l}_i$ is easily derived classically. Quantum-mechanically, it is obtained by considering the generalization of the Schrödinger equation for a particle moving in an electromagnetic field.* The spin magnetic moment $\boldsymbol{\mu}_i^{\text{sp}}$, which has no classical analogue, was originally introduced empirically to explain atomic spectra. In particular this necessitates the gyromagnetic ratio (i.e. the ratio of magnetic moment to angular momentum) $-e/m$ rather than $-e/(2m)$ which one would have expected by analogy with the orbital magnetic moment. This correct gyromagnetic ratio for the spin comes out 'automatically' from Dirac's relativistic equation for the electron.[†]

Combining Eqs. (7.59) to (7.61) and introducing the resultant orbital and spin angular momentum operators $\mathbf{L}$ and $\mathbf{S}$, the magnetic interaction operator becomes

$$V_{\text{mag}} = \frac{e}{2m}(\mathbf{L} + 2\mathbf{S}) \cdot \mathbf{B}. \qquad (7.62)$$

The order of magnitude of V_{mag} is easily estimated. The angular momenta are of the order of $\hbar$ and the Bohr magneton μ_B is

$$\mu_B = \frac{e\hbar}{2m} = 9.27 \times 10^{-24} \text{ J/T}. \qquad (7.63)$$

Hence V_{mag} is of the order

$$V_{\text{mag}} \sim 10^{-23} B \text{ J} = 0.6 \times 10^{-4} B \text{ eV} \qquad (7.64)$$

where the magnetic field B is measured in tesla. For $B = 1\text{T}$, which is a very strong applied field, $V_{\text{mag}} \sim 10^{-4}$ eV. This is to be compared with our earlier estimate of the spin–orbit interaction, Eq. (7.44): $V_{\text{SO}} \sim Z^2 \times 10^{-4}$ eV. Except for very strong fields one deals with a situation in which $V_{\text{mag}} \ll V_{\text{SO}}$. This is the usual regime and we shall deal with it first.

---

* See, for example, Sakurai, pp. 307–309.

[†] For a derivation see, for example, Bethe and Jackiw, pp. 376–378. Quantum electrodynamics leads to corrections to the electron's magnetic moment. These are fully borne out by experiment, the agreement between theory and experiment being to an accuracy better than one part in a million! A success as impressive as that for the Lamb shift.

7.5.1   The weak field case: $V_{mag} \ll V_{SO}$

In this situation, the complete Hamiltonian (7.58) is divided into $(H_0 + V_{SO})$ and V_{mag}, with V_{mag} treated by perturbation theory. The unperturbed energy levels and states are

$$E(\lambda LSJ) = E_0(\lambda LS) + \Delta E_{FS}(\lambda LSJ), \tag{7.65a}$$

$$|\lambda LSJM\rangle, \quad M = J, J - 1, \ldots, -J, \tag{7.65b}$$

discussed in the last section. The magnetic interaction (7.62) causes a shifting and splitting of the levels (7.65a), given by

$$\Delta E_{mag}(\lambda LSJM) = \langle \lambda LSJM | V_{mag} | \lambda LSJM \rangle. \tag{7.66}$$

Taking the direction of **B** as z-axis, we substitute Eqs. (7.62) and (7.63), with $J_z = L_z + S_z$, in (7.66) and obtain

$$\Delta E_{mag}(\lambda LSJM) = \frac{\mu_B B}{\hbar} [\hbar M + \langle \lambda LSJM | S_z | \lambda LSJM \rangle]. \tag{7.67}$$

The expectation value of S_z will be derived in section 7.5.3. Here we only quote the result:

$$\langle JM | S_z | JM \rangle = \hbar M \frac{\langle JM | \mathbf{S} \cdot \mathbf{J} | JM \rangle}{\hbar^2 J(J + 1)} \tag{7.68}$$

where $|JM\rangle$ is any eigenstate of $\mathbf{J}^2$ and J_z. For the eigenstates $|\lambda LSJM\rangle$ and with

$$\mathbf{S} \cdot \mathbf{J} = \tfrac{1}{2}(\mathbf{J}^2 + \mathbf{S}^2 - \mathbf{L}^2),$$

Eq. (7.68) gives

$$\langle \lambda LSJM | S_z | \lambda LSJM \rangle = \hbar M \frac{J(J + 1) + S(S + 1) - L(L + 1)}{2J(J + 1)} \tag{7.69}$$

and Eq. (7.67) becomes

$$\Delta E_{mag}(\lambda LSJM) = M \mu_B g B \tag{7.70a}$$

where the Landé g factor is defined by

$$g = 1 + \frac{J(J + 1) + S(S + 1) - L(L + 1)}{2J(J + 1)}. \tag{7.70b}$$

Eqs. (7.70) represent our final result: the magnetic field splits the level $E(\lambda LSJ)$ into $2J + 1$ equidistant levels with the spacing between neighbouring levels given by

$$\delta = \mu_B g B. \tag{7.71}$$

Determining δ from atomic spectra, for a known field B, gives the Landé factor and hence information about the angular momentum quantum numbers of the states involved. For singlet states ($S = 0$, $J = L$) the magnetic moment is entirely due to the orbital motion of the electrons, and $g = 1$. For S-states ($L = 0$, $J = S$) the magnetic moment comes from the spins only, and $g = 2$. For historical reasons, and rather misleadingly, the case with $S = 0$ and $g = 1$ is known as the normal Zeeman effect; the general case with $g \neq 1$ as the anomalous Zeeman effect, i.e. it appeared anomalous before spin had been thought of.

Fig. 7.2 illustrates the Zeeman effect in sodium for the ground state $^2S_{1/2}$ and the ^2P first excited state. The latter is split into two levels, $^2P_{1/2}$ and $^2P_{3/2}$, by the spin–orbit interaction. Figure (a) shows the levels with no magnetic field; figure (b) shows the Zeeman splitting in a field. The vertical lines represent the atomic transitions allowed by the electric dipole selection rules (6.24). The

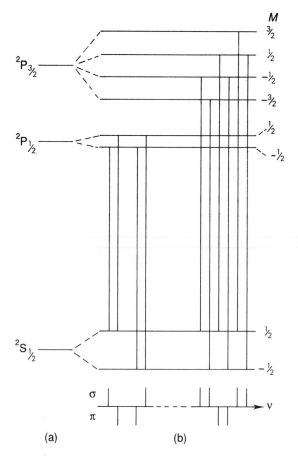

Fig. 7.2   The Zeeman effect for the ^2S and ^2P levels of sodium. (*a*) The levels without magnetic field. (*b*) The Zeeman splitting of the levels, the electric dipole transitions and the line spectrum in a weak magnetic field.

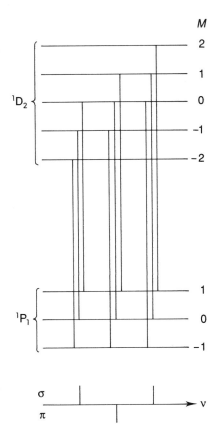

Fig. 7.3   The normal Zeeman effect for the transitions $^1D_2 \rightarrow {}^1P_1$.

selection rule (6.24c) now comes into play. As discussed in section 6.2.2, transitions with $\Delta M = 0$ correspond to photons polarized parallel to the field **B**, transitions with $\Delta M = \pm 1$ to photons with polarization in a plane perpendicular to **B**. Spectral lines with $\Delta M = 0$ and $\Delta M = \pm 1$ are called π and σ components respectively. The resulting line spectrum is shown at the bottom of figure (b).

Fig. 7.3 similarly illustrates the normal Zeeman effect for a transition $^1D_2 \rightarrow {}^1P_1$. (For example, this could be the transition $(1s)(3d)^1D_2 \rightarrow (1s)(2p)^1P_1$ in helium.) Both levels have the same level spacing $\delta = \mu_B B$, so that only three spectral lines are observed.

The reader should remember that the above analysis assumed LS coupling for the atomic states. For heavy atoms, where LS coupling is not a good approximation, the experimental g values differ greatly from the LS-coupling values but are in good agreement with values calculated assuming jj coupling.

7.5.2   The strong field case: $V_{\text{mag}} \gg V_{\text{SO}}$

The strong-field limit of the Zeeman effect is also known as the Paschen–Back effect. In the first instant we neglect the spin orbit interaction V_{SO} in the Hamiltonian (7.58). The unperturbed levels are the $(2L + 1)(2S + 1)$-fold degenerate energies $E_0(\lambda LS)$ of the Hamiltonian H_0 with the eigenstates $|\lambda LSM_L M_S\rangle$, $M_L = L, \ldots, -L$, $M_S = S, \ldots, -S$. The magnetic interaction V_{mag}, Eq. (7.62), causes a shifting and splitting of the levels $E_0(\lambda LS)$, given by

$$\Delta E_{\text{mag}}(\lambda LSM_L M_S) = \langle \lambda LSM_L M_S | V_{\text{mag}} | \lambda LSM_L M_S \rangle$$

$$= \mu_B B(M_L + 2M_S). \tag{7.72}$$

[We have again taken $\mathbf{B} = (0, 0, B)$.] The electric dipole selection rules $\Delta M_S = 0$ and $\Delta M_L = 0, \pm 1$ [see Eqs. (6.24)] always lead to three spectral lines with equal spacing of $\mu_B B$, provided the strong-field limit is valid for both initial and final levels. This situation is illustrated in Fig. 7.4a which shows the spectral lines for the $^2P \rightarrow \,^2S$ transitions in sodium.

The observed spectrum is more complicated than that shown in Fig. 7.4a, due to the spin–orbit interaction V_{SO} which we have neglected. V_{SO} causes a shifting of levels which is given by

$$\Delta E_{\text{SO}}(\lambda LSM_L M_S) = \langle \lambda LSM_L M_S | V_{\text{SO}} | \lambda LSM_L M_S \rangle.$$

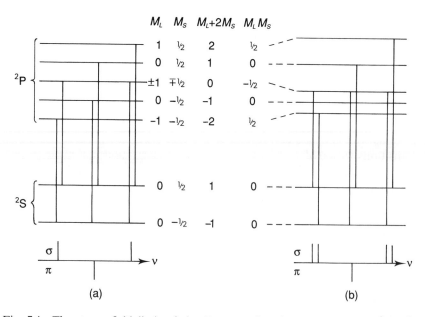

Fig. 7.4   The strong-field limit of the Zeeman effect for the transitions $^2P \rightarrow \,^2S$ in sodium: (a) neglecting the spin–orbit interaction, (b) including the spin–orbit interaction.

This expectation value can be reduced to the form

$$\Delta E_{SO}(\lambda LSM_LM_S) = A(\lambda LS)\hbar^2 M_LM_S.$$

I shall not give the derivation of this expression. Taking it into account leads to level schemes and line spectra in agreement with experiment. The effects of the shift ΔE_{SO} on the ^2S and ^2P levels in sodium are shown in Fig. 7.4b. The shift ΔE_{SO} is proportional to M_LM_S. The four levels with $M_L = 0$ are not shifted. All other levels are shifted up or down by the same amount ($M_LM_S = \pm\frac{1}{2}$ for all of them). The middle ^2P level remains two-fold degenerate. The effect on the spectrum is shown at the bottom of the figure. The lines connecting two states with $M_L = 0$ (and $\Delta M = 0$: π components) are not shifted and remain degenerate. The other two lines (for which $\Delta M = \pm 1$: σ components) are each split into two lines with equal spacing.

7.5.3 Derivation of $\langle JM|S_z|JM\rangle$, Eq. (7.68)

I shall take as starting point the operator identity

$$[J^2, [J^2, A]] = 2\hbar^2(J^2A + AJ^2) - 4\hbar^2(A \cdot J)J \qquad (7.73)$$

which can be shown to hold for any vector operator $A = (A_x, A_y, A_z)$ which satisfies the commutation relations

$$[J_x, A_x] = 0, \quad [J_x, A_y] = i\hbar A_z, \quad [J_x, A_z] = -i\hbar A_y, \quad \text{etc.} \qquad (7.74)$$

where etc. stands for cyclic permutations of x, y and z.* In particular, taking the spin operator S for A satisfies these commutation relations. Hence Eq. (7.73) holds for $A = S$.

If $|JM\rangle$ is an eigenstate of J^2 and J_z, with the eigenvalues $\hbar^2J(J + 1)$ and $\hbar M$, and X is any operator, then

$$\langle JM|[J^2, X]|JM\rangle = \langle JM|J^2X - XJ^2|JM\rangle = 0. \qquad (7.75)$$

With $X = [J^2, S_z]$, Eq. (7.75) becomes

$$\langle JM|[J^2, [J^2, S_z]]|JM\rangle = 0.$$

Using the identity (7.73) to evaluate this expression, we obtain

$$0 = 2\hbar^2\langle JM|J^2S_z + S_zJ^2|JM\rangle - 4\hbar^2\langle JM|(S \cdot J)J_z|JM\rangle$$

$$= 4\hbar^4J(J + 1)\langle JM|S_z|JM\rangle - 4\hbar^3M\langle JM|S \cdot J|JM\rangle$$

whence Eq. (7.68) follows at once.

---

* For a derivation of this identity, which is straightforward but laborious, see, for example, Slater, volume II, pp. 369–371.

PROBLEMS 7

7.1 Consider the proton as a uniformly charged sphere of radius R, much smaller than the Bohr radius a_0. Derive the shift in the energy of the ground state of the hydrogen atom due to the proton's finite size. Estimate the order of magnitude of this shift, given that $R \approx 10^{-15}$ m.

7.2 Consider the one-dimensional motion of a particle of mass m and charge e in the oscillator potential

$$V(x) = \tfrac{1}{2}m\omega^2 x^2.$$

A weak electric field E, which is constant in space and time, is applied in the x-direction. Show that to first order in E the oscillator energy levels are not shifted, and calculate the level shifts to second order in E.

   Show that the second-order result is, in fact, exact.

   (Helpful relation: With $|n\rangle$ denoting the normalized oscillator eigenstate with energy $E_n^{(0)} = \hbar\omega(n + \tfrac{1}{2})$, $n = 0, 1, \ldots$:

$$\langle p|x|n\rangle = \left(\frac{\hbar}{2m\omega}\right)^{1/2}[\delta_{p,n-1}\sqrt{n} + \delta_{p,n+1}\sqrt{(n+1)}], \qquad n, p = 0, 1, \ldots . \quad (7.76)$$

This relation follows from the properties of the oscillator wave functions; see section 2.7 and in particular Eq. (2.160).)

7.3 A particle moves in the central potential $V_0(r)$. If a small central potential $V(r)$ is added to $V_0(r)$, show that, to first order in $V(r)$, the shifts in the energy levels are given by Eqs. (7.41a–b).

7.4 The relativistic mass correction to the energy levels of the hydrogen atom can be found by writing the Hamiltonian in the form

$$H = (\mu^2 c^4 + c^2 \mathbf{p}^2)^{1/2} - \mu c^2 + V_c(r) \approx H_0 - \frac{1}{2\mu c^2}\left(\frac{\mathbf{p}^2}{2\mu}\right)^2,$$

where $V_c(r) = -e^2/(4\pi\varepsilon_0 r)$, $H_0 = \mathbf{p}^2/(2\mu) + V_c(r)$ and μ is the rest mass of the electron, and treating

$$V = \frac{-1}{2\mu c^2}\left(\frac{\mathbf{p}^2}{2\mu}\right)^2 = \frac{-1}{2\mu c^2}(H_0 - V_c)^2$$

in perturbation theory. Show that, to first order in V, the energy E_n of the hydrogenic state $|nlm\rangle$ (where n, l, m are the conventional quantum numbers) is shifted by

$$E_{\text{mass}}^{(1)}(nl) = -\frac{|E_n|}{n}\left(\frac{e^2}{4\pi\varepsilon_0 \hbar c}\right)^2\left(\frac{2}{2l+1} - \frac{3}{4n}\right). \quad (7.77)$$

(Helpful expectation values: With $\langle nlm|nlm\rangle = 1$,

$$\langle nlm|\frac{1}{r}|nlm\rangle = \frac{1}{n^2 a_0}, \qquad \langle nlm|\frac{1}{r^2}|nlm\rangle = \frac{2}{(2l+1)n^3 a_0^2}, \quad (7.78)$$

where a_0 is the Bohr radius.)

7.5 The D_1 and D_2 spectral lines of sodium are due to the transitions $(3p)\,^2P_{1/2} \to (3s)\,^2S_{1/2}$ and $(3p)\,^2P_{3/2} \to (3s)\,^2S_{1/2}$ respectively. In a weak magnetic field B, these lines are split into 4 and 6 lines respectively. Calculate the frequencies of these lines, given that v_1 and v_2 are the frequencies of the D_1 and D_2 lines.

7.6 Consider the splitting by a weak magnetic field B of the spectral line, of frequency v, which is due to an atomic transition $^3P_2 \rightarrow {}^3S_1$, i.e. construct a diagram for this case analogous to Fig. 7.2, and calculate the frequencies of the resulting spectral lines.

7.7 The 1850 Å spectral line of mercury arises from the transition from an excited state to the $^1S_0$ ground state. A magnetic field of 0.2 T splits this line into three components with separation 0.0032 Å between neighbouring lines. What information can be deduced about the excited state?

7.8 An atom, described in LS coupling, is in a ^4P state. Calculate how this energy level is shifted and split as a result of the spin–orbit interaction, assuming the latter can be represented by the potential

$$V_{SO} = A\mathbf{L} \cdot \mathbf{S}$$

where A is a positive constant.

What is the effect of a weak magnetic field B (i.e. $\mu_B B \ll A\hbar^2$) on this level scheme? Draw the resulting level diagram.

What is the corresponding level scheme if the atom, in a ^4P state, is placed in a strong magnetic field (i.e. $\mu_B B \gg A\hbar^2$) and the spin–orbit interaction is neglected in first approximation? How is this level scheme altered if the spin–orbit interaction is taken into account in second approximation? Draw energy level diagrams for the ^4P multiplet in the strong-field limit in both first and second approximations, analogous to the level schemes for the ^2P multiplet in Figs. 7.4(a) and (b).

The variational method

There are many stationary state problems which cannot be solved exactly and for which perturbation theory is not satisfactory because in first order it is not sufficiently accurate and in higher orders it involves enormous calculations. The helium atom, discussed in section 7.2, illustrates this situation. In the present chapter we shall consider a variational method—also known as the Rayleigh–Ritz method—which does not presuppose a knowledge of the solutions of simpler problems and is therefore very versatile. The method is particularly suitable for calculating the ground state energy of a system, and we shall develop it for this case in section 8.1 and apply it to some simple examples in section 8.2. The application of the variational method to excited states is more difficult and we shall consider it in section 8.3.

8.1 THE GROUND STATE

Consider a system described by a Hamiltonian H, which possesses the complete set of orthonormal eigenstates $u_1, u_2, \ldots$ and corresponding energy eigenvalues $E_1, E_2, \ldots$ ordered in increasing sequence

$$E_1 \leqslant E_2 \cdots. \tag{8.1}$$

Any state ψ of the system can be expanded in terms of these eigenstates,

$$\psi = \sum_{n=1}^{\infty} c_n u_n \tag{8.2}$$

and the expectation value of the energy in the state ψ is then given by

$$\frac{\langle \psi | H | \psi \rangle}{\langle \psi | \psi \rangle} = \frac{\sum\limits_{n=1}^{\infty} |c_n|^2 E_n}{\sum\limits_{n=1}^{\infty} |c_n|^2}. \tag{8.3}$$

From Eq. (8.1) we have $E_n \geqslant E_1$ for all n. It follows from Eq. (8.3) that

$$\frac{\langle \psi | H | \psi \rangle}{\langle \psi | \psi \rangle} \geqslant E_1. \tag{8.4}$$

Eq. (8.4) is our basic result. It states that the expectation value of H for any normalized state ψ is an upper bound to the ground state energy E_1, i.e. the left-hand side of the inequality (8.4) is greater than or equal to E_1, and the minimum value E_1 is only attained if ψ is an eigenfunction of H belonging to the eigenvalue E_1.

Eq. (8.4) forms the basis for approximate calculations of the ground state energy. One chooses a trial wave function ψ_T which depends on some parameters $\alpha_1, \alpha_2, \ldots, \alpha_s$,

$$\psi_T = \psi_T(\alpha_1, \alpha_2, \ldots, \alpha_s), \tag{8.5}$$

and calculates

$$E(\alpha_1, \alpha_2, \ldots, \alpha_s) = \frac{\langle \psi_T | H | \psi_T \rangle}{\langle \psi_T | \psi_T \rangle}. \tag{8.6}$$

One then minimizes $E(\alpha_1, \alpha_2, \ldots, \alpha_s)$ with respect to the variational parameters $\alpha_1, \alpha_2, \ldots, \alpha_s$ by solving the equations

$$\frac{\partial E(\alpha_1, \alpha_2, \ldots, \alpha_s)}{\partial \alpha_i} = 0, \qquad i = 1, \ldots, s.$$

The resulting minimum value of $E(\alpha_1, \alpha_2, \ldots, \alpha_s)$ represents the best estimate of the ground state energy for a trial function of the form (8.5). It gives an upper bound to the ground state energy, i.e. (8.4) is a minimum principle. Of two estimates, obtained with two different trial functions, the lower estimate is always the better one.

The success of the variational method depends on choosing a trial function which incorporates the correct qualitative features of the state. The advent of large computers allows the use of trial functions of great complexity, and therefore of great flexibility, leading to very accurate results. The variational method optimizes the energy. On the other hand, the trial function is not necessarily a good approximation to the true wave function and may give poor results when used in calculations of quantities other than the energy.

8.2 SIMPLE EXAMPLES

8.2.1 The ground state of the hydrogen atom

For our first example, we take the ground state of the hydrogen atom. This is an unrealistic example—after all we know the exact solutions—but it illustrates in a simple case how one chooses the trial function and how one handles such a problem.

We know from the general considerations of section 2.5 that the hydrogenic ground state must be an s-state, since there is no centrifugal repulsive potential in an s-state. For $l = 0$, the wave function depends on the radial coordinate only, $\psi = \psi(r)$, with $\psi(0) \neq 0$ [see Eq. (2.84)]. At large distances, a bound-state wave function must vanish. A simple trial wave function satisfying both these requirements is

$$\psi_T = C\,e^{-\alpha r} \qquad (8.7)$$

where α is our variational parameter. The normalization constant C is found from

$$\langle \psi_T | \psi_T \rangle = C^2 \int_0^\infty e^{-2\alpha r} 4\pi r^2 \, dr = 1, \qquad (8.8)$$

whence*

$$C = (\alpha^3/\pi)^{1/2}. \qquad (8.10)$$

The Hamiltonian of the hydrogen atom is

$$H = -\frac{\hbar^2}{2m}\nabla^2 + \frac{-e^2}{4\pi\varepsilon_0 r}. \qquad (8.11)$$

For the expectation value of the potential energy one obtains

$$\left\langle \psi_T \left| \frac{-e^2}{4\pi\varepsilon_0 r} \right| \psi_T \right\rangle = \frac{-e^2}{4\pi\varepsilon_0}\alpha. \qquad (8.12)$$

For our simple trial function we could easily evaluate the expectation value of the kinetic energy directly. Instead I would like to illustrate a trick which frequently simplifies this calculation. It depends on the vector identity, true for any function $\psi(\mathbf{r})$,

$$\int \nabla \cdot (\psi^* \nabla \psi)\, d^3r = \int \psi^* \nabla^2 \psi \, d^3r + \int \nabla \psi^* \cdot \nabla \psi \, d^3r \qquad (8.13)$$

---

* The evaluation of this integral and of some of the integrals below follows from

$$\int_0^\infty dr\, r^n\, e^{-\alpha r} = \frac{n!}{a^{n+1}}, \qquad a > 0, \qquad n = 0, 1, 2, \ldots. \qquad (8.9)$$

where the integrals are over all space. Using Gauss's theorem, the integral on the left-hand side can be converted to a surface integral at infinity. If ψ tends to zero sufficiently rapidly as $r \to \infty$, this surface integral vanishes. This is certainly the case if for ψ we take the exponentially decaying trial function (8.7), and Eq. (8.13) then gives

$$\langle \psi_T | -\frac{\hbar^2}{2m} \nabla^2 | \psi_T \rangle = -\frac{\hbar^2}{2m} \int \psi_T^* \nabla^2 \psi_T \, d^3\mathbf{r} = \frac{\hbar^2}{2m} \int |\nabla \psi_T|^2 \, d^3\mathbf{r}. \quad (8.14)$$

The expression for the kinetic energy on the right-hand side of this equation involves only $\nabla \psi_T$, which for a complicated trial function is often simpler to handle than $\nabla^2 \psi_T$. For the trial function (8.7)

$$|\nabla \psi_T|^2 = |\partial \psi_T / \partial r|^2 = \alpha^2 |\psi_T|^2, \quad (8.15)$$

so that (8.14) reduces to

$$\langle \psi_T | -\frac{\hbar^2}{2m} \nabla^2 | \psi_T \rangle = \frac{\hbar^2 \alpha^2}{2m}. \quad (8.16)$$

From Eq. (8.6) we obtain, on substituting Eqs. (8.8), (8.11), (8.12) and (8.16),

$$E(\alpha) = \langle \psi_T | H | \psi_T \rangle = \frac{\hbar^2 \alpha^2}{2m} - \frac{e^2}{4\pi\varepsilon_0} \alpha. \quad (8.17)$$

From $\partial E(\alpha)/\partial \alpha = 0$, one finds that $E(\alpha)$ has a minimum for

$$\alpha = \frac{e^2 m}{4\pi\varepsilon_0 \hbar^2} = \frac{1}{a_0}, \quad (8.18)$$

where a_0 is the Bohr radius, giving the upper bound

$$E(1/a_0) = -\frac{1}{2} \frac{e^2}{4\pi\varepsilon_0 a_0} \equiv -1 \text{ Ry} = -13.6 \text{ eV}$$

for the ground-state energy. This is the exact ground-state energy due to the fact that our trial function (8.7), with $\alpha = 1/a_0$, is the exact hydrogenic (1s) wave function.

8.2.2 The ground state of the helium atom

For our second, more realistic example we apply the variational method to the ^{1}S ground state of the helium atom or of helium-like ions. The simplest trial function describes each electron by a hydrogenic (1s) orbital with an effective nuclear charge which allows for screening by the other electron:

$$\psi_T(r_1, r_2) = \frac{\alpha^3}{\pi a_0^3} e^{-\alpha(r_1 + r_2)/a_0}. \quad (8.19)$$

No screening corresponds to $\alpha = $ atomic number Z. This trial function is correctly normalized and, when multiplied by the singlet spin function, is antisymmetric in the coordinates of the two electrons. The variational parameter α is to be determined by minimizing

$$E(\alpha) = \langle \psi_T | H | \psi_T \rangle, \tag{8.20}$$

with the Hamiltonian H of the helium atom or helium-like ion given by Eqs. (7.23).

With the trial function (8.19) we obtain for the kinetic energy term in (7.23b), on using the result (8.16) appropriately modified,

$$\langle \psi_T | -\frac{\hbar^2}{2m} (\nabla_1^2 + \nabla_2^2) | \psi_T \rangle = \frac{\hbar^2 \alpha^2}{m a_0^2}, \tag{8.21a}$$

and for the potential energy term in (7.23b), on using Eq. (8.12) appropriately modified,

$$\langle \psi_T | \frac{-Ze^2}{4\pi\varepsilon_0} \left(\frac{1}{r_1} + \frac{1}{r_2} \right) | \psi_T \rangle = -\frac{2Ze^2}{4\pi\varepsilon_0 a_0} \alpha. \tag{8.21b}$$

The electron–electron interaction term (7.23c) gives, from Eq. (7.28),

$$\langle \psi_T | \frac{e^2}{4\pi\varepsilon_0 r_{12}} | \psi_T \rangle = \frac{5}{4} \alpha \text{ Ry.} \tag{8.21c}$$

Expressing Eqs. (8.21a) and (8.21b) in terms of Rydbergs $[1 \text{ Ry} = \frac{1}{2}e^2/(4\pi\varepsilon_0 a_0) = \hbar^2/(2m a_0^2)]$ and substituting Eqs. (8.21a) to (8.21c) in (8.20) we obtain

$$E(\alpha) = [\alpha^2 - (2Z - \tfrac{5}{8})\alpha]2 \text{ Ry.} \tag{8.22}$$

From $\partial E(\alpha)/\partial \alpha = 0$ it follows that for

$$\alpha = Z_{\text{eff}} \equiv (Z - \tfrac{5}{16}) \tag{8.23}$$

$E(\alpha)$ attains its minimum value

$$E(\alpha = Z_{\text{eff}} = Z - \tfrac{5}{16}) = -2Z_{\text{eff}}^2 \text{ Ry} = -2(Z - \tfrac{5}{16})^2 \text{ Ry.} \tag{8.24}$$

For helium we have $Z = 2$, $Z_{\text{eff}} = \frac{27}{16}$, and for the upper bound to the ground-state energy we obtain from Eq. (8.24)

$$E(\alpha = Z_{\text{eff}} = \tfrac{27}{16}) = -2(\tfrac{27}{16})^2 \text{ Ry} = -5.7 \text{ Ry.} \tag{8.25}$$

This value agrees with the experimental value of -5.81 Ry to better than 2 per cent. In section 7.2 we found that perturbation theory gives the value -5.5 Ry (see Eq. (7.29) and the comments following it). The latter value is just $E(\alpha = 2)$, i.e. the improved variational value (8.25) is obtained by replacing in the wave function (8.19) $\alpha = Z = 2$ by $\alpha = Z_{eff} = Z - \frac{5}{16} = \frac{27}{16}$; the actual nuclear charge $Ze = 2e$ is replaced by the smaller effective nuclear charge $Z_{eff}e = (Z - \frac{5}{16})e = \frac{27}{16}e$, which allows for the fact that each electron is partly shielded from the nucleus by the other electron.

To improve further on this variational result, one must use a more elaborate trial function. Allowing different screening parameters for the two electrons, i.e. a trial function of the form

$$\psi_T(r_1, r_2) = C[e^{-(\alpha r_1 + \beta r_2)/a_0} + e^{-(\alpha r_2 + \beta r_1)/a_0}], \tag{8.26}$$

with α and β variational parameters, leads to the improved upper bound -5.75 Ry. To get a significantly better value, one must allow for the fact that the electrons repel each other and prefer to be on opposite sides of—and shielded from each other by—the nucleus. A trial function depending on r_1 and r_2 only cannot allow for such correlations. One can do so with a trial function which also depends on the inter-electron distance r_{12}: $\psi_T(r_1, r_2, r_{12})$. Such a trial function leads to much heavier calculations but gives excellent results. A trial function of this type, containing 3 variational parameters, gives the upper bound -5.8048 Ry to the ground-state energy of helium, within less than a tenth of a per cent of the experimental value of -5.808 Ry.

The variational result (8.24) also applies to the two-electron ions H^-, Li^+, Be^{2+}. For Li^+ and Be^{2+} the agreement with experiment is to better than one per cent, but for H^- the simple trial function (8.19) is not adequate. The negative ion of atomic hydrogen, H^-, is stable, albeit with the small binding energy 0.76 eV ($= 0.056$ Ry). Eq. (8.24) predicts a binding energy greater or equal to

$$[-1 \text{ Ry}] - [-2(\tfrac{11}{16})^2 \text{ Ry}] = -0.055 \text{ Ry},$$

i.e. it does not even predict with certainty the existence of a bound state. A trial function of the form (8.26) does predict a bound state but to obtain quantitatively reasonable results one must go to a trial function of the form $\psi_T(r_1, r_2, r_{12})$; with 3, 6 or 11 variational parameters, it leads to upper bounds to the binding energy of H^-, within respectively 10, 6 and 2 per cent of the experimental value.*

★8.3   EXCITED STATES

The variational method for the ground state of a system, developed in section 8.1, can be adapted to excited states. Let $E_n, u_n, n = 1, 2, \ldots$, be the eigenvalues and orthonormal eigenfunctions of a Hamiltonian H. Suppose the ground state is p-fold degenerate so that for the energies ordered in increasing sequence

$$E_1 = E_2 = \cdots = E_p < E_{p+1} \leqslant E_{p+2} \cdots. \tag{8.27}$$

---

* For further discussion of these results, see Moiseiwitsch, pp. 172–181.

We choose a trial function ψ_T orthogonal to the degenerate ground-state eigenfunctions $u_1, u_2, \ldots, u_p$, by imposing the conditions

$$\langle u_1|\psi_T\rangle = \langle u_2|\psi_T\rangle = \cdots = \langle u_p|\psi_T\rangle = 0 \tag{8.28}$$

on ψ_T. The expansion of ψ_T in terms of the eigenfunctions u_n then takes the form

$$\psi_T = \sum_{n=p+1}^{\infty} c_n u_n, \tag{8.29}$$

i.e. the expansion does not contain terms in $u_1, u_2, \ldots, u_p$. From Eqs. (8.27) and (8.29) we obtain

$$\frac{\langle \psi_T|H|\psi_T\rangle}{\langle \psi_T|\psi_T\rangle} = \frac{\displaystyle\sum_{n=p+1}^{\infty} |c_n|^2 E_n}{\displaystyle\sum_{n=p+1}^{\infty} |c_n|^2} \geqslant E_{p+1}. \tag{8.30}$$

Eq. (8.30) is a minimum principle for the first excited energy level E_{p+1}. It differs from the minimum principle for the ground state in that the trial function ψ_T must be orthogonal to all eigenfunctions of H belonging to the ground-state energy E_1, i.e. ψ_T must satisfy the subsidiary conditions (8.28). With this proviso, the variational method is applied as previously: for a trial function $\psi_T(\alpha_1, \ldots, \alpha_n)$, satisfying the subsidiary conditions (8.28), the minimum of

$$E(\alpha_1, \alpha_2, \ldots, \alpha_n) = \frac{\langle \psi_T|H|\psi_T\rangle}{\langle \psi_T|\psi_T\rangle} \tag{8.31}$$

with respect to $\alpha_1, \alpha_2, \ldots, \alpha_n$ is an upper bound to the first excited energy level E_{p+1}. The extension of this procedure for higher excited states should be obvious to the reader.

The need to impose the subsidiary conditions on the trial function severely limits the usefulness of the variational method for excited energy levels. Usually one does not know the exact energy eigenstates to which the trial function must be orthogonal, and if one uses approximate eigenfunctions one cannot even be sure that of two estimates, obtained with different trial functions, the lower estimate is the better one. These difficulties do not arise if one looks for an excited energy level whose symmetries differ from those of all lower-lying levels. In this case, choosing a trial function with the correct symmetries automatically ensures orthogonality to all lower-lying states and the variational method will give an upper bound to the energy of the lowest-lying level with those symmetries.

Consider as a first example a particle moving in an attractive central potential. The ground state will be an s-state. A function of the form

$$\psi_T(\mathbf{r}) = f_t(r, \alpha_1, \ldots, \alpha_n) \cos \theta \tag{8.32}$$

represents a p-state and is necessarily orthogonal to all states with angular momentum quantum number $l \neq 1$. Using a function of the type (8.32) as trial function in the variational method gives an upper bound to the lowest-lying p-state.

For our second example, consider the helium atom. Its ground state is a ^1S state. Its first four excited states are ^3S, ^1S, ^3P and ^1P states (see section 6.1.2). Of these excited states, all but the ^1S state have symmetries different from those of the ground state. Upper bounds to their energies are found by a straightforward application of the variational method, employing trial functions with the appropriate symmetries. These will ensure orthogonality to the ground state and to each other: singlet and triplet states are necessarily orthogonal, so are states with orbital angular momentum quantum numbers $L = 0$ and $L = 1$. Simple trial functions, meeting these requirements, are:

(a) for the ^3S state:

$$\psi_T(r_1, r_2) = C[f(r_1)g(r_2) - f(r_2)g(r_1)]; \tag{8.33a}$$

(b) for the ^1P and ^3P states:

$$\psi_T^\pm(\mathbf{r}_1, \mathbf{r}_2) = C_\pm[f(r_1)h(r_2)\cos\theta_2 \pm f(r_2)h(r_1)\cos\theta_1] \tag{8.33b}$$

where the $+$ and $-$ signs go with ^1P and ^3P states respectively. The functions $f(r)$, $g(r)$ and $h(r)$ in Eqs. (8.33) are suitable trial functions. Simple choices, suggested by the configurational descriptions $(1s)(2s)\,^3S$ and $(1s)(2p)\,^{1,3}P$ of these states and by the hydrogenic wave functions (2.100), are

$$f(r) = e^{-\alpha r/a_0}, \quad g(r) = (1 - \tfrac{1}{2}\beta'r)\,e^{-\beta r/2a_0}, \quad h(r) = r\,e^{-\gamma r/2a_0} \tag{8.34}$$

with α, β, β' and γ variational parameters. Even these simple trial functions produce upper bounds for the energies which exceed the experimental values by less than one per cent. For further discussion of these results and of a simple technique for obtaining upper bounds for excited states with the same symmetries as lower-lying states, see Moiseiwitsch, pp. 182–183 and pp. 166-168, and Bransden and Joachain (1983), pp. 284–286 and pp. 120–122.

PROBLEMS 8

8.1 Use a simple trial function to obtain a variational upper bound to the ground state energy of a particle of mass m moving in the one-dimensional potential

$$V(x) = Ax^4, \quad -\infty < x < \infty, \quad A > 0.$$

8.2 The harmonic oscillator Hamiltonian

$$H = -\frac{\hbar^2}{2m}\frac{d^2}{dx^2} + \frac{1}{2}m\omega^2 x^2$$

posseses the ground state wave function

$$\psi_0(x) = \text{const. } \exp(-m\omega x^2/2\hbar).$$

Use simple trial functions to obtain variational upper bounds to the energies of the first and second excited states.

8.3 A particle of mass m is bound in the ground state of the exponential potential

$$V(r) = -\frac{4\hbar^2}{3ma^2} e^{-r/a}.$$

Use a simple trial function to obtain an upper bound for the ground-state energy.

8.4 E_1 and E_2 are the ground-state energies of a particle of mass m moving in the attractive potentials $V_1(\mathbf{r})$ and $V_2(\mathbf{r})$ respectively. If $V_1(\mathbf{r}) \leqslant V_2(\mathbf{r})$, one intuitively expects $E_1 \leqslant E_2$. Use a variational argument to derive this result.

Time dependence

9.1 INTRODUCTION

When discussing selection rules for electric dipole transitions in atoms, in sections 4.1 and 6.2, I introduced the idea of transitions between different states of a system. We shall now study how transitions come about and, more generally, how the state of a system changes with time.

Consider a system, described by the time-independent Hamiltonian H_0, which at time $t = 0$ is in an energy eigenstate $|u_1\rangle$ of H_0 with the energy E_1. At later times, the system will be in the state

$$|\psi_1(t)\rangle = |u_1\rangle \, e^{-iE_1 t/\hbar} \; ; \tag{9.1}$$

$|\psi_1(t)\rangle$ and $|u_1\rangle$ differ only by a phase factor, i.e. the state of the system has not really changed.

On the other hand, suppose at $t = 0$ the system is in a state $|\psi_0\rangle$ which is *not* a stationary state of H_0. We can expand $|\psi_0\rangle$ in terms of the complete set of time-independent orthonormal energy eigenstates $|u_1\rangle, |u_2\rangle, \ldots$ of H_0 belonging to the eigenvalues $E_1, E_2, \ldots$:

$$|\psi_0\rangle = \sum_n c_n |u_n\rangle \tag{9.2}$$

with

$$c_n = \langle u_n | \psi_0 \rangle. \tag{9.3}$$

The state of the system at later times is then given by

$$|\psi(t)\rangle = \sum_n c_n |u_n\rangle \, e^{-iE_n t/\hbar}, \tag{9.4}$$

as we saw in section 1.3, Eq. (1.58). The state $|\psi(t)\rangle$ will usually differ from $|\psi_0\rangle$. For example, if at time $t = 0$, the system is in the state

$$|a\rangle = |u_1\rangle + |u_2\rangle$$

then at time $t = \pi\hbar/(E_1 - E_2)$ it will be in the state

$$|b\rangle = \exp\left(\frac{i\pi E_1}{E_2 - E_1}\right)(|u_1\rangle - |u_2\rangle)$$

which is clearly different from $|a\rangle$; in fact $|b\rangle$ is orthogonal to $|a\rangle$. We say a transition has taken place from the state $|a\rangle$ to the state $|b\rangle$. We conclude that a system, described by a time-independent Hamiltonian, will undergo transitions if initially it is not in an eigenstate of the Hamiltonian.

More generally, a time-dependent Hamiltonian $H(t)$ does not possess stationary states, and the state $|\psi(t)\rangle$ of such a system necessarily changes with time. Its time-dependence follows from the time-dependent Schrödinger equation

$$i\hbar \frac{\partial}{\partial t} |\psi(t)\rangle = H(t)|\psi(t)\rangle. \tag{9.5}$$

If at an initial time $t = t_0$ the system is in the state $|\psi_0\rangle$, i.e.

$$|\psi(t_0)\rangle = |\psi_0\rangle, \tag{9.6}$$

then the state $|\psi(t)\rangle$ of the system at later times, $t \geqslant t_0$, is given by the solution of the time-dependent Schrödinger equation (9.5) with (9.6) as initial condition. [A time-independent Hamiltonian is of course a special case of this. With $H(t)$ replaced by H_0, Eq. (9.4) is the solution of Eq. (9.5) which at $t = 0$ reduces to the state (9.2).]

We shall now consider a time-dependent Hamiltonian of the form

$$H(t) = H_0 + V(t) \tag{9.7}$$

i.e. consisting of a time-independent part H_0 and a time-dependent part $V(t)$. Systems with Hamiltonians of this form occur frequently. An example is an atom placed in an applied oscillating electromagnetic field: H_0 is the atomic Hamiltonian in the absence of the field, and $V(t)$ is the interaction of the atom with the field. It is this system we considered when deriving the selection rules for electric dipole transitions. In section 9.5 we shall obtain the matrix element for these transitions, on which our earlier analysis was based.

To solve the time-dependent Schrödinger equation for a time-dependent Hamiltonian is anything but trivial. The state $|\psi(t)\rangle$ of the system at any instant of time can, of course, be expanded in terms of a complete set of states, e.g. in terms of the energy eigenstates $|u_1\rangle$, $|u_2\rangle$, ... of H_0. Eq. (9.4) suggests that the most convenient form of the expansion is

$$|\psi(t)\rangle = \sum_n c_n(t)|u_n\rangle\, e^{-iE_nt/\hbar}. \tag{9.8}$$

For a time-independent Hamiltonian ($V = 0$, $H = H_0$) this reduces to the form (9.4) with constant coefficients c_n; thus, the time dependence of the coefficients $c_n(t)$ in Eq. (9.8) results exclusively from the term V in the Hamiltonian (9.7).

With $|\psi(t)\rangle$ normed to unity, the coefficients $c_n(t)$ in Eq. (9.8) are probability amplitudes: $|c_n(t)|^2$ is the probability that at time t the system is in the state $|u_n\rangle$. These probabilities are time-dependent: the interaction V effects transitions between different states, and a system which is initially in the state $|\psi_0\rangle = |u_i\rangle$ will subsequently be found in other states. This is in contrast to the case $V = 0$: the probability amplitudes c_n in Eq. (9.4) are independent of time.

To determine the time-dependent amplitudes $c_n(t)$, we substitute Eqs. (9.7) and (9.8) in the Schrödinger equation (9.5), giving

$$i\hbar \sum_n \dot{c}_n(t)|u_n\rangle\, e^{-iE_nt/\hbar} = \sum_n V(t)|u_n\rangle\, e^{-iE_nt/\hbar}c_n(t) \tag{9.9}$$

where we used $H_0|u_n\rangle = E_n|u_n\rangle$ and $\dot{c}_n(t)$ stands for $dc_n(t)/dt$. Taking the scalar product of Eq. (9.9) with $|u_p\rangle$, we obtain

$$i\hbar\dot{c}_p(t) = \sum_n V_{pn}(t)\, e^{i\omega_{pn}t}c_n(t) \tag{9.10}$$

where the matrix element $V_{pn}(t)$ is defined by

$$V_{pn}(t) = \langle u_p|V(t)|u_n\rangle \tag{9.11}$$

and the Bohr angular frequency ω_{pn} by

$$\omega_{pn} = (E_p - E_n)/\hbar. \tag{9.12}$$

Eqs. (9.10) are an infinite set of coupled equations for the amplitudes $c_1(t)$, $c_2(t)$, ... For $V = 0$, Eqs. (9.10) reduce to $\dot{c}_p = 0$ and hence to time-independent amplitudes c_p, as expected. The initial condition (9.6) for $|\psi(t)\rangle$ now is translated into initial conditions for amplitudes $c_n(t)$ at $t = t_0$; from Eqs. (9.6) and (9.8) we obtain

$$c_n(t_0) = \langle u_n|\psi_0\rangle\, e^{iE_nt_0/\hbar}. \tag{9.13}$$

Solving the coupled equations (9.10) with the initial conditions (9.13) is equivalent to solving the time-dependent Schrödinger equation (9.5) with the initial condition (9.6). This formulation of the time-dependent problem in terms of the coupled amplitudes $c_n(t)$ is due to Dirac and is known as the method of the variations of the constants.

In general it is not possible to solve the coupled equations (9.10) exactly. Later in this chapter, we shall develop a time-dependent perturbation theory which, like its time-independent counterpart, is often extremely useful. For certain simple systems, exact solutions are possible. In the next section, we shall study one such system, as it gives insight into the way the state of a system changes with time.

★9.2   AN EXACTLY SOLUBLE TWO-STATE SYSTEM*

In this section we shall discuss an exactly soluble time-dependent two-state system. Consider a spin $\frac{1}{2}$ system possessing a magnetic moment. The Hamiltonian for this system placed in a time-dependent magnetic field $\mathbf{B}(t)$ is

$$H(t) = -\boldsymbol{\mu} \cdot \mathbf{B}(t). \tag{9.14}$$

Here the magnetic moment operator $\boldsymbol{\mu}$ is related to the spin operator $\mathbf{s}$ by

$$\boldsymbol{\mu} = \gamma \mathbf{s} = \gamma \tfrac{1}{2}\hbar\boldsymbol{\sigma}$$

where γ is the gyromagnetic ratio (i.e. ratio of magnetic moment to angular momentum) of the system and $\boldsymbol{\sigma} = (\sigma_x, \sigma_y, \sigma_z)$ are the Pauli matrices (5.43b):

$$\sigma_x = \begin{bmatrix} 0 & 1 \\ 1 & 0 \end{bmatrix}, \quad \sigma_y = \begin{bmatrix} 0 & -i \\ i & 0 \end{bmatrix}, \quad \sigma_z = \begin{bmatrix} 1 & 0 \\ 0 & -1 \end{bmatrix}. \tag{9.15}$$

For a static uniform magnetic field in the z-direction,

$$\mathbf{B}_0 = (0, 0, B_0), \tag{9.16a}$$

the Hamiltonian (9.14) becomes

$$H_0 = -\tfrac{1}{2}\gamma\hbar B_0\sigma_z = -\tfrac{1}{2}\gamma\hbar B_0 \begin{bmatrix} 1 & 0 \\ 0 & -1 \end{bmatrix}. \tag{9.17}$$

For this time-independent Hamiltonian, the eigenstates $|u_1\rangle$ and $|u_2\rangle$ are just the spin up and spin down eigenstates

$$|u_1\rangle = |\alpha\rangle = \begin{bmatrix} 1 \\ 0 \end{bmatrix}, \quad |u_2\rangle = |\beta\rangle = \begin{bmatrix} 0 \\ 1 \end{bmatrix} \tag{9.18a}$$

---

* Although not required later in this book, the reader is encouraged to read this section for the insight it provides.

with the energy eigenvalues

$$E_1 = -\tfrac{1}{2}\gamma\hbar B_0, \quad E_2 = +\tfrac{1}{2}\gamma\hbar B_0 \tag{9.18b}$$

respectively.

We now superpose on the static field $\mathbf{B}_0$ the time-dependent magnetic field

$$\mathbf{B}_1(t) = (B_1 \cos \omega t, B_1 \sin \omega t, 0). \tag{9.16b}$$

$\mathbf{B}_1(t)$ is a field of magnitude B_1 rotating with constant angular frequency ω in the (x, y) plane; for $\omega > 0$, the field rotates anticlockwise as seen from the positive z-direction. The effect of the time-dependent field is to cause transitions between the spin up and the spin down states, as discussed handwavingly in the last section. For the field

$$\mathbf{B}(t) = \mathbf{B}_0 + \mathbf{B}_1(t) \tag{9.16c}$$

and the resulting Hamiltonian (9.14), the time-dependent Schrödinger equation

$$i\hbar \frac{\partial}{\partial t} |\psi(t)\rangle = H(t)|\psi(t)\rangle \tag{9.19}$$

can be solved exactly, allowing an exact quantitative discussion of this case. Since any spin state can be expressed as a superposition of $|\alpha\rangle$ and $|\beta\rangle$, we write the general spin state $|\psi(t)\rangle$ in (9.19) as

$$|\psi(t)\rangle = c_1(t)|\alpha\rangle\, e^{-iE_1 t/\hbar} + c_2(t)|\beta\rangle\, e^{-iE_2 t/\hbar} \tag{9.20}$$

which is just Eq. (9.8) for the present case. The Hamiltonian (9.14) for the fields (9.16) becomes

$$H(t) = H_0 + V(t) \tag{9.21a}$$

where H_0 is given by Eq. (9.17) and

$$V(t) = -\tfrac{1}{2}\gamma\hbar B_1(\sigma_x \cos \omega t + \sigma_y \sin \omega t) = -\tfrac{1}{2}\gamma\hbar B_1 \begin{bmatrix} 0 & e^{-i\omega t} \\ e^{i\omega t} & 0 \end{bmatrix}. \tag{9.21b}$$

The above equations are easily related to the general case treated in the last section. With

$$V_{pn}(t) = \langle u_p | V(t) | u_n \rangle, \quad p, n = 1, 2, \tag{9.22}$$

we have at once from Eqs. (9.18a) and (9.21b) that

$$V_{11}(t) = V_{22}(t) = 0, \quad V_{12}(t) = V_{21}^*(t) = -\tfrac{1}{2}\gamma\hbar B_1\, e^{-i\omega t}, \tag{9.23}$$

and Eqs. (9.10) for the coupled amplitudes $c_1(t)$ and $c_2(t)$ become

$$\left. \begin{aligned} i\hbar\dot{c}_1(t) &= -\tfrac{1}{2}\gamma\hbar B_1\, e^{i(\omega_{12} - \omega)t} c_2(t) \\ i\hbar\dot{c}_2(t) &= -\tfrac{1}{2}\gamma\hbar B_1\, e^{i(\omega_{21} + \omega)t} c_1(t) \end{aligned} \right\} \tag{9.24}$$

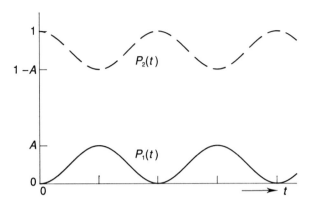

Fig. 9.1   The probabilities $P_1(t)$ and $P_2(t)$, Eqs. (9.27–28), that a system which is in the spin down state at $t = 0$ will at time t be in the spin up or spin down states respectively.

where

$$\omega_{12} = -\omega_{21} = (E_1 - E_2)/\hbar = -\gamma B_0. \tag{9.25}$$

Eqs. (9.24) admit an exact solution. In particular, if initially at $t = 0$ the system is in the spin down state $|\beta\rangle$, i.e.

$$c_1(0) = 0, \quad c_2(0) = 1, \tag{9.26}$$

then the probability that at a later time t the system is in the spin up state $|\alpha\rangle$ is given by (see problem 9.3)

$$P_1(t) = |c_1(t)|^2 = A \sin^2\{\tfrac{1}{2}[(\gamma B_1)^2 + (\omega + \gamma B_0)^2]^{1/2} t\} \tag{9.27a}$$

where

$$A = \frac{(\gamma B_1)^2}{[(\gamma B_1)^2 + (\omega + \gamma B_0)^2]}. \tag{9.27b}$$

Correspondingly, the probability that at time t the system is in the spin down state $|\beta\rangle$ is given by

$$P_2(t) = |c_2(t)|^2 = 1 - |c_1(t)|^2. \tag{9.28}$$

These probabilities are shown in Fig. 9.1. They are periodic; $P_1(t)$ oscillates between 0 (when $P_2 = 1$) and its maximum value A (when P_2 has its minimum value $1 - A$).

The transitions between the spin up and spin down states are brought about by the rotating field $\mathbf{B}_1$. In practice one usually has $B_1 \ll B_0$. The amplitude A and hence the probability $P_1(t)$ of a spin flip from spin down to spin up are

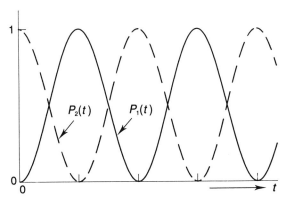

Fig. 9.2   The probabilities $P_1(t)$ and $P_2(t)$, Eqs. (9.30), i.e. at resonance when the frequency ω of the rotating field equals the Bohr frequency ω_{12} of the transition:

$$\omega = \omega_{12} = (E_1 - E_2)/\hbar = -\gamma B_0. \tag{9.29}$$

At resonance $A = 1$, from Eq. (9.27b), and Fig. 9.1 goes over into this figure.

then of order $(B_1/B_0)^2$, i.e. quite small. We shall now assume that the gyromagnetic ratio γ is negative.* For $\gamma < 0$, the system shows a remarkable phenomenon if the frequency ω of the rotating field is chosen equal to the Bohr frequency ω_{12} of the transition

$$\omega = \omega_{12} = (E_1 - E_2)/\hbar = -\gamma B_0. \tag{9.29}$$

The amplitude A now has the value 1, and the probabilities (9.27)–(9.28) become

$$P_1(t) = \sin^2 \left(\tfrac{1}{2}\gamma B_1 t\right), \quad P_2(t) = \cos^2 \left(\tfrac{1}{2}\gamma B_1 t\right), \tag{9.30}$$

and are large, oscillating between 0 and 1 (see Fig. 9.2) even for $B_1/B_0 \ll 1$. This is typical of a resonance phenomenon: if the frequency ω of the rotating field equals the natural frequency ω_{12} of the system, a very weak rotating field (classically one would think of this as a weak driving torque) produces large effects. It is interesting to note, but not unexpected, that at or near resonance one cannot treat the rotating field as a small perturbation and retain only terms linear in B_1^2 in (9.27), even if $B_1 \ll B_0$. Eq. (9.27a) reduces to

$$P_1(t) = \frac{(\gamma B_1)^2}{(\omega + \gamma B_0)^2} \sin^2 \left[\tfrac{1}{2}(\omega + \gamma B_0)t\right], \tag{9.31}$$

---

* $\gamma < 0$ means that the magnetic moment of the system is antiparallel to its spin, as for the spin magnetic moment of the electron, Eq. (7.61). Our analysis is quite general and applies to $\gamma > 0$ or $\gamma < 0$. We shall come back to the case $\gamma > 0$ below.

only if $|\gamma B_1| \ll |\omega + \gamma B_0|$, and this condition is violated for frequencies ω sufficiently near the resonance frequency, $\omega \approx -\gamma B_0$, however weak the rotating field B_1. (See problem 9.2.)

If we consider Eqs. (9.27) for $\gamma > 0$, a resonance would occur for $\omega = -\gamma B_0 < 0$, i.e. it would occur for a clockwise rotating field $\mathbf{B}_1$. In practice one usually employs an oscillating field, $\mathbf{B}_1 = (B_1 \cos \omega t, 0, 0)$, which can be thought of as a superposition of clockwise and anticlockwise rotating fields. For a given sign of γ and resonance conditions, i.e. $|\omega| \approx |\gamma B_0|$, the effect of one of the rotating components dominates; the other can be neglected or, if need be, treated in perturbation theory.

The analysis given in this section, or generalizations of it, describe many situations of practical importance. It forms the basis of high-precision methods (paramagnetic resonance, nuclear magnetic resonance (NMR), Rabi's molecular beam techniques) for measuring magnetic moments of particles (atoms, molecules, elementary particles like the muon) and for studying the properties of condensed matter. With appropriate reinterpretation of the symbols, the formalism also applies to non-magnetic systems; for example, the ammonia maser is a two-state system which can be described in this way.* An example from elementary particle physics is given in problem 9.6.

9.3 TIME-DEPENDENT PERTURBATION THEORY

In section 9.1 we saw that the time-dependent Schrödinger equation (9.5) is equivalent to the coupled equations (9.10) for the probability amplitudes $c_n(t)$. We shall now treat the time-dependent part $V(t)$ of the Hamiltonian $H = H_0 + V(t)$ as small, and solve Eqs. (9.10) in first-order perturbation theory. We shall take as initial condition that at time $t = 0$ the system is in the eigenstate $|u_i\rangle$ of H_0, i.e. the amplitudes $c_n(t)$ have the initial values

$$c_n(0) = \delta_{ni}. \qquad (9.32)$$

For $V = 0$, the amplitudes $c_n(t)$ retain their initial values at all times. To find the amplitudes $c_n^{(1)}(t)$ in first-order perturbation theory we need only substitute their initial values (9.32) in the right-hand sides of Eqs. (9.10), since these right-hand sides are already of first order in $V(t)$, giving

$$i\hbar \dot{c}_p^{(1)}(t) = V_{pi}(t) \, e^{i\omega_{pi}t} \qquad (9.33)$$

---

* For an elementary excellent discussion of the ammonia maser and other two-state systems, see Feynman III.

which holds for all values of p, including $p = i$. We obtain the first-order amplitudes $c_p^{(1)}(\tau)$ by integrating these equations (9.33) from $t = 0$ to $t = \tau$, subject to the initial conditions (9.32):

$$c_i^{(1)}(\tau) = 1 + \frac{1}{i\hbar} \int_0^\tau V_{ii}(t)\, dt \tag{9.34a}$$

$$c_p^{(1)}(\tau) = \frac{1}{i\hbar} \int_0^\tau V_{pi}(t)\, e^{i\omega_{pi}t}\, dt, \qquad p \neq i. \tag{9.34b}$$

$|c_p^{(1)}(\tau)|^2$, $p \neq i$, is the probability, in first-order perturbation theory, that at time τ the system is in the state $|u_p\rangle$, i.e. that a transition $i \to p$ has occurred. For first-order perturbation theory to be valid, the probability for transitions from the initial state i to have occurred must be very small, i.e. the initial state i is only slightly depleted:

$$|c_i^{(1)}(\tau)| \approx 1 \tag{9.35a}$$

and

$$1 - |c_i^{(1)}(\tau)|^2 = \sum_{p \neq i} |c_p^{(1)}(\tau)|^2 \ll 1, \tag{9.35b}$$

so that the transition probability for any final state $p(\neq i)$ must satisfy

$$P_{i \to p}^{(1)}(\tau) = |c_p^{(1)}(\tau)|^2 \ll 1, \qquad p \neq i. \tag{9.35c}$$

In order to exploit these results, one must know $V(t)$ so that the integrals in Eqs. (9.34) can be calculated. In the following two sections, we shall consider two important cases.

9.4 TIME-INDEPENDENT PERTURBATIONS

The perturbation theory for time-dependent perturbations $V(t)$, developed in the last section, can also be applied to time-independent perturbations V. For the latter, one could employ the time-independent perturbation theory, derived in Chapter 7, and it depends on the situation one is studying which formalism is appropriate. As we saw at the beginning of this chapter, for a system not in a stationary state, transitions will occur and the time-dependent formalism is the appropriate one. In particular, if a system with the Hamiltonian

$$H = H_0 + V$$

is initially in the energy eigenstate $|u_i\rangle$ of H_0, then transitions to other eigenstates $|u_p\rangle$ of H_0 will occur and the transition probabilities for these are given by the results of the last section.* Most collision problems are of this kind. If we consider the scattering of an electron by an atom, the Hamiltonian of the system (atom plus projectile electron) is time-independent. Long before and long after the collision, electron and atom are far apart and the interaction between them is negligible. The scattering process can be described in terms of transitions from an initial state of the system, in which the electron has a definite momentum and the atom is in a given state, to a final state of the system, in which the electron has a different momentum and the atom may have changed its state. We shall consider collision processes in this way in the next chapter.

For a time-independent perturbation V, Eq. (9.34b) can be integrated to give the amplitude for a transition from the initial state i to a final state p

$$c_p^{(1)}(\tau) = \frac{V_{pi}}{\hbar \omega_{pi}} (1 - e^{i\omega_{pi}\tau}). \tag{9.36}$$

Hence the probability that a transition from the initial state i to the final state p has occurred in the time interval τ is given by

$$P_{i \to p}^{(1)}(\tau) = |c_p^{(1)}(\tau)|^2 = \frac{4|V_{pi}|^2}{\hbar^2} \frac{\sin^2 (\frac{1}{2}\omega_{pi}\tau)}{\omega_{pi}^2}. \tag{9.37}$$

The function

$$D(\omega_{pi}, \tau) = \frac{\sin^2 (\frac{1}{2}\omega_{pi}\tau)}{\omega_{pi}^2} \tag{9.38}$$

is shown in Fig. 9.3. For sufficiently large values of τ, it consists essentially of a large peak, centred at $\omega_{pi} = 0$, of height $\tau^2/4$ and width $4\pi/\tau$. It follows that for ω_{pi} appreciably different from zero the transition probability (9.37) remains small for all values of τ. $P_{i \to p}^{(1)}(\tau)$ becomes large only if $\omega_{pi} \approx 0$, i.e. if

$$\hbar |\omega_{pi}| = |E_p - E_i| \lesssim 2\pi\hbar/\tau. \tag{9.39}$$

---

* I am adopting the approach used by Dirac in his classic treatise (Dirac, section 46). Many authors employ the following alternative formulation. The problem for the time-independent perturbation V, which we are considering here, can be thought of as one for the explicitly time-dependent perturbation

$$V(t) = \begin{cases} 0, & \text{for } t < t_0, \\ V, & \text{for } t > t_0, \end{cases}$$

with the system being in the stationary state

$$|\psi(t)\rangle = |u_i\rangle \exp[-iE_i(t - t_0)/\hbar]$$

of the unperturbed Hamiltonian H_0 for $t < t_0$.

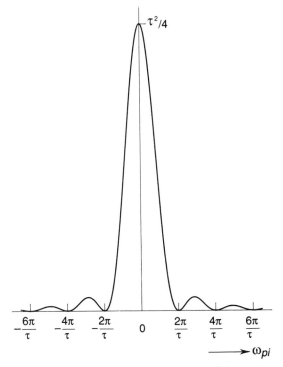

Fig. 9.3 The function $D(\omega_{pi}, \tau) = \dfrac{\sin^2(\frac{1}{2}\omega_{pi}\tau)}{\omega_{pi}^2}$.

A frequently occurring situation is one in which the final states form a continuum. In this case it is not the transition probability to a particular final state, with energy $E_p \approx E_i$, but to a group G of final states with energy in the interval

$$E_i - \Delta E \leqslant E_p \leqslant E_i + \Delta E \qquad (9.40)$$

which is of interest. This probability is obtained by summing the probabilities (9.37) over this group of states. In order to count states, it is convenient to convert a system with a continuum of states into a system with a discrete set of states lying closely together. We know from section 2.6.1 that this is achieved by enclosing the system in a large box and imposing suitable boundary conditions on the wave functions at the surface of the box. With this box normalization, it is easy—as we shall see later—to specify the number $\rho(E_p)\,dE_p$ of final states with energy in the interval E_p to $E_p + dE_p$. $\rho(E_p)$ is known as the density of states per unit energy. The probability that in the time interval τ a transition has occurred from the initial state i to one of the group G of final states is then given by

$$P_{i \to G}^{(1)}(\tau) = \int_{E_i - \Delta E}^{E_i + \Delta E} dE_p \rho(E_p) \frac{4|V_{pi}|^2}{\hbar^2} \frac{\sin^2 (\tfrac{1}{2}\omega_{pi}\tau)}{\omega_{pi}^2}. \tag{9.41}$$

For τ sufficiently large and

$$\Delta E \gg 2\pi\hbar/\tau, \tag{9.42}$$

it follows from Eq. (9.38) and Fig. 9.3 that the contributions to the integral (9.41) originate from the range of energies (9.39). Within this narrow energy range, the functions $\rho(E_p)$ and $|V_{pi}|^2$ can be taken as constant. They can therefore be taken outside the integral, being evaluated at $E_p = E_i$, and Eq. (9.41) becomes

$$P_{i \to G}^{(1)}(\tau) = \left[\frac{4|V_{pi}|^2}{\hbar^2} \rho(E_p) \right]_{E_p = E_i} I \tag{9.43}$$

where

$$I = \int_{E_i - \Delta E}^{E_i + \Delta E} dE_p \frac{\sin^2 (\tfrac{1}{2}\omega_{pi}\tau)}{\omega_{pi}^2} = \tfrac{1}{2}\tau\hbar \int_{-\tau\Delta E/2\hbar}^{\tau\Delta E/2\hbar} d\xi \frac{\sin^2 \xi}{\xi^2}, \tag{9.44}$$

with $\xi = \tfrac{1}{2}\omega_{pi}\tau$. On account of (9.42), the range of integration in the last integral can be extended over all values of ξ, from $\xi = -\infty$ to $\xi = +\infty$. The resulting integral can be shown to have the value π, so that $I = \tfrac{1}{2}\tau\hbar\pi$, and Eq. (9.43) becomes

$$P_{i \to G}^{(1)}(\tau) = \tau \frac{2\pi}{\hbar} \left[|V_{pi}|^2 \rho(E_p) \right]_{E_p = E_i}. \tag{9.45}$$

This transition probability is proportional to the time interval τ. Hence the transition probability per unit time $\dot{P}_{i \to G}$ ($= dP_{i \to G}^{(1)}(\tau)/d\tau$) is independent of time in this case:

$$\dot{P}_{i \to G}^{(1)} = \frac{2\pi}{\hbar} [|V_{pi}|^2 \rho(E_p)]_{E_p = E_i}. \tag{9.46}$$

Eq. (9.46) for the transition rate is so useful in applications that Fermi called it 'Golden Rule No. 2.'*

In Chapter 10 we shall use the golden rule to solve collision problems. In the next section, I shall first generalize it so as to apply to time-dependent perturbations.

---

* See Fermi, p. 142. Somewhat misleadingly, it is often referred to as Fermi's golden rule. Actually, time-dependent perturbation theory and the above results are due to Dirac. Fermi is not in need of 'borrowed feathers'.

The golden rule is conveniently expressed in terms of the Dirac delta function.* One sees from Fig. 9.3 and the discussion following Eq. (9.38) that the function $D(\omega_{pi}, \tau)$, defined in Eq. (9.38), possesses the essential properties of the delta function for large values of τ. Moreover with $E_p - E_i = \hbar\omega_{pi}$ one has

$$\int_{-\infty}^{\infty} D\left(\frac{E_p - E_i}{\hbar}, \tau\right) dE_p = \int_{-\infty}^{\infty} \frac{\sin^2\left(\frac{1}{2}\omega_{pi}\tau\right)}{\omega_{pi}^2} \hbar \, d\omega_{pi} = \frac{1}{2}\pi\hbar\tau,$$

independently of the value of τ. Hence we can define the delta function by

$$\delta(E_p - E_i) = \delta(\hbar\omega_{pi}) = \underset{\tau \to \infty}{\mathrm{Lim}} \frac{2}{\pi\hbar\tau} D\left(\frac{E_p - E_i}{\hbar}, \tau\right).$$

In deriving the golden rule (9.46), we took the limit $\tau \to \infty$. Anticipating this limit, we replace $D(\omega_{pi}, \tau)$ in Eq. (9.37) by $\frac{1}{2}\pi\hbar\tau\delta(E_p - E_i)$ giving

$$P_{i \to p}^{(1)}(\tau) = \tau\frac{2\pi}{\hbar} |V_{pi}|^2 \delta(E_p - E_i).$$

Differentiating this equation with respect to τ, we obtain the transition probability per unit time

$$\dot{P}_{i \to p}^{(1)} = \frac{2\pi}{\hbar} |V_{pi}|^2 \delta(E_p - E_i). \tag{9.46a}$$

This is the golden rule stated for transitions between two individual states. This is in contrast to Eq. (9.46) which gave the golden rule for transitions to a group of final states in the continuum. The occurrence of the delta function in Eq. (9.46a) means, of course, that we must integrate this equation over a continuum of states to obtain a useful result. If $\rho(E_p) \, dE_p$ is the number of final states with energy in the range E_p to $E_p + dE_p$, then it follows from Eq. (9.46a) and the basic property (2.122) of the delta function that

$$\dot{P}_{i \to G}^{(1)} = \int \dot{P}_{i \to p}^{(1)}\rho(E_p) \, dE_p = \frac{2\pi}{\hbar}\left[|V_{pi}|^2\rho(E_p)\right]_{E_p = E_i}, \tag{9.46}$$

i.e. we have recovered our earlier form of the golden rule. The two forms are, of course, equivalent but (9.46a), in which the integration to eliminate the delta function has not yet been performed, is more versatile and it is frequently found in the literature.

---

* In the rest of this section, a knowledge of the basic properties of the Dirac delta function is presupposed. These are derived in the starred section 2.6.2.

★9.5   EMISSION AND ABSORPTION OF RADIATION BY ATOMS

The results of the last section for time-independent perturbations are easily generalized to the case of a perturbation with harmonic time-dependence. In this section this will be illustrated for the important case of electromagnetic transitions in atoms.

We consider an atom placed in the path of an electromagnetic wave, specified by the electric field

$$\mathscr{E}(\mathbf{r}, t) = \mathscr{E}_0 \mathbf{n} \cos (\mathbf{k} \cdot \mathbf{r} - \omega t), \tag{9.47a}$$

where the unit vector $\mathbf{n}$ defines the polarization of the wave, of wavelength $\lambda = 2\pi/|\mathbf{k}|$ and angular frequency $\omega = c|\mathbf{k}|$. The magnetic interaction of the radiation with the atoms is usually small compared to the interaction of the electric field and we shall neglect it. We shall also approximate the electric field (9.47a) by

$$\mathscr{E}(t) = \mathscr{E}_0 \mathbf{n} \cos \omega t \tag{9.47b}$$

i.e. we shall neglect the spatial variation of the electric field over the size of the atom which is justified provided λ is large compared to the linear dimensions of the atom. The interaction of the atom with the radiation field is then just the electric dipole interaction

$$V(t) = \mathbf{D} \cdot \mathscr{E}(t) = \tfrac{1}{2} D_{\mathbf{n}} \mathscr{E}_0 (e^{-i\omega t} + e^{i\omega t}) \tag{9.48}$$

where $D_{\mathbf{n}} = \mathbf{D} \cdot \mathbf{n}$ is the component, in the direction $\mathbf{n}$, of the electric dipole operator

$$\mathbf{D} = -e \sum_{j=1}^{Z} \mathbf{r}_j \tag{9.49}$$

of the atom.

The electromagnetic interaction is weak and can be treated as a perturbation of H_0, the atomic Hamiltonian in the absence of the radiation field. If the atom (i.e. H_0) possesses the stationary states $|u_1\rangle, |u_2\rangle, \ldots$, with energies $E_1, E_2, \ldots$, and at the initial time $t = 0$ the atom is in the state $|u_i\rangle$, then the probability amplitude that at time τ the atom has made a transition to the state $|u_p\rangle$ follows from Eqs. (9.34b) and (9.48), and is given by

$$c_p^{(1)}(\tau) = \frac{1}{i\hbar} \langle u_p | \tfrac{1}{2} D_{\mathbf{n}} \mathscr{E}_0 | u_i \rangle \left\{ \frac{e^{i(\omega_{pi} - \omega)\tau} - 1}{i(\omega_{pi} - \omega)} + \frac{e^{i(\omega_{pi} + \omega)\tau} - 1}{i(\omega_{pi} + \omega)} \right\}. \tag{9.50}$$

Eq. (9.50) is similar to Eq. (9.36) for a time-independent perturbation, except that the amplitude (9.50) contains two terms. We know from the discussion in the last section that the first of these terms is large only if $\omega = \omega_{pi}$ and the second term can then be neglected; on the other hand, the second term is large only if $\omega = -\omega_{pi}$ and the first term can then be neglected. We therefore

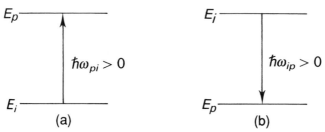

Fig. 9.4   Level diagrams for the atomic transition $i \to p$: (a) absorption of a photon of frequency $\omega = \omega_{pi} > 0$, (b) emission of a photon of frequency $\omega = \omega_{ip} > 0$.

distinguish two situations in which the transition probabilities are large:

(a) $$\omega = \omega_{pi} = (E_p - E_i)/\hbar > 0 \tag{9.51a}$$

corresponds to the absorption by the atom of a photon of angular frequency ω from the electromagnetic wave, the atom absorbing energy $\hbar\omega$ in the transition $i \to p$ to a more highly excited state (Fig. 9.4a). The probability that this transition has occurred in the time interval τ follows from the first term in (9.50):

$$P_{i \to p}^{(1)}(\tau) = |c_p^{(1)}(\tau)|^2$$
$$= \frac{1}{\hbar^2} |\langle u_p | D_n | u_i \rangle|^2 \mathscr{E}_0^2 \frac{\sin^2[\tfrac{1}{2}(\omega_{pi} - \omega)\tau]}{(\omega_{pi} - \omega)^2}. \tag{9.52a}$$

(b) $$\omega = -\omega_{pi} = \omega_{ip} = (E_i - E_p)/\hbar > 0 \tag{9.51b}$$

corresponds to emission of a photon with the same frequency ω as the electromagnetic wave, the atom making a transition $i \to p$ in which it loses energy $\hbar\omega$ (Fig. 9.4b). For this case

$$P_{i \to p}^{(1)}(\tau) = |c_p^{(1)}(\tau)|^2$$
$$= \frac{1}{\hbar^2} |\langle u_p | D_n | u_i \rangle|^2 \mathscr{E}_0^2 \frac{\sin^2[\tfrac{1}{2}(\omega_{pi} + \omega)\tau]}{(\omega_{pi} + \omega)^2}. \tag{9.52b}$$

We have so far considered a monochromatic electromagnetic wave. Except in the case of laser sources, one deals with an *incoherent* superposition of waves of different frequencies, and we must next adapt our results for this situation. The energy density of the electromagnetic wave (9.47a) is* $\tfrac{1}{2}\varepsilon_0 \mathscr{E}_0^2$, whereas the energy density of the incoherent wave is of the form

$$\tfrac{1}{2}\varepsilon_0 \int_0^\infty \mathscr{E}_0^2(\omega) \, d\omega.$$

---

* The energy density of the electric field of the wave is the time average of $\tfrac{1}{2}\varepsilon_0 \mathscr{E}^2(t)$ which is $\tfrac{1}{4}\varepsilon_0 \mathscr{E}_0^2$. The energy density of the magnetic field of the wave has the same value, leading to the above result. For a detailed derivation, see Grant and Phillips, section 11.4.

(This is typical of incoherent waves: we add squares of amplitudes, not amplitudes; the interference terms cancel out.) For the incoherent case, the waves in each frequency range ω to $\omega + d\omega$ lead to transition probabilities which are independent of other frequencies and which are given by Eqs. (9.52) with $\mathscr{E}_0^2$ replaced by $\mathscr{E}_0^2(\omega)\,d\omega$. Finally, we must add the contributions from all frequencies, giving the transition probabilities*

$$P^{(1)}_{i \to p}(\tau) = \frac{1}{\hbar^2} |\langle u_p | D_n | u_i \rangle|^2 \int_0^\infty d\omega \, \mathscr{E}_0^2(\omega) \left\{ \frac{\sin^2 \left[\frac{1}{2}(\omega_{pi} \mp \omega)\tau \right]}{(\omega_{pi} \mp \omega)^2} \right\} \qquad (9.53)$$

where the upper and lower signs refer to absorption and emission processes respectively. Since the matrix element in (9.53) does not depend on ω, we were able to take its square modulus outside the integral.

We evaluated an integral similar to that in Eq. (9.53) in the last section. For sufficiently large values of τ, the factor in curly parentheses in Eq. (9.53) is essentially zero except when $\omega \approx \omega_{pi}\,(> 0)$ for the absorption process, and when $\omega \approx -\omega_{pi}\,(> 0)$ for the emission process. Hence we can replace $\mathscr{E}_0^2(\omega)$ in Eq. (9.53) by $\mathscr{E}_0^2(|\omega_{pi}|)$. In the remaining integral, the range of integration can be extended over all values of ω, i.e. $-\infty \leqslant \omega \leqslant \infty$, and as in the last section one obtains the value $\frac{1}{2}\pi\tau$ for it. Hence Eq. (9.53) becomes

$$P^{(1)}_{i \to p}(\tau) = \tau \frac{\pi}{2} \frac{1}{\hbar^2} |\langle u_p | D_n | u_i \rangle|^2 \mathscr{E}_0^2(|\omega_{pi}|). \qquad (9.54)$$

It is convenient to express $\mathscr{E}_0^2$ in this equation in terms of the photon flux. If $N(\omega)\,d\omega$ is the number of photons in the frequency range ω to $\omega + d\omega$ striking unit area normal to the direction of wave propagation per unit time, then the energy density of photons in this frequency range is given by

$$\hbar\omega \frac{1}{c} N(\omega)\,d\omega. \qquad (9.55)$$

(This follows since each photon possesses energy $\hbar\omega$, and the number of photons striking unit area per second is contained within a column of unit cross-sectional area and length c.) In terms of the electric field, the same energy density is given by $\frac{1}{2}\varepsilon_0 \mathscr{E}_0^2(\omega)\,d\omega$, whence

$$\mathscr{E}_0^2(\omega) = \frac{2}{\varepsilon_0 c} \hbar\omega N(\omega) \qquad (9.56)$$

---

* I have obtained this result in a handwaving way, giving the underlying physics. For a more formal derivation see, for example, Bethe and Jackiw, pp. 199–203.

and Eq. (9.54) becomes

$$P^{(1)}_{i \to p}(\tau) = \tau \frac{\pi}{\varepsilon_0 c\hbar} |\omega_{pi}| N(|\omega_{pi}|) |\langle u_p | D_n | u_i \rangle|^2. \tag{9.57}$$

Since this transition probability is proportional to the time interval τ, it leads to a transition probability per unit time which is independent of time

$$\dot{P}^{(1)}_{i \to p} = \frac{\pi}{\varepsilon_0 c\hbar} |\omega_{pi}| N(|\omega_{pi}|) |\langle u_p | D_n | u_i \rangle|^2. \tag{9.58}$$

One further step is required to reach the final result. The matrix element in Eq. (9.58) contains the operator $D_n = \mathbf{D} \cdot \mathbf{n}$, which depends on the orientation of the atomic dipole moment, i.e. of the atom, relative to the polarization vector $\mathbf{n}$ of the electric field. Usually the atoms will be randomly oriented and one must average (9.58) over the random orientations, i.e. over an isotropic distribution of orientations. Averaging over random orientations of $\mathbf{D}$ relative to a fixed direction $\mathbf{n}$ is equivalent to averaging over randomly oriented unit vectors $\mathbf{n}$, keeping $\mathbf{D}$ fixed. To carry out this averaging, we write $\mathbf{n} = (\cos \theta_x, \cos \theta_y, \cos \theta_z)$, where $\theta_x, \theta_y, \theta_z$ are the angles which $\mathbf{n}$ makes with the Cartesian axes x, y, z. Then

$$\langle u_p | D_n | u_i \rangle = \sum_{k=x,y,z} \cos \theta_k \langle u_p | D_k | u_i \rangle$$

and

$$|\langle u_p | D_n | u_i \rangle|^2 = \sum_{\substack{k=x,y,z \\ l=x,y,z}} \cos \theta_k \cos \theta_l \langle u_p | D_k | u_i \rangle \langle u_p | D_l | u_i \rangle^*. \tag{9.59}$$

But

$$\overline{\cos^2 \theta_k} = \tfrac{1}{3}, \quad \overline{\cos \theta_k \cos \theta_l} = \overline{\cos \theta_k} \, \overline{\cos \theta_l} = 0 \quad \text{for } k \neq l,$$

where the bar denotes averaging over the isotropic distribution of unit vectors $\mathbf{n}$. Hence the average over orientations of (9.59) becomes

$$\overline{|\langle u_p | D_n | u_i \rangle|^2} = \frac{1}{3} \sum_{k=x,y,z} |\langle u_p | D_k | u_i \rangle|^2 = \tfrac{1}{3} |\langle u_p | \mathbf{D} | u_i \rangle|^2 \tag{9.60}$$

where the last step defines $|\langle u_p | \mathbf{D} | u_i \rangle|^2$. Writing the electric dipole operator (9.49) as

$$\mathbf{D} = -e\mathbf{R}, \quad \mathbf{R} = \sum_{j=1}^{z} \mathbf{r}_j, \tag{9.61}$$

and introducing the fine structure constant $\alpha = e^2/(4\pi\varepsilon_0 \hbar c) \approx 1/137$, we can write Eq. (9.60)

$$\overline{|\langle u_p | D_n | u_i \rangle|^2} = \tfrac{1}{3} 4\pi\varepsilon_0 \hbar c\alpha |\langle u_p | \mathbf{R} | u_i \rangle|^2. \tag{9.62}$$

Replacing the square modulus of the matrix element in Eq. (9.58) by the average (9.62), we obtain the probability per unit time for an atomic transition $i \rightarrow p$, averaged over randomly oriented atoms; it is given by

$$W_{i \rightarrow p} = \overline{\dot{P}^{(1)}_{i \rightarrow p}} = \tfrac{4}{3} \pi^2 \alpha |\omega_{pi}| N(|\omega_{pi}|) |\langle u_p | \mathbf{R} | u_i \rangle|^2. \tag{9.63}$$

For $E_p > E_i$, i.e. $\omega_{pi} > 0$, Eq. (9.63) refers to absorption by the atom of a photon from the radiation field; for $E_p < E_i$, i.e. $\omega_{pi} < 0$, to emission of a photon by the atom.

Eq. (9.63) is our final result. That $W_{i \rightarrow p}$ is proportional to the fine structure constant α is typical of processes involving the absorption or emission of one photon. The smallness of $\alpha \approx 1/137$ is the reason why perturbation theory describes successfully the interaction of radiation with charged particles. The matrix element which occurs in Eq. (9.63) is the electric dipole transition matrix element which we already met in sections 4.1 and 6.2 and from which the selection rules for electric dipole transitions follow.

9.5.1   Spontaneous emission

It follows from Eq. (9.63) that the probabilities for radiative transitions are proportional to the photon flux $N(|\omega_{pi}|)$ of the incident radiation, and this is so for both absorption and emission of photons. This implies that the transition probability for photon emission vanishes if no radiation is incident on the atom. For this reason the emission process described by Eq. (9.63) is called *stimulated* or *induced emission*; it is brought about by radiation of the right frequency falling on the atom. But we know that excited states of atoms usually decay to less excited states by photon emission, even when the atoms are not placed in an applied electromagnetic field. This *spontaneous emission* process is not contained within a theory which treats only the atoms quantum mechanically, but the electromagnetic field classically, as we have done. Quantization of the electromagnetic field, i.e. quantum electrodynamics (QED) is required to account for the process of spontaneous photon emission.* According to QED, the probability per unit time, $W_{i \rightarrow p}(\text{spont.})$, for the spontaneous emission of an electric dipole photon in the atomic transition $i \rightarrow p$, is given by

$$W_{i \rightarrow p}(\text{spont.}) = \frac{4\alpha}{3c^2} \omega_{ip}^3 |\langle u_p | \mathbf{R} | u_i \rangle|^2. \tag{9.64}$$

The transition probability (9.64) for spontaneous emission involves the same electric dipole matrix element which occurs in Eq. (9.63) for induced emission and for absorption, so that all three processes obey the same selection rules.

---

* For a simple derivation of the probability for spontaneous photon emission from QED, see Mandl and Shaw, section 1.3.

Although QED is required to derive Eq. (9.64), one can adapt an ingenious argument, due to Einstein, to show that spontaneous and induced emission processes are needed to balance photon absorption, in order to be consistent with Planck's law for radiation in thermal equilibrium.*

9.5.2   Justifying the electric dipole approximation

The electric dipole interaction (9.48) was obtained by neglecting the spatial variation of the electric field and approximating Eq. (9.47a) by Eq. (9.47b). To justify this, consider, for simplicity of writing, a one-electron atom or ion for which $\mathbf{D} = -e\mathbf{r}$. Retaining the full space-dependent field (9.47a) would have led to the transition matrix element

$$\langle u_p | \tfrac{1}{2} D_n \mathscr{E}_0 \, \mathrm{e}^{\pm i\mathbf{k}\cdot\mathbf{r}} | u_i \rangle \tag{9.65}$$

in Eq. (9.50), i.e. in that equation we made the approximation

$$\mathrm{e}^{\pm i\mathbf{k}\cdot\mathbf{r}} = 1 \pm i\mathbf{k}\cdot\mathbf{r} + \cdots = 1. \tag{9.66}$$

The matrix element (9.65) involves the integrals

$$\int \mathrm{d}^3 r \, u_p^*(\mathbf{r}) \mathbf{r} \, \mathrm{e}^{\pm i\mathbf{k}\cdot\mathbf{r}} u_i(\mathbf{r})$$

whose effective range of integration is determined by the extension of the atomic wave functions u_i and u_p, i.e. the range of integration is restricted to $r \leqslant a$. Here a is a measure of the radius of the atom and is of the order of the Bohr radius $a_0 = 0.5 \times 10^{-10}$ m. Hence the approximation (9.66) is justified provided $ka_0 \ll 1$. For visible light the wavelength λ varies from 4×10^{-7} m to 7×10^{-7} m, so that $ka_0 < 10^{-3} \ll 1$ and the electric dipole approximation is justified for optical spectra and even for UV spectra for which $\lambda \gtrsim 10^{-8}$ m.

We have considered the electric interactions only and have neglected the magnetic interactions. A proper treatment describes the interaction of the atoms with the radiation field in terms of the vector potential of the field and this formulation automatically allows for both electric and magnetic interactions. For electric dipole transitions it leads precisely to the results and validity criterion which we have obtained here. Other multipole transitions may become significant if electric dipole transitions are forbidden by selection rules, as discussed in section 6.2.4.†

---

* See, for example, Bethe and Jackiw, pp. 206–208, or Schiff, pp. 414–415.

† A simple treatment in terms of the vector potential and of higher multipoles is given in Davydov, sections 78 and 79.

9.5.3 Line width

In discussing radiative processes, we have treated them as transitions between stationary atomic states. The possibility of excited states decaying by spontaneous photon emission means that they cannot be truly stationary states. Treating them as such is a good approximation since the electromagnetic interaction, which is responsible for their decay, is a weak one so that they have long lifetimes. In the next approximation, the decay of the excited state is taken into account in the perturbation theory treatment.

The probability per unit time for spontaneous emission, quoted in Eq. (9.64), is independent of time. One easily sees that this implies an exponential decay law: if $P(t)$ is the probability that at time t a system is in an excited state for which the decay probability per unit time is γ, then

$$P(t + dt) = P(t)(1 - \gamma\, dt)$$

and

$$P(t) = P(0)\, e^{-\gamma t}. \tag{9.67}$$

Whereas a stationary state is of the form

$$|\psi(t)\rangle = |u\rangle\, e^{-iEt/\hbar} \tag{9.68}$$

with

$$\langle \psi(t)|\psi(t)\rangle = \langle u|u\rangle = 1,$$

a decaying state will be of the form

$$|\psi(t)\rangle = |u\rangle\, e^{-iEt/\hbar - \gamma t/2}, \tag{9.69}$$

leading to the exponential decay law

$$P(t) = \langle \psi(t)|\psi(t)\rangle = e^{-\gamma t}, \tag{9.70}$$

in agreement with Eq. (9.67). Eq. (9.70) means that a system, which is in the excited state (9.69) at time $t = 0$, has a probability $1/e$ of still being in that state at $t = 1/\gamma$. The time interval $T = 1/\gamma$ is known as the lifetime of the excited state. The only truly stationary state is the ground state which, by definition, cannot decay. For it $\gamma = 0$ and the lifetime $T = 1/\gamma = \infty$.

In our approximate stationary state treatment of radiative transitions we obtained infinitely sharp spectral lines whose frequencies are determined by the Bohr frequency condition. Replacing stationary states of the form (9.68) by decaying states of the form (9.69) means that these states no longer have an infinitely sharp energy: they can be represented as a superposition of stationary states and so have an energy spread. It can be shown that $\hbar\gamma$ is a measure of this spread. (Mathematically, this amounts to writing the decaying state (9.69) as a Fourier transform with respect to t, but we shall not carry this through.) Because the electromagnetic interaction is weak, this spread is very small:

typically, the lifetime $1/\gamma$ is of the order of 10^{-8} s giving $\hbar\gamma \approx 10^{-7}$ eV. As a consequence of this energy spread, the spectral line emitted in a transition $|\psi_1(t)\rangle \to |\psi_2(t)\rangle$, where the atomic states are given by

$$|\psi_n(t)\rangle = |u_n\rangle\, e^{-iE_nt/\hbar - \gamma_nt/2}, \qquad n = 1, 2,$$

acquires a finite spectral width. One can show that its intensity distribution is given by

$$I(\omega) = I_0 \frac{\frac{1}{4}(\gamma_1 + \gamma_2)^2}{(\omega - \omega_{12})^2 + \frac{1}{4}(\gamma_1 + \gamma_2)^2}. \tag{9.71}$$

(For the absorption process $2 \to 1$, Eq. (9.71) gives the shape of the absorption line.)

The intensity distribution (9.71), known as a Lorentzian distribution, is illustrated in Fig. 9.5. It consists of a single peak, centred at $\omega = \omega_{12}$. For

$$\omega = \omega_{12} \pm \tfrac{1}{2}(\gamma_1 + \gamma_2) \tag{9.72}$$

the intensity has dropped to half its maximum value, and

$$\frac{\Gamma_{12}}{\hbar} = \gamma_1 + \gamma_2 = \frac{1}{T_1} + \frac{1}{T_2} \tag{9.73}$$

is a measure of the width of the peak. Γ_{12} is called the *width* of the spectral line emitted in the transition $1 \to 2$.

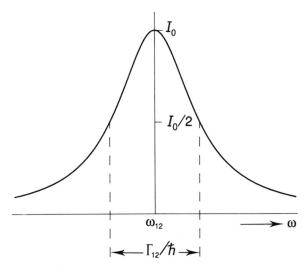

Fig. 9.5 The Lorentzian line shape (9.71).

An interesting aspect of Eq. (9.73) is that the width Γ_{12} depends on the lifetimes of both initial and final states. Sharp spectral lines result only if both γ_1 and γ_2 are small: both initial and final states must have sharply defined energies and long lifetimes; both are nearly stationary states. On the other hand broad spectral lines occur when one or other or both states are shortlived and so have poorly defined energies.

The line width we have here discussed is called the *natural line width*. It arises necessarily as a result of the spontaneous emission process and represents a theoretical lower limit to the width of a spectral line. There are other processes which cause broadening of spectral lines; such as Doppler broadening, i.e. the frequency shifts of the emitted radiation due to the thermal motion of the atoms of the emitting source; or collisional broadening, e.g. in a gaseous source at high densities, excited atoms are deexcited through collisions thereby shortening the lifetimes of the excited states. The line broadening due to these effects is usually much greater than the natural line width.

PROBLEMS 9

9.1 An electron is placed in a uniform static magnetic field B which points in the z-direction. At the initial time $t = 0$ the electron spin points in the positive x-direction. Assuming that the interaction of the electron spin with the magnetic field is given by the Hamiltonian $H = -\gamma \mathbf{B} \cdot \mathbf{s}$, where $\mathbf{s}$ is the spin operator and the constant γ the electron's gyromagnetic ratio, calculate the probability that at a later time t the electron spin points (a) in the positive x-direction, (b) in the negative x-direction, and (c) in the positive z-direction.

9.2 The Hamiltonian

$$H(t) = H_0 + V(t), \tag{9.74a}$$

with

$$H_0 = -\tfrac{1}{2}\gamma\hbar B_0 \sigma_z, \quad V(t) = -\tfrac{1}{2}\gamma\hbar B_1(\sigma_x \cos \omega t + \sigma_y \sin \omega t), \tag{9.74b}$$

describes a spin $\tfrac{1}{2}$ system which has the gyromagnetic ratio γ and which is subject to the static magnetic field $(0, 0, B_0)$ and the rotating magnetic field $(B_1 \cos \omega t, B_1 \sin \omega t, 0)$. B_0 and B_1 are uniform throughout space. Initially, at time $t = 0$, the spin of the system is pointing in the direction of the negative z-axis. Assuming that the field B_1 is weak and the interaction $V(t)$ can be treated in first-order perturbation theory, calculate the probability that at time t the spin points in the positive z-direction.

Show that in the case of resonance, i.e. the rotational frequency ω equals the energy difference of the spin-up and spin-down eigenstates of H_0 divided by $\hbar$, perturbation theory must break down for sufficiently large t, however weak B_1, and suggest a criterion for it to hold.

9.3 Eqs. (9.24), derived in section 9.2, are the general equations of motion for the system specified by the Hamiltonian (9.74a–b) of the last problem. Obtain the solution of Eqs. (9.24) which satisfy the initial conditions (9.26). Hence derive Eqs. (9.27a–b) for the probability that at time t the system is in the spin-up eigenstate of σ_z, if at $t = 0$ it was in the spin-down eigenstate of σ_z. From your solution for $c_2(t)$, verify Eq. (9.28).

9.4 A hydrogen atom is placed in a spatially homogeneous time-dependent electric field

$$\mathscr{E} = 0 \quad \text{for } t < 0, \quad \mathscr{E} = \mathscr{E}_0 e^{-t/\tau} \quad \text{for } t > 0, \quad (\tau > 0).$$

At time $t = 0$, the atom is in the (1s) ground state. Find the probability, in lowest order of perturbation theory, that after a sufficiently long time the atom is in the (2p) state in which the component of the orbital angular momentum in the direction of the field has the value zero.

9.5 A one-dimensional harmonic oscillator of mass m and angular frequency ω in its ground state is subject to a small constant force F acting for a time interval τ. What value of τ gives the greatest chance that the oscillator will be found in its first excited state thereafter?

(Note: This problem can be solved either by using the oscillator wave functions (see section 2.7) or, more simply, by using the matrix elements (7.76) of the position coordinate with respect to the oscillator eigenstates.)

9.6 The K_1 and K_2 mesons have slightly different masses m_1 and m_2 and the lifetimes $\tau_1 = 0.9 \times 10^{-10}$ s and $\tau_2 = 0.5 \times 10^{-7}$ s respectively. The K^0 and $\overline{K}^0$ mesons are the superpositions

$$|K^0\rangle = \frac{1}{\sqrt{2}}(|K_1\rangle + |K_2\rangle), \qquad |\overline{K}^0\rangle = \frac{1}{\sqrt{2}}(|K_1\rangle - |K_2\rangle). \tag{9.75}$$

A K^0 meson is produced in the process $\pi^- + p \to \Lambda^0 + K^0$ at time $t = 0$. What is the probability that at a later time t the K^0 meson has turned into a $\overline{K}^0$ meson? [Since the K_1 and K_2 mesons are unstable, their states have the time-dependence (9.69).]

CHAPTER

10

Scattering I:
time-dependent approach

The quantum theory of collisions is a huge subject. The diversity of systems, processes and conditions leads to many different theoretical approaches and approximation methods, which usually require large-scale computations for the solution of realistic problems. I shall develop the basic ideas of scattering theory and illustrate them for simple systems. It turns out that first-order perturbation theory is both easy to apply to scattering problems and under suitable conditions represents a reasonable approximation. This situation is similar to that found, in earlier chapters, for time-independent and time-dependent perturbation theory. In this chapter I shall use the golden rule to derive the time-dependent formulation of scattering theory. I shall apply this method to potential scattering and, more realistically, to the scattering of electrons by atoms. This last problem will enable me to introduce the very useful concept of a form factor for scattering by a complex target.

The time-dependent approach affords perhaps the easiest introduction to scattering theory, but it has much wider importance. For example, it is one starting point for the formulation of quantum field theory—and systematic perturbation expansions generally—in terms of Feynman graphs. However, there are many situations where the complementary time-independent formulation of scattering theory is most appropriate, and some of its basic ideas will be introduced in Chapter 11.

10.1   THE CROSS-SECTION

In a typical scattering experiment, a collimated beam of particles of well-defined energy is incident on a target, and the particles deflected into different directions are counted by means of suitable detectors. We shall assume that the density of particles in the incident beam is sufficiently low for interactions between them to be negligible. We shall also assume that we are dealing with a thin target so that multiple collisions of a projectile with several scattering centres in the target can be neglected. The collision process can then be treated as involving one projectile and one scattering centre only.

If the projectile or target particle possess internal structures, their internal states may change during a collision. Collisions are called elastic, if the internal states do not change during the collision. For example, in the scattering of electrons by atoms in their ground state, the atoms will remain in their ground state in elastic collisions; they will make transitions to excited states in inelastic collisions. For the time being, we shall restrict ourselves to elastic collisions.

We shall assume that the scattering centre is very massive compared with the mass of the projectile. (For example, this holds for electron-atom scattering.) In this case, the recoil which the scattering centre experiences during the collision is negligible and for an elastic collision the kinetic energy of the projectile does not change in the collision. Our assumption of an infinitely massive scattering centre is not a restriction, since the case of a scattering centre of finite mass is easily reduced to that of one of infinite mass. As in classical mechanics, one achieves this by analysing the collision in a frame of reference in which the centre of mass of the whole system (projectile plus scattering centre) is at rest. In this frame of reference, our two-body collision problem reduces to the scattering of a projectile, whose actual mass is replaced by it reduced mass, by an infinitely massive scattering centre (see problem 1.5).

We shall now consider the elastic scattering of a beam of particles of mass m and momentum $\mathbf{p}$ by an infinitely massive scattering centre, located at the origin of coordinates $\mathbf{r} = 0$. In the absence of the scattering centre, the system consists of a plane wave with wave vector $\mathbf{k} = \mathbf{p}/m$ and flux I. I specifies the number of particles in the beam crossing unit area normal to $\mathbf{k}$ per unit time, i.e. I is the current density of the beam. The effect of the scattering centre is to deflect particles. In terms of waves, these scattered particles are represented by a spherical outgoing wave, originating from the scattering centre. We shall take the direction of propagation of the incident plane wave as the z-axis of our system of polar coordinates (r, θ, ϕ), so that $\mathbf{p}$ has the Cartesian components $(0, 0, p)$. In a scattering experiment one will count the number of particles scattered into a given direction per unit time by placing a counter at $\mathbf{r}$, i.e. at a large distance r in the direction (θ, ϕ) from the scattering centre. If the counter is oriented so that its aperture, of area dS, is normal to $\mathbf{r}$, it will subtend the solid angle $d\Omega = dS/r^2$ at the scattering centre. We shall denote by $n(\theta, \phi)\, d\Omega$

the number of particles scattered into this solid angle $d\Omega$ per unit time, i.e. $n(\theta, \phi)\, d\Omega$ is the current of particles crossing dS.

It is important to be clear about the meaning of this quantity $n(\theta, \phi)\, d\Omega$. We represented the incident monoenergetic beam of particles by a plane wave, i.e. by a wave train of infinite length and infinitely wide wave fronts, but in reality the incident beam of course has finite length and finite width. (Its representation as a plane wave will be a very good approximation because these dimensions will be very large compared with the appropriate deBroglie wavelength of the particles.) In the forward direction, this incident wave will be present as well as the spherical outgoing wave, which represents the scattered particles, and there will be interference between them. Away from the forward direction, only the scattered particles will be present. This is illustrated schematically in Fig. 10.1. Except for scattering through very small angles, only scattered particles will be observed sufficiently far from the scattering centre. $n(\theta, \phi)\, d\Omega$ denotes the current of these scattered particles into $d\Omega$.

Clearly the current $n(\theta, \phi)\, d\Omega$ through dS is proportional to the incident flux I: if we double the latter, we double the former. To interpret the counts in a scattering experiment, we require some sort of normalization. The appropriate

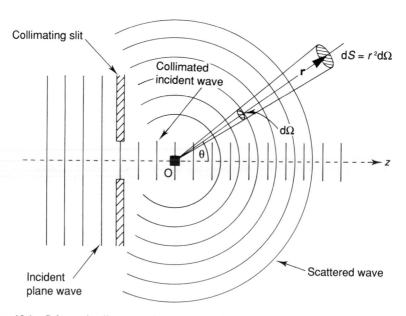

Fig. 10.1   Schematic diagram of scattering of a plane wave incident on a scattering centre at O. θ is the angle of scattering. The shadow of the target in the forward direction and the diffraction of the collimated incident wave near its edges are not shown.

normalized quantity is the cross-section. The differential cross-section for scattering into the element of solid angle $d\Omega$ is defined by

$$\sigma(\theta, \phi)\, d\Omega = \frac{n(\theta, \phi)\, d\Omega}{I}. \tag{10.1}$$

It follows from this definition that the cross-section has the dimensions of an area.

As indicated, the differential cross-section can depend on both the scattering angle θ and on the angle ϕ which defines the plane of scattering. In many situations, it depends on θ only. For example, this is so for the scattering of particles by a spherically symmetric potential $V(r)$. In this case, the system is axially symmetric about the direction of the incident wave vector $\mathbf{k}$, i.e. about the z-axis, and as a consequence the scattering cannot depend on ϕ.

From the differential cross-section the total cross-section σ_{tot} is obtained by integrating over all directions

$$\sigma_{tot} = \int \sigma(\theta, \phi)\, d\Omega = \int_0^{2\pi} d\phi \int_0^{\pi} \sin\theta\, d\theta\, \sigma(\theta, \phi). \tag{10.2}$$

10.2 POTENTIAL SCATTERING

We shall now consider potential scattering, i.e. we shall represent the scattering centre, fixed at the origin, by a potential $V(\mathbf{r})$. We shall see in section 10.3 how a complex target (e.g. the atom in electron-atom scattering) can be approximately described in this way. Only elastic scattering is now possible. Our object is to calculate the differential cross-section for potential scattering using first-order time-dependent perturbation theory, i.e. we shall consider the potential $V(\mathbf{r})$ as causing transitions between states describing the incident and the scattered particles. The initial state corresponds to a plane wave of momentum $\mathbf{p} = (0, 0, p)$. For the scattering of particles into the element of solid angle $d\Omega$ in the direction (θ, ϕ), we must take into account the group of final states describing particles with momentum $\mathbf{p}'$, pointing into the solid angle $d\Omega$ in the direction (θ, ϕ) and of magnitude $|\mathbf{p}'| = |\mathbf{p}| = p$. In order to apply time-dependent perturbation theory, and in particular the golden rule (9.46), we must count the final states comprising this group, i.e. we must first find the density of momentum eigenstates per unit energy interval.

10.2.1 Density of states

For particles in infinite space, the momentum eigenstates form a continuum and defining their density of states is a rather contrived exercise. In section 2.6.1 we saw how to avoid this difficulty by enclosing the system in a very large but finite cubic box and imposing periodic boundary conditions on the wave functions at the walls of the box. This trick converts the continuous spectrum of momentum eigenvalues into a completely discrete spectrum, with eigenstates

$$u_{\mathbf{p}}(\mathbf{r}) = \frac{1}{L^{3/2}} e^{i\mathbf{p}\cdot\mathbf{r}/\hbar} \tag{10.3}$$

where L is the length of an edge of the box and the momentum eigenvalues are given by

$$\mathbf{p} = \frac{2\pi\hbar}{L}(n_1, n_2, n_3), \qquad n_1, n_2, n_3 = 0, \pm 1, \pm 2, \ldots . \tag{10.4}$$

(Compare Eqs. (2.105) and (2.106) for these results.)

We introduce momentum space, whose rectangular coordinates are the components of momentum (p_x, p_y, p_z). Any momentum vector $\mathbf{p}$ is represented by a point in this space, and the momentum eigenvalues (10.4) form a cubic lattice of points with lattice spacing $2\pi\hbar/L$. It follows that the volume in momentum space per lattice point, i.e. per state, is $(2\pi\hbar/L)^3$, and the number of states in a volume element $d^3\mathbf{p}$ of momentum space is

$$\rho(\mathbf{p})\, d^3\mathbf{p} = \left(\frac{L}{2\pi\hbar}\right)^3 d^3\mathbf{p}. \tag{10.5}$$

In spherical polar coordinates (p, θ, ϕ), the volume element $d^3\mathbf{p}$ in momentum space becomes

$$d^3\mathbf{p} = p^2\, dp\, d\Omega = p^2\, dp \sin\theta\, d\theta\, d\phi \tag{10.6}$$

and

$$\rho(\mathbf{p})\, d^3\mathbf{p} = \left(\frac{L}{2\pi\hbar}\right)^3 p^2\, dp\, d\Omega \tag{10.7}$$

is the number of states for which the momentum lies in an element of solid angle $d\Omega$ in the direction (θ, ϕ) and its magnitude lies in the interval $(p, p + dp)$.

Eq. (10.7) gives the density of states in momentum space. We require $\rho(E_p)$, the density of states per unit energy, where

$$E_p = \frac{p^2}{2m} \tag{10.8}$$

is the energy of the incident particle. If the momentum interval $p \leqslant |\mathbf{p}| \leqslant p + dp$ corresponds to the energy range $(E_p, E_p + dE_p)$, then

$$\rho(E_p)\, dE_p = \rho(\mathbf{p})\, d^3\mathbf{p} = \left(\frac{L}{2\pi\hbar}\right)^3 p^2\, dp\, d\Omega \tag{10.9}$$

is the number of states with energy in the interval $(E_p, E_p + dE_p)$ and momentum $\mathbf{p}$ pointing into the solid angle $d\Omega$ in the direction (θ, ϕ). From Eqs. (10.8) and (10.9) follows the density of states per unit energy

$$\rho(E_p) = \left(\frac{L}{2\pi\hbar}\right)^3 pm\, d\Omega. \tag{10.10}$$

10.2.2   Born approximation

We now derive the differential cross-section for the potential scattering problem. We consider transitions from the initial state with momentum $\mathbf{p} = (0, 0, p)$ to the group of final states with momentum $\mathbf{p}'$, of magnitude $|\mathbf{p}'| = p' = p$ and lying within the solid angle $d\Omega$ in the direction (θ, ϕ). The transition probability per unit time for a particle to be scattered into $d\Omega$ is found by combining the golden rule (9.46) and the density of states (10.10). In this way, we obtain for the number of particles $n(\theta, \phi)\, d\Omega$, scattered into $d\Omega$ per unit time,

$$n(\theta, \phi)\, d\Omega = \frac{2\pi}{\hbar} \left[|\langle u_{\mathbf{p}'}| V(\mathbf{r})|u_{\mathbf{p}}\rangle|^2 \left(\frac{L}{2\pi\hbar}\right)^3 p'm\, d\Omega \right]_{p'=p}$$

$$= \frac{d\Omega}{(2\pi\hbar^2)^2} \frac{pm}{L^3} \left[\left| \int d^3r\, e^{i(\mathbf{p}-\mathbf{p}')\cdot\mathbf{r}/\hbar} V(\mathbf{r}) \right|^2 \right]_{p'=p}. \tag{10.11}$$

Taking $p' = p$ in the evaluation of this expression corresponds to the condition $E_p = E_i$ in Eq. (9.46).

Next we calculate the incident flux I. The wave function (10.3) represents a density of one particle in a volume L^3. Since the velocity of the incident particles is p/m, a volume p/m of beam crosses unit area normal to $\mathbf{p}$ in unit time, so that

$$I = (p/m)/L^3. \tag{10.12}$$

Substituting the last two equations in Eq. (10.1), we obtain the differential cross-section for scattering into the element of solid angle $d\Omega$ in the direction (θ, ϕ):

$$\sigma(\theta, \phi)\, d\Omega = d\Omega \left(\frac{m}{2\pi\hbar^2}\right)^2 \left[\left| \int d^3r\, e^{i(\mathbf{p}-\mathbf{p}')\cdot\mathbf{r}/\hbar} V(\mathbf{r}) \right|^2 \right]_{p'=p}. \tag{10.13}$$

We note first of all that this result is independent of the normalization volume L^3. This is an essential requirement, as discussed in section 2.6.1: a physically meaningful quantity, such as a cross-section, cannot depend on the size of a normalization volume which is very much larger than the system.

Eq. (10.13) is our basic result: it is the differential cross-section for potential scattering, calculated in first-order perturbation theory. It was first derived by Max Born and is known as the Born approximation. It shares with other first-order perturbation theory results the virtue of great simplicity but its validity is rather restricted. For a potential $V(\mathbf{r})$ limited to a region of linear dimension a and mean strength V_0, the Born approximation can be shown to be valid if

$$V_0 a \ll \hbar v \tag{10.14}$$

where v is the velocity of the incident particle. (This result will be derived in section 11.3.1.) Thus the Born approximation is a high-energy approximation which improves with increasing energy of the bombarding particles.

We can write the matrix element in Eq. (10.13) as

$$\tilde{V}(\mathbf{K}) = \int d^3\mathbf{r} V(\mathbf{r}) e^{i\mathbf{K}\cdot\mathbf{r}} \tag{10.15}$$

where $\hbar\mathbf{K}$ is the momentum transfer in the collision:

$$\hbar\mathbf{K} = \mathbf{p} - \mathbf{p}', \tag{10.16}$$

Thus the Born approximation cross-section depends only on the Fourier transform $\tilde{V}(\mathbf{K})$ of the scattering potential $V(\mathbf{r})$.

For a central potential $V(r)$, Eq. (10.15) simplifies. We take the momentum transfer vector $\hbar\mathbf{K}$ as polar axis of a set of polar coordinates with angles α and β. Referred to these coordinates, the vectors $\mathbf{r}$ and $\mathbf{K}$ have Cartesian components $(r\sin\alpha\cos\beta, r\sin\alpha\sin\beta, r\cos\alpha)$ and $(0, 0, K)$ respectively, so that

$$\mathbf{K}\cdot\mathbf{r} = Kr\cos\alpha.$$

The matrix element (10.15) then becomes

$$\begin{aligned}
\tilde{V}(K) &= \int_0^\infty r^2\,dr V(r) \int_0^{2\pi} d\beta \int_0^\pi \sin\alpha\,d\alpha\, e^{iKr\cos\alpha} \\
&= \frac{4\pi}{K} \int_0^\infty V(r)\sin(Kr)\,r\,dr. \tag{10.17}
\end{aligned}$$

From the last expression one sees that, for a central potential $V(r)$, the Fourier transform $\tilde{V}$ is a function of $K = |\mathbf{K}|$ only.* We anticipated this result and wrote $\tilde{V}(K)$ on the left-hand side of Eq. (10.17). Similarly, the differential cross-section also only depends on K for a central potential. Since

$$(\hbar\mathbf{K})^2 = (\mathbf{p} - \mathbf{p}')^2 = 2p^2(1 - \cos\theta) = 4p^2\sin^2\frac{\theta}{2}, \tag{10.18a}$$

---

* This can be seen from Eq. (10.15) directly. For a central potential $V(r)$, the right-hand side of Eq. (10.15) is invariant under rotations and the only functions of $\mathbf{K}$ with this property are functions of $K = |\mathbf{K}|$ only.

i.e.

$$\hbar K = 2p \sin \frac{\theta}{2}, \tag{10.18b}$$

it follows that for a central potential the differential cross-section depends on the scattering angle θ and the momentum p of the incident particle through the combination $p \sin(\theta/2)$ only. This result does not hold in general but only in Born approximation and can be used to test the validity of the Born approximation.

10.2.3   Scattering by a screened Coulomb potential

We conclude this section with a simple example: the Born approximation for the scattering of a particle of mass m by the potential

$$V(r) = \frac{-Ze^2}{4\pi\varepsilon_0 r} e^{-\mu r}. \tag{10.19}$$

This potential is known as a Yukawa potential or a screened Coulomb potential. The parameter $1/\mu$ is a length which characterizes the screening distance. We can think of this scattering problem as an extremely simplified description of electron–atom scattering. For $r \ll 1/\mu$, i.e. when the bombarding electron has penetrated well inside the atom, it experiences the Coulomb field of the full nuclear charge Ze. At larger distances, r of the order of $1/\mu$ or greater, the incident electron sees the nucleus shielded by the surrounding atomic electrons and experiences a weakened field only.

Substituting Eq. (10.19) in (10.17), we obtain the Fourier transform of the screened Coulomb potential

$$\tilde{V}(K) = \frac{-Ze^2}{4\pi\varepsilon_0} \int d^3 r \, \frac{e^{-\mu r}}{r} e^{i\mathbf{K}\cdot\mathbf{r}} = \frac{-Ze^2}{\varepsilon_0(K^2 + \mu^2)}, \tag{10.20}$$

and hence from Eq. (10.13) the differential cross-section per unit solid angle

$$\sigma(\theta) = \left(\frac{mZe^2}{2\pi\varepsilon_0\hbar^2}\right)^2 \frac{1}{(K^2 + \mu^2)^2}$$

$$= \left(\frac{mZe^2}{2\pi\varepsilon_0}\right)^2 \frac{1}{[4p^2 \sin^2(\tfrac{1}{2}\theta) + \hbar^2\mu^2]^2}, \tag{10.21}$$

where we have substituted for K from Eq. (10.18b). From Eq. (10.21) the total cross-section (10.2) is easily obtained (see problem 10.1).

For large momentum transfer, $\hbar K = 2p \sin(\tfrac{1}{2}\theta) \gg \hbar\mu$, we can neglect μ^2 compared to K^2 in Eq. (10.21). Taking $\mu = 0$ in this equation, it reduces to

$$\sigma(\theta) = \left(\frac{Ze^2}{8\pi\varepsilon_0 mv^2}\right)^2 \frac{1}{\sin^4(\tfrac{1}{2}\theta)} \tag{10.22}$$

where $v = p/m$ is the velocity of the electron. The limiting case, $\mu = 0$, corresponds to scattering by a pure Coulomb potential $-Ze^2/(4\pi\varepsilon_0 r)$. For large momentum transfers, the electron must penetrate well inside the atom where it sees the unscreened nuclear charge. Eq. (10.22) is Rutherford's celebrated formula for Coulomb scattering. Rutherford derived it from classical mechanics. We obtained the *same* result using quantum mechanics and the Born approximation. Even more curiously, it can be shown that the Rutherford formula (10.22) is also the exact quantum-mechanical result for Coulomb scattering.*

The Coulomb differential cross-section (10.22) diverges in the forward direction $\theta = 0$. This is due to the fact that the Coulomb potential is a long range potential which decreases only slowly with distance. Hence electrons passing at even a great distance from the scattering centre will experience a small deflection. In reality one never deals with pure Coulomb scattering. There is always some screening mechanism present and cross-sections do not diverge.

★10.3   ELECTRON–ATOM SCATTERING AT HIGH ENERGIES

The results of the last section are easily generalized to the scattering of particles by complex targets. To illustrate this, we shall derive the Born approximation for electron scattering by atoms.

We write the total Hamiltonian of the electron–atom system as

$$H = H_0 + V \qquad (10.23)$$

where

$$H_0 = -\frac{\hbar^2}{2m}\nabla^2 + H_A, \qquad V = \frac{-Ze^2}{4\pi\varepsilon_0 r} + \sum_{j=1}^{Z} \frac{e^2}{4\pi\varepsilon_0 |\mathbf{r} - \mathbf{r}_j|}. \qquad (10.24)$$

H_0 consists of the kinetic energy operator of the projectile electron plus the Hamiltonian H_A of the target atom which has the atomic number Z [H_A was stated explicitly in Eq. (4.23)], and V is the electrostatic interaction of the projectile electron with the target atom. We shall treat V as a perturbation causing transitions between eigenstates of H_0. The normalized eigenstates and the corresponding eigenvalues of H_A will be denoted by

$$|a\rangle = \phi_a(\mathbf{r}_1, \ldots, \mathbf{r}_Z) \quad \text{and} \quad \varepsilon_a, \qquad a = 1, 2, \ldots, \qquad (10.25)$$

respectively. In writing ϕ_a we are not showing explicitly the spin variables of the electrons.

We shall consider electron–atom collisions in which the electron's momentum changes from $\mathbf{p}$ to $\mathbf{p}'$ and the atom makes a transition from the state $|a\rangle$ to the state $|b\rangle$. Usually $|a\rangle$ will be the ground state of the target atom. If the

---

* For a derivation see, for example, Davydov, pp. 400–404, or Merzbacher, pp. 245–250.

atomic state does not change, $|b\rangle = |a\rangle$, we are dealing with elastic scattering for which $p' = p$ (where $p = |\mathbf{p}|$, etc.). The case when $|a\rangle$ and $|b\rangle$ are different states usually corresponds to inelastic collisions with $\varepsilon_b \neq \varepsilon_a$; the magnitude p' of the momentum of the scattered electron follows from conservation of energy:

$$E_b \equiv \frac{p'^2}{2m} + \varepsilon_b = E_a \equiv \frac{p^2}{2m} + \varepsilon_a. \qquad (10.26)$$

[E_a and E_b are the energies of the system (atom plus projectile electron) in the initial and final states.] We shall not be considering ionization processes, so that $|b\rangle$ as well as $|a\rangle$ are bound states of the atom.

For the transitions we are considering, the initial and final states are given by

$$|a, \mathbf{p}\rangle = |a\rangle \frac{e^{i\mathbf{p}\cdot\mathbf{r}/\hbar}}{L^{3/2}}, \qquad |b, \mathbf{p}'\rangle = |b\rangle \frac{e^{i\mathbf{p}'\cdot\mathbf{r}/\hbar}}{L^{3/2}} \qquad (10.27)$$

where we are again using box normalization within a cube of volume L^3 for the momentum eigenstates.

We know from the discussion in section 4.4, that the wave functions of many-electron systems must be antisymmetric. The wave functions (10.27) are not antisymmetrized with respect to the coordinates of the projectile electron and of the atomic electrons (although the atomic wave functions are anti-symmetrized). If the projectile electron and the atomic electrons could have distinguishing labels attached, we could distinguish between two types of processes: 'direct' processes in which the incident and the scattered electron have the same label, and 'exchange' scattering in which the incident electron has exchanged places with one of the atomic electrons. For identical particles, such a distinction is not possible. One should use properly antisymmetrized wave functions, instead of (10.27), and these lead to exchange effects which are neglected in our treatment. At high energies, where the Born approximation is valid, one can show that these exchange effects are negligible. It is as though a fast electron retains its identity in the collision. In addition, we are also neglecting all effects of spin.

To apply the golden rule (9.46), we require the density of final states. This is obtained as in section 10.2.1. For the transition

$$|a, \mathbf{p}\rangle \rightarrow |b, \mathbf{p}'\rangle,$$

with $\mathbf{p}'$ lying in the solid angle $d\Omega$ in the direction (θ, ϕ) and of magnitude p' given by Eq. (10.26), one finds instead of Eq. (10.10),

$$\rho(E_b) = \left(\frac{L}{2\pi\hbar}\right)^3 p'm \, d\Omega. \qquad (10.28)$$

Using this result in the golden rule (9.46), we obtain for the number of electrons scattered into $d\Omega$ per unit time, accompanied by the atomic transition $|a\rangle \rightarrow |b\rangle$,

$$n_{ba}(\theta, \phi) \, d\Omega = \frac{2\pi}{\hbar} \left[|\langle b, \mathbf{p}'|V|a, \mathbf{p}\rangle|^2 \left(\frac{L}{2\pi\hbar}\right)^3 p'm \, d\Omega \right]_{E_b = E_a}. \qquad (10.29)$$

The incident electron flux is given by Eq. (10.12) as before,

$$I = (p/m)/L^3. \qquad (10.30)$$

Combining the last two equations, we obtain the differential cross-section for the electron to be scattered into the solid angle $d\Omega$ in the direction (θ, ϕ), accompanied by the atomic transition $|a\rangle \rightarrow |b\rangle$:

$$\sigma_{ba}(\theta, \phi) \, d\Omega = n_{ba}(\theta, \phi) \, d\Omega/I$$

$$= d\Omega \left(\frac{m}{2\pi\hbar^2}\right)^2 \frac{p'}{p} \left| \int d^3r \, e^{i\mathbf{K}\cdot\mathbf{r}} U_{ba}(\mathbf{r}) \right|^2 \qquad (10.31)$$

where $\hbar\mathbf{K} = \mathbf{p} - \mathbf{p}'$, and where we have written

$$U_{ba}(\mathbf{r}) = \langle b|V|a\rangle, \qquad (10.32)$$

to show that the matrix element $\langle b|V|a\rangle$ is a function of $\mathbf{r}$ only. Eq. (10.31), like (10.29) before it, is to be evaluated for $E_b = E_a$, i.e. with p' given by Eq. (10.26); in Eq. (10.31) and in what follows this is always implied.

We must next calculate the matrix element (10.32). Using the definition (10.24) of V, we write

$$U_{ba}(\mathbf{r}) = \frac{-Ze^2}{4\pi\varepsilon_0 r} \delta_{ba} + U'_{ba}(\mathbf{r}) \qquad (10.33)$$

where

$$U'_{ba}(\mathbf{r}) = \langle b| \sum_{j=1}^{Z} \frac{e^2}{4\pi\varepsilon_0 |\mathbf{r} - \mathbf{r}_j|} |a\rangle$$

$$= \int \phi_b^*(\mathbf{r}_1, \ldots, \mathbf{r}_Z) \phi_a(\mathbf{r}_1, \ldots, \mathbf{r}_Z) \left[\sum_{j=1}^{Z} \frac{e^2}{4\pi\varepsilon_0 |\mathbf{r} - \mathbf{r}_j|} \right] \prod_{k=1}^{Z} d^3r_k$$

$$= \sum_{j=1}^{Z} \int \phi_b^*(\mathbf{r}_1, \ldots, \mathbf{r}_Z) \phi_a(\mathbf{r}_1, \ldots, \mathbf{r}_Z) \frac{e^2}{4\pi\varepsilon_0 |\mathbf{r} - \mathbf{r}_j|} \prod_{k=1}^{Z} d^3r_k. \qquad (10.34)$$

In each of the multiple integrals in the last expression we relabel the variable of integration $\mathbf{r}_j$, calling it $\mathbf{s}$ instead of $\mathbf{r}_j$; Eq. (10.34) then becomes

$$U'_{ba}(\mathbf{r}) = \int d^3s \, \frac{e^2}{4\pi\varepsilon_0 |\mathbf{r} - \mathbf{s}|} \rho_{ba}(\mathbf{s}) \qquad (10.35)$$

where

$$\rho_{ba}(\mathbf{s}) = \sum_{j=1}^{Z} \int \phi_b^*(\mathbf{r}_1, \ldots, \mathbf{r}_{j-1}, \mathbf{s}, \mathbf{r}_{j+1}, \ldots, \mathbf{r}_Z)$$

$$\times \phi_a(\mathbf{r}_1, \ldots, \mathbf{r}_{j-1}, \mathbf{s}, \mathbf{r}_{j+1}, \ldots, \mathbf{r}_Z) \prod_{\substack{k=1 \\ k \neq j}}^{Z} d^3\mathbf{r}_k. \qquad (10.36)$$

Combining Eqs. (10.33) and (10.35) gives

$$U_{ba}(\mathbf{r}) = \langle b | V | a \rangle = \frac{-Ze^2}{4\pi\varepsilon_0 r} \delta_{ba} + \int d^3\mathbf{s} \, \frac{e^2}{4\pi\varepsilon_0 |\mathbf{r} - \mathbf{s}|} \rho_{ba}(\mathbf{s}). \qquad (10.37)$$

Before completing the evaluation of the cross-section (10.31), we consider elastic electron scattering with the atom remaining in the state $|a\rangle$, i.e. $|b\rangle = |a\rangle$ and $p' = p$. The cross-section (10.31) now becomes (writing σ_a for σ_{aa})

$$\sigma_a(\theta, \phi) \, d\Omega = d\Omega \left(\frac{m}{2\pi\hbar^2} \right)^2 \left| \int d^3\mathbf{r} \, e^{i\mathbf{K}\cdot\mathbf{r}} U_{aa}(\mathbf{r}) \right|^2. \qquad (10.38)$$

Comparison of this equation with Eq. (10.13) shows that Eq. (10.38) is the cross-section for potential scattering by the potential

$$U_{aa}(\mathbf{r}) = \frac{-Ze^2}{4\pi\varepsilon_0 r} + \int d^3\mathbf{s} \, \frac{e^2}{4\pi\varepsilon_0 |\mathbf{r} - \mathbf{s}|} \rho_a(\mathbf{s}) \qquad (10.39)$$

where

$$\rho_a(\mathbf{s}) = \rho_{aa}(\mathbf{s}) = \sum_{j=1}^{Z} \int |\phi_a(\mathbf{r}_1, \ldots, \mathbf{r}_{j-1}, \mathbf{s}, \mathbf{r}_{j+1}, \ldots, \mathbf{r}_Z)|^2 \prod_{\substack{k=1 \\ k \neq j}}^{Z} d^3\mathbf{r}_k. \qquad (10.40)$$

This equation admits a simple interpretation. $\rho_a(\mathbf{s})$ is the density of atomic electrons in the state $|a\rangle$.* Hence $U_{aa}(\mathbf{r})$ is the electrostatic potential of the projectile electron in the field of the atom in the state $|a\rangle$, i.e. due to the point nuclear charge Ze and to the atomic electron density $\rho_a(\mathbf{s})$, and Eq. (10.38) gives the elastic scattering cross-section in this potential. Indeed we could have guessed the result (10.38) in this way. In this description, the projectile electron

---

* One easily sees this from Eq. (10.40). The term

$$\int |\phi_a(\mathbf{s}, \mathbf{r}_2, \ldots, \mathbf{r}_Z)|^2 \prod_{k=2}^{Z} d^3\mathbf{r}_k$$

is the probability density that the electron labelled 1 in the wave function $\phi_a(\mathbf{r}_1, \ldots, \mathbf{r}_z)$ is at the point $\mathbf{s}$, irrespective of the positions of the electrons with other labels. Hence the right-hand side of (10.40) is the sum of the probability densities that in the state $\phi_a(\mathbf{r}_1, \ldots, \mathbf{r}_z)$ the electron labelled 1 is at $\mathbf{s}$, the electron labelled 2 is at $\mathbf{s}$, etc.

appears to see the atomic electrons 'frozen' in the state $|a\rangle$. This is an oversimplification which results from the use of first-order perturbation theory. In reality, the projectile electron distorts the electron distribution in the atom. For fast incident electrons, the comparatively slowly moving atomic electrons cannot adjust their distribution during the short time interval which the incident electron spends in their vicinity. Hence for high energy scattering this distortion of the atomic electron distribution is small and can be neglected.

To complete the evaluation of the cross-section (10.31), we require the Fourier transform of $U_{ba}(\mathbf{r})$, defined in Eq. (10.37). Taking the limit $\mu \to 0$ in Eq. (10.20), one obtains

$$\int d^3\mathbf{r} \, e^{i\mathbf{K}\cdot\mathbf{r}} \frac{1}{r} = \frac{4\pi}{K^2}, \tag{10.41}$$

from which the Fourier transform of the first term in $U_{ba}(\mathbf{r})$ follows at once:

$$\int d^3\mathbf{r} \, e^{i\mathbf{K}\cdot\mathbf{r}} \frac{-Ze^2}{4\pi\varepsilon_0 r} \delta_{ba} = \frac{-Ze^2}{\varepsilon_0 K^2} \delta_{ba}. \tag{10.42}$$

The Fourier transform of the second term in $U_{ba}(\mathbf{r})$, i.e. of $U'_{ba}(\mathbf{r})$ defined in Eq. (10.35), we write

$$\int d^3\mathbf{r} \, e^{i\mathbf{K}\cdot\mathbf{r}} U'_{ba}(\mathbf{r}) = \int d^3\mathbf{r} \, e^{i\mathbf{K}\cdot\mathbf{r}} \int d^3\mathbf{s} \frac{e^2}{4\pi\varepsilon_0|\mathbf{r}-\mathbf{s}|} \rho_{ba}(\mathbf{s})$$

$$= \int d^3\mathbf{s} \frac{e^2}{4\pi\varepsilon_0} \rho_{ba}(\mathbf{s}) I, \tag{10.43}$$

where

$$I = \int d^3\mathbf{r} \frac{e^{i\mathbf{K}\cdot\mathbf{r}}}{|\mathbf{r}-\mathbf{s}|}. \tag{10.44}$$

In this integral, we change the variable of integration from $\mathbf{r}$ to $\mathbf{u} = \mathbf{r} - \mathbf{s}$, so that

$$I = \int d^3\mathbf{u} \frac{e^{i\mathbf{K}\cdot(\mathbf{u}+\mathbf{s})}}{u} = e^{i\mathbf{K}\cdot\mathbf{s}} \frac{4\pi}{K^2}; \tag{10.45}$$

hence Eq. (10.43) reduces to

$$\int d^3\mathbf{r} \, e^{i\mathbf{K}\cdot\mathbf{r}} U'_{ba}(\mathbf{r}) = \frac{e^2}{\varepsilon_0 K^2} F_{ba}(\mathbf{K}) \tag{10.46}$$

where

$$F_{ba}(\mathbf{K}) = \int d^3\mathbf{s} \, e^{i\mathbf{K}\cdot\mathbf{s}} \rho_{ba}(\mathbf{s}). \tag{10.47}$$

From Eqs. (10.33), (10.42) and (10.46), we obtain

$$\int d^3r\, e^{i\mathbf{K}\cdot\mathbf{r}} U_{ba}(\mathbf{r}) = \frac{-e^2}{\varepsilon_0 K^2}[Z\delta_{ba} - F_{ba}(\mathbf{K})], \qquad (10.48)$$

and substituting this expression in Eq. (10.31) leads to our final result for the differential cross-section in the Born approximation

$$\sigma_{ba}(\theta, \phi)\, d\Omega = d\Omega \left(\frac{me^2}{2\pi\varepsilon_0\hbar^2 K^2}\right)^2 \frac{p'}{p} |Z\delta_{ba} - F_{ba}(\mathbf{K})|^2. \qquad (10.49)$$

This equation gives the cross-section for inelastic scattering if $\varepsilon_b \neq \varepsilon_a$; and for elastic scattering if $|b\rangle = |a\rangle$ (and consequently $p' = p$).

In the case of elastic scattering, Eq. (10.47) becomes

$$F_a(\mathbf{K}) \equiv F_{aa}(\mathbf{K}) = \int d^3r\, e^{i\mathbf{K}\cdot\mathbf{r}} \rho_a(\mathbf{r}). \qquad (10.50)$$

$F_a(\mathbf{K})$ is known as the atomic form factor of the state $|a\rangle$. It is the Fourier transform of the electron density $\rho_a(\mathbf{r})$ in the state $|a\rangle$. We shall assume that the state $|a\rangle$ has definite parity so that from Eq. (10.40)

$$\rho_a(-\mathbf{r}) = \rho_a(\mathbf{r}), \qquad (10.51)$$

and it follows from Eq. (10.50) that $F_a(\mathbf{K})$ is real. The Born approximation expression (10.49) for the elastic cross-section can now be written

$$\sigma_a(\theta, \phi)\, d\Omega = d\Omega \left(\frac{me^2}{2\pi\varepsilon_0\hbar^2 K^2}\right)^2 [Z - F_a(\mathbf{K})]^2. \qquad (10.52)$$

We see from Eqs. (10.50) and (10.52) that the form factor relates the cross-section to the electron density. It plays the same useful role in other situations, connecting the cross-section to the particle density or the charge density of a complex target; for example, in the scattering of electrons by nuclei or by nucleons in order to probe the charge distributions within them. This requires very high electron energies. Our treatment can be generalized to apply at these relativistic energies. This leads to modifications of detail but does not alter the essential feature: the cross-section is proportional to the square of the modulus of the form factor of the charge distribution of the target.

Finally, we consider two limiting cases of the elastic cross-section (10.52).

(i) We consider large values of K, i.e. $KR \gg 1$ where R is the radius of the charge distribution $\rho_a(\mathbf{r})$. The form factor $F_a(\mathbf{K})$ then is very small, since the exponential factor in (10.50) will oscillate very rapidly compared with the spatial variation of $\rho_a(\mathbf{r})$. Hence

$$|F_a(\mathbf{K})| = \left|\int d^3r\, e^{i\mathbf{K}\cdot\mathbf{r}} \rho_a(\mathbf{r})\right| \ll \left|\int d^3r \rho_a(\mathbf{r})\right| = Z \qquad (10.53)$$

and $F_a(\mathbf{K})$ can be neglected compared with Z in Eq. (10.52) which reduces to the Rutherford scattering formula (10.22). This is as expected: to achieve large momentum transfer, the incident electron must penetrate deeply inside the atom where the atomic electrons do not screen it.

(ii) For small-angle scattering, it follows from $\hbar K = 2p \sin(\tfrac{1}{2}\theta)$, Eq. (10.18b), that we can have small values of K even though the energy is sufficiently high for the Born approximation to be valid. In this case we expand the exponential in Eq. (10.50) in a power series in K. Taking the direction of $\mathbf{K}$ as z-axis, so that $\mathbf{K} = (0, 0, K)$, we have $\exp(i\mathbf{K} \cdot \mathbf{r}) = \exp(iKz)$, and Eq. (10.50) becomes

$$F_a(\mathbf{K}) = \int d^3r\rho_a(\mathbf{r}) + iK \int d^3r\rho_a(\mathbf{r})z + \tfrac{1}{2}(iK)^2 \int d^3r\rho_a(\mathbf{r})z^2 + \cdots. \qquad (10.54)$$

In this series, the first term has the value Z. The term linear in K vanishes with our assumption that the state $|a\rangle$ has a definite parity, as do all terms in odd powers of K. Hence

$$F_a(\mathbf{K}) = Z - \tfrac{1}{2}K^2 \int d^3r\rho_a(\mathbf{r})z^2 + O(K^4) \qquad (10.55)$$

where $O(K^4)$ denotes the next term in the expansion and is of order K^4. Substituting Eq. (10.55) in (10.52), we obtain the Born approximation cross-section for small-angle scattering

$$\sigma_a(\theta, \phi) \, d\Omega = d\Omega \left(\frac{me^2}{4\pi\varepsilon_0\hbar^2} \right)^2 \left[\int d^3r\rho_a(\mathbf{r})z^2 + O(K^2) \right]^2. \qquad (10.56)$$

We see from this formula that for $K = 0$, i.e. in the forward direction $\theta = 0$, the differential cross-section is finite, in contrast to the infinite result for Coulomb scattering: the nucleus is now screened by the atomic electrons. Furthermore, for small K, i.e. for small-angle scattering, the cross-section is independent of the angle. For spherically symmetric atomic states, ρ_a depends on r only, so that

$$\int d^3r\rho_a(r)z^2 = \tfrac{1}{3}\overline{r^2}, \qquad (10.57)$$

and the form factor and the differential cross-section depend on the mean-square-radius of the charge distribution only.

PROBLEMS 10

10.1 Derive the total cross-section, in Born approximation, for the scattering of particles of mass m by the screened Coulomb potential (10.19).

10.2 Obtain, in Born approximation, the differential cross-section for the scattering of particles of mass m by the potential

$$V(r) = V_0 e^{-ar}.$$

10.3 Obtain, in Born approximation, the differential and total cross-sections for the elastic scattering of electrons by atomic hydrogen in its ground state.

10.4 Derive the Born approximation differential and total cross-sections for elastic electron-helium scattering. You may treat the helium atom as in the 'frozen' ground state described by the wave function

$$\psi(r_1, r_2) = \frac{1}{\pi b^3} \exp\left[-(r_1 + r_2)/b\right] \tag{10.58}$$

where $b = \frac{16}{27} a_0$. (The wave function (10.58) was obtained in section 8.2.2 as a simple variational approximation to the helium ground state wave function.)

10.5 Derive the Born approximation expression for the differential cross-section for the inelastic scattering of electrons by atomic hydrogen, the hydrogen atoms being excited from the ground state to the (2s) state.

CHAPTER

Scattering II: time-independent approach

In this chapter I shall introduce the basic ideas of time-independent scattering theory. This approach is of great importance both for the analysis and the computational treatment of collision phenomena. We shall only consider potential scattering but the theory can of course be applied to more complicated realistic situations.

11.1 THE SCATTERING AMPLITUDE

We shall consider potential scattering, i.e. the scattering of particles of mass m incident on a fixed scattering centre which is represented by the potential $V(\mathbf{r})$. In the absence of the scattering centre, the incident particles, with energy E and travelling in the z-direction, are represented by the plane wave

$$\psi_0(\mathbf{r}) = e^{ikz} \tag{11.1}$$

where the wave number k is given by

$$k = (2mE/\hbar^2)^{1/2} \tag{11.2}$$

and $\psi_0(\mathbf{r})$ is normalized to one particle per unit volume. $\psi_0(\mathbf{r})$ is a solution of the free-particle Schrödinger equation

$$(\nabla^2 + k^2)\psi_0(\mathbf{r}) = 0. \tag{11.3}$$

In the presence of the potential $V(\mathbf{r})$, the plane wave $\psi_0(\mathbf{r})$ is distorted. The wave function describing the particles is now a solution of the Schrödinger equation

$$(\nabla^2 + k^2)\psi(\mathbf{r}) = \frac{2m}{\hbar^2} V(\mathbf{r})\psi(\mathbf{r}), \tag{11.4}$$

satisfying the appropriate boundary conditions. These are that

$$\psi(\mathbf{r}) \to \psi_0(\mathbf{r}) = e^{ikz} \qquad \text{as } V(\mathbf{r}) \to 0, \tag{11.5}$$

i.e. as the strength of the potential is decreased, the solution $\psi(\mathbf{r})$ goes over into the plane wave $\psi_0(\mathbf{r})$. Secondly, we require a solution which represents the plane incident wave $\psi_0(\mathbf{r})$ together with a scattered outgoing wave. We shall assume that $V(\mathbf{r})$ is a short-range potential; more precisely, either that it has a finite range a (i.e. $V(\mathbf{r}) = 0$ for $r > a$) or that it tends to zero sufficiently rapidly so that

$$rV(\mathbf{r}) \to 0 \qquad \text{as } r \to \infty \tag{11.6}$$

in all directions.* In the next section, we shall show that, for a potential satisfying this condition, the Schrödinger equation (11.4) possesses solutions which are of the form

$$\frac{1}{r} e^{\pm ikr} f(\theta, \phi), \qquad \text{as } r \to \infty. \tag{11.7}$$

These wave functions represent spherical waves, centred on $\mathbf{r} = 0$; outgoing waves for the positive exponent, ingoing waves for the negative exponent. To see this, we calculate the radial component of the current density $\mathbf{j}$. For a wave function $\psi(\mathbf{r})$ this is defined by (see problem 1.6)

$$\mathbf{j} = \frac{-i\hbar}{2m} (\psi^* \nabla \psi - \psi \nabla \psi^*) \tag{11.8}$$

with radial component in the outward direction given by

$$j_r = \frac{-i\hbar}{2m} \left(\psi^* \frac{\partial \psi}{\partial r} - \psi \frac{\partial \psi^*}{\partial r} \right). \tag{11.9}$$

For the wave functions (11.7) this expression gives the radial current densities

$$j_r = \pm \frac{\hbar k}{m} |f(\theta, \phi)|^2 \frac{1}{r^2}, \tag{11.10}$$

---

* This condition is not satisfied for the Coulomb potential $1/r$ and the following analysis does not apply. An exact analytic solution of the Coulomb scattering problem is possible; see, for example, Dayvdov, section 100, or Schiff, section 21.

justifying the interpretation of (11.7) as spherical outgoing and ingoing waves.

In our scattering problem we require a wave function $\psi(\mathbf{r})$ which corresponds to the plane incident wave (11.1) together with a scattered outgoing wave, i.e. with the asymptotic form

$$\psi(\mathbf{r}) \simeq e^{ikz} + \frac{1}{r} e^{ikr} f(\theta, \phi), \qquad \text{as } r \to \infty. \tag{11.11}$$

$f(\theta, \phi)$ is called the scattering amplitude.

From Eq. (11.11) the cross-section is easily found. The number of particles $n(\theta, \phi)\, d\Omega$ scattered into an element of solid angle $d\Omega$ in the direction (θ, ϕ) per unit time is given by

$$n(\theta, \phi)\, d\Omega = j_r r^2\, d\Omega = \frac{\hbar k}{m} |f(\theta, \phi)|^2\, d\Omega, \tag{11.12}$$

where j_r refers only to the scattered particles, i.e. only to the second term on the right-hand side of Eq. (11.11).* With the incident wave (11.1) normed to one particle per unit volume, the incident flux I equals $\hbar k/m$ [compare Eq. (10.12)]. Hence the differential cross-section (10.1) becomes

$$\sigma(\theta, \phi)\, d\Omega = |f(\theta, \phi)|^2\, d\Omega. \tag{11.13}$$

As discussed at the end of section 10.1, for a potential with axial symmetry about the direction of travel of the incident particles the scattering cannot depend on ϕ, so that the scattering amplitude and the cross-section become functions of θ (and the energy) only:

$$\sigma(\theta)\, d\Omega = |f(\theta)|^2\, d\Omega. \tag{11.14}$$

Eqs. (11.13) and (11.14) are exact. Comparison of Eq. (11.13) with the Born approximation cross-section (10.13), derived from time-dependent perturbation theory, shows that the latter gives only the modulus of the Born approximation to the scattering amplitude but not its phase. The complete expression for the scattering amplitude in Born approximation is given by

$$f_{\mathrm{BA}}(\theta, \phi) = \frac{-m}{2\pi\hbar^2} \int d^3 r\, e^{i\mathbf{K}\cdot\mathbf{r}} V(\mathbf{r}) \tag{11.15}$$

where $\mathbf{K} = (\mathbf{p} - \mathbf{p}')/\hbar$, Eq. (10.16), and $p' = p$. To derive this result, we shall first, in section 11.2, obtain an exact expression for the scattering amplitude $f(\theta, \phi)$ in terms of the exact solution $\psi(\mathbf{r})$ of the Schrödinger equation for the scattering problem. From this exact expression for $f(\theta, \phi)$, the Born approximation result (11.15) will then be derived in section 11.3.

---

* As discussed in section 10.1, the scattered particles are usually observed in a region, away from the forward direction, where only scattered particles are present.

11.2   THE INTEGRAL EQUATION FOR POTENTIAL SCATTERING

In the last section we saw that solving the potential scattering problem amounts to solving the Schrödinger equation (11.4) with the wave function satisfying the boundary conditions (11.5) and (11.11). In this section we shall show how to transform this problem into an integral equation. The great advantage of this formulation lies in the fact that the integral equation already incorporates the boundary conditions (11.5) and (11.11). It therefore readily forms the starting point for different approximation procedures, such as perturbation expansions or variational methods.

We shall start by solving a problem with which the reader may be more familiar; namely, to obtain the electrostatic potential $\psi(\mathbf{r})$ due to a density distribution $\rho(\mathbf{r})$ of electric charges. $\psi(\mathbf{r})$ is the solution of Poisson's equation*

$$\nabla^2 \psi(\mathbf{r}) = -\frac{1}{\varepsilon_0} \rho(\mathbf{r}). \tag{11.16}$$

If we consider a unit point charge situated at the point $\mathbf{r} = \mathbf{s}$, then Eq. (11.16) becomes

$$\nabla^2 \psi(\mathbf{r}) = -\frac{1}{\varepsilon_0} \delta^{(3)}(\mathbf{r} - \mathbf{s}) \tag{11.17}$$

where $\delta^{(3)}(\mathbf{r} - \mathbf{s})$ is the three-dimensional Dirac delta function which was defined in section 2.6.2.[†] The solution of Eq. (11.17) is the Coulomb potential

$$\psi(\mathbf{r}) = \frac{1}{4\pi\varepsilon_0 |\mathbf{r} - \mathbf{s}|}. \tag{11.20}$$

To obtain the potential due to the charge distribution $\rho(\mathbf{r})$, we go back to Eq. (11.16). Since this equation is linear, the potential it generates will be a

---

* For Poisson's equation and its solution see, for example, Grant and Phillips, section 3.1.

[†] For the benefit of readers who omitted section 2.6.2, I briefly restate the relevant properties of $\delta^{(3)}(\mathbf{r} - \mathbf{s})$. It is defined by

(i) $\delta^{(3)}(\mathbf{r} - \mathbf{s}) = 0,$ if $\mathbf{r} \neq \mathbf{s}$;   (11.18a)

(ii) $\delta^{(3)}(\mathbf{r} - \mathbf{s}) \to \infty$ as $\mathbf{r} \to \mathbf{s}$ in such a way that

$$\int_R \delta^{(3)}(\mathbf{r} - \mathbf{s}) \, d^3\mathbf{r} = 1 \tag{11.18b}$$

for any three-dimensional region of integration R for which the point $\mathbf{r} = \mathbf{s}$ lies inside R.

From these properties it follows at once that for any function $f(\mathbf{r})$, continuous at $\mathbf{r} = \mathbf{s}$, and any three-dimensional region of integration R

$$\int_R d^3\mathbf{r} f(\mathbf{r}) \delta^{(3)}(\mathbf{r} - \mathbf{s}) = \begin{cases} f(\mathbf{s}), & \text{if } \mathbf{s} \text{ lies inside the region } R \\ 0, & \text{if } \mathbf{s} \text{ lies outside the region } R. \end{cases} \tag{11.19}$$

superposition of the potentials due to the individual elements of charge $\rho(s)\,d^3s$ in the different volume elements d^3s, i.e.

$$\psi(\mathbf{r}) = \int \frac{1}{4\pi\varepsilon_0 |\mathbf{r} - \mathbf{s}|}\,\rho(s)\,d^3s, \tag{11.21}$$

the integration being over all space.

We now apply the same procedure to the inhomogeneous wave equation

$$(\nabla^2 + k^2)\psi(\mathbf{r}) = F(\mathbf{r}). \tag{11.22}$$

We can think of $F(\mathbf{r})$ as a source term which generates or absorbs waves, and the solutions of the inhomogeneous wave equation (11.22) consist of two parts: waves generated or absorbed by the source term, to which may be added any solution of the homogeneous wave equation (11.3), representing waves generated and absorbed at infinity, i.e. at very large distances.

We again start by considering the wave equation with a unit point source at the origin

$$(\nabla^2 + k^2)\psi(\mathbf{r}) = \delta^{(3)}(\mathbf{r}). \tag{11.23}$$

One easily verifies by direct substitution that this equation possesses the solutions

$$\psi(r) = \frac{-e^{\pm ikr}}{4\pi r} \tag{11.24}$$

which are functions of the radial coordinate r only (see problem 11.1). We saw in the last section that these solutions correspond to spherical outgoing and ingoing waves respectively. To represent scattered waves, we shall require the outgoing wave solutions only.

Analogously to our treatment of Poisson's equation, we can at once write down a solution of the inhomogeneous wave equation (11.22) containing the source term $F(\mathbf{r})$, as a linear superposition of waves produced by different volume elements of the source. The solution corresponding to outgoing waves only is given by

$$\psi(\mathbf{r}) = -\frac{1}{4\pi} \int \frac{e^{ik|\mathbf{r}-\mathbf{s}|}}{|\mathbf{r} - \mathbf{s}|}\,F(s)\,d^3s. \tag{11.25}$$

Other solutions of the inhomogeneous wave equation (11.22) are obtained by adding any solution of the homogeneous wave equation (11.3) to the particular solution (11.25).

We apply these results to the scattering problem, defined by the Schrödinger equation (11.4) together with the boundary conditions (11.5) and (11.11). Comparing Eqs. (11.4) and (11.22), we see that for the source term $F(\mathbf{s})$ in Eq. (11.25) we must take

$$F(\mathbf{s}) = \frac{2m}{\hbar^2} V(\mathbf{s})\psi(\mathbf{s}), \qquad (11.26)$$

and in order to satisfy the boundary condition (11.11), we must add the solution

$$\psi_0(\mathbf{r}) = e^{ikz}$$

of the homogeneous wave equation (11.3) to (11.25). In this way we finally obtain

$$\psi(\mathbf{r}) = e^{ikz} - \frac{m}{2\pi\hbar^2} \int \frac{e^{ik|\mathbf{r}-\mathbf{s}|}}{|\mathbf{r}-\mathbf{s}|} V(\mathbf{s})\psi(\mathbf{s})\, d^3\mathbf{s}. \qquad (11.27)$$

This equation is an integral equation. It contains the unknown solution $\psi(\mathbf{r})$ of the scattering problem in the integrand. Consequently this equation is not a solution of the scattering problem, unlike Eq. (11.21) which is a solution of Poisson's equation. However, Eq. (11.27) is equivalent to Eqs. (11.4), (11.5) and (11.11) as a complete statement of the scattering problem, and the fact that it incorporates the boundary conditions makes it a most useful starting point for different approximation methods of solving the scattering problem.

We derive the scattering amplitude from Eq. (11.27) by considering the asymptotic form of $\psi(\mathbf{r})$ for large values of r. For $r \gg s$ we can expand

$$|\mathbf{r}-\mathbf{s}| = r\left[1 - \frac{2}{r^2}\mathbf{r}\cdot\mathbf{s} + \frac{s^2}{r^2}\right]^{1/2} \simeq r\left[1 - \frac{1}{r^2}\mathbf{r}\cdot\mathbf{s} + O\left(\frac{s^2}{r^2}\right)\right],$$

so that

$$\frac{e^{ik|\mathbf{r}-\mathbf{s}|}}{|\mathbf{r}-\mathbf{s}|} \simeq \frac{e^{ikr}}{r}\exp(-i\mathbf{k}'\cdot\mathbf{s}), \qquad \text{for } r \gg s, \qquad (11.28)$$

where

$$\mathbf{k}' = k\mathbf{r}/r. \qquad (11.29)$$

The range of integration in (11.27) is in effect limited since $V(\mathbf{s})$ is a short-range potential. For sufficiently large values of r we can therefore substitute the approximation (11.28) in Eq. (11.27), giving for $\psi(\mathbf{r})$ the asymptotic form (11.11) with

$$f(\theta, \phi) = \frac{-m}{2\pi\hbar^2} \int \exp(-i\mathbf{k}'\cdot\mathbf{s})V(\mathbf{s})\psi(\mathbf{s})\, d^3\mathbf{s}. \qquad (11.30)$$

Eq. (11.30) is an exact expression for the scattering amplitude $f(\theta, \phi)$ in terms of the exact solution $\psi(\mathbf{r})$ of the scattering problem.

11.3   BORN APPROXIMATION

The Born approximation is easily derived from Eq. (11.30). If the potential $V(\mathbf{r})$ is weak so that it distorts the incident plane wave only slightly, then we can, in (11.30), approximate the exact wave function $\psi(\mathbf{s})$ by the incident plane wave

$$\psi_0(\mathbf{s}) = e^{i\mathbf{k}\cdot\mathbf{s}} \tag{11.31}$$

where $\mathbf{k} = (0, 0, k)$. With this replacement, Eq. (11.30) at once reduces to

$$f_{\mathrm{BA}}(\theta, \phi) = \frac{-m}{2\pi\hbar^2}\int d^3r\, e^{i\mathbf{K}\cdot\mathbf{r}}V(\mathbf{r}), \qquad (\mathbf{K} = \mathbf{k} - \mathbf{k}'), \tag{11.32}$$

which is the Born approximation to the scattering amplitude quoted in Eq. (11.15), and the Born approximation for the cross-section becomes

$$\sigma_{\mathrm{BA}}(\theta, \phi)\, d\Omega = |f_{\mathrm{BA}}(\theta, \phi)|^2\, d\Omega = d\Omega\left(\frac{m}{2\pi\hbar^2}\right)^2\left|\int d^3r\, e^{i\mathbf{K}\cdot\mathbf{r}}V(\mathbf{r})\right|^2, \tag{11.33}$$

in agreement with our earlier result (10.13).

★11.3.1   Validity of Born approximation

Eq. (11.27) can serve to investigate the validity of the Born approximation. The condition that the actual wave function $\psi(\mathbf{r})$ shall differ only slightly from the plane wave (11.31) can, from Eq. (11.27), be written

$$\left|\frac{-m}{2\pi\hbar^2}\int\frac{e^{ik|\mathbf{r}-\mathbf{s}|}}{|\mathbf{r}-\mathbf{s}|}V(\mathbf{s})\psi(\mathbf{s})\, d^3s\right| \ll 1. \tag{11.34}$$

Assuming that the distortion of the plane wave is largest at $\mathbf{r} = 0$ and approximating $\psi(\mathbf{s})$ by the plane wave (11.31), we can rewrite the condition (11.34) at $\mathbf{r} = 0$ as

$$\left|\frac{-m}{2\pi\hbar^2}\int\frac{e^{i(ks+\mathbf{k}\cdot\mathbf{s})}}{s}V(\mathbf{s})\, d^3s\right| \ll 1.$$

For a spherically symmetric potential $V(s)$, this reduces to

$$\frac{m}{\hbar^2 k}\left|\int_0^\infty (e^{2iks} - 1)V(s)\, ds\right| \ll 1. \tag{11.35}$$

We illustrate this validity criterion for the square-well potential

$$V(r) = \begin{cases} V_0, & r < a, \\ 0, & r > a. \end{cases} \tag{11.36}$$

Substituting (11.36) in (11.35) and performing the integration, one obtains

$$\frac{m|V_0|}{2\hbar^2 k^2} (\xi^2 - 2\xi \sin \xi + 2 - 2 \cos \xi)^{1/2} \ll 1 \qquad (11.37)$$

where $\xi = 2ka$.

In the limit of high energies, i.e. $ka \gg 1$, Eq. (11.37) gives as the criterion for the validity of the Born approximation

$$|V_0|a \ll \hbar v \qquad (11.38)$$

where $v = \hbar k/m$ is the velocity of the incident particle. More generally, one can show that Eq. (11.38) is the criterion for the validity of the Born approximation at high energies for a short-range potential $V(r)$, of range a and mean strength V_0.*

11.4   SCATTERING OF IDENTICAL PARTICLES

The elastic collisions of two particles interacting through a potential $V(r)$, which depends on their distance of separation r only, is easily reduced to the potential scattering problem which we have discussed so far.† As in classical mechanics, one introduces the centre-of-mass coordinate **R** and the relative coordinate **r**, defined by

$$\mathbf{R} = \frac{m_1 \mathbf{r}_1 + m_2 \mathbf{r}_2}{m_1 + m_2}, \qquad \mathbf{r} = \mathbf{r}_1 - \mathbf{r}_2, \qquad (11.39)$$

where m_1 and m_2 are the masses of the two particles, and $\mathbf{r}_1$ and $\mathbf{r}_2$ their position coordinates. Expressed in terms of **R** and **r**, the Schrödinger equation of the two-particle system is separable (see problem 1.5). It decouples into an equation in **R** which describes the motion of the centre of mass of the system and an equation in **r** which describes the motion of the particles in a frame of reference in which their centre of mass is at rest. (We can also think of this equation as describing the relative motion of the two particles.) This latter equation is just the Schrödinger equation of a particle of mass m, equal to the reduced mass of the particles, i.e.

$$m = \frac{m_1 m_2}{m_1 + m_2}, \qquad (11.40)$$

---

* For further discussion of criteria for the validity of the Born approximation, see Messiah, vol. II, pp. 812–815.

† The restriction to an interaction of the form $V(r)$ means that we are ignoring spin-dependent forces in the case of particles with spin.

in the potential $V(r)$. Hence our analysis and results for potential scattering at once apply to the elastic collisions of two particles in their centre-of-mass system.

For the collisions of identical particles, our analysis must be modified. The states for identical particles must satisfy certain symmetry conditions under the interchange of particle labels and we must now take these into account.

Interchanging particle labels $(1 \leftrightarrow 2)$ corresponds to replacing $\mathbf{r} = (r, \theta, \phi)$ by $-\mathbf{r} = (r, \pi - \theta, \phi + \pi)$. From the scattering state $\psi(\mathbf{r})$ with the asymptotic form (11.11) one obtains the symmetric and antisymmetric scattering states $\psi(\mathbf{r}) \pm \psi(-\mathbf{r})$ with the asymptotic forms

$$\left.\begin{array}{c} \psi_S(\mathbf{r}) \\ \psi_A(\mathbf{r}) \end{array}\right\} = \psi(\mathbf{r}) \pm \psi(-\mathbf{r})$$

$$\simeq (e^{ikz} \pm e^{-ikz}) + \frac{e^{ikr}}{r} [f(\theta, \phi) \pm f(\pi - \theta, \phi + \pi)]. \tag{11.41}$$

Let us start with the case of spinless particles. For these we know, from section 4.4, that the wave functions must be symmetric under interchange of particle labels, and we must take the upper (positive) signs in Eq. (11.41). The two plane wave terms represent the colliding beams travelling in the positive and negative z-directions; in particular, the positive exponential (e^{ikz}) represents a beam incident from $z = -\infty$ and a density of one particle per unit volume, i.e. an incident flux $I = \hbar k/m$. The term proportional to e^{ikr}/r in Eq. (11.41) describes a spherical outgoing wave. We know from Eqs. (11.11) and (11.12) that the amplitudes $f(\theta, \phi)$ and $f(\pi - \theta, \phi + \pi)$ in Eq. (11.41) are the scattering amplitudes for the scattering of particles, incident from $z = -\infty$, into the directions (θ, ϕ) and $(\pi - \theta, \phi + \pi)$, respectively (see Figs. 11.1a and b). But in the latter case, the other colliding particle, incident from $z = +\infty$, is scattered into the direction (θ, ϕ). Thus, in both cases, a particle is scattered into the direction (θ, ϕ). Analogously to Eq. (11.12), one finds for the number of particles scattered into the element of solid angle $d\Omega$ in the direction (θ, ϕ) per unit time

$$n(\theta, \phi) \, d\Omega = \frac{\hbar k}{m} |f(\theta, \phi) + f(\pi - \theta, \phi + \pi)|^2 \, d\Omega. \tag{11.42}$$

Dividing this expression by the incident flux I gives the differential cross-section

$$\sigma(\theta) \, d\Omega = |f(\theta) + f(\pi - \theta)|^2 \, d\Omega$$

$$= \{|f(\theta)|^2 + |f(\pi - \theta)|^2 + 2 \, \mathcal{R}e[f^*(\theta)f(\pi - \theta)]\} \, d\Omega. \tag{11.43}$$

In this equation we have written $f(\theta)$ instead of $f(\theta, \phi)$, etc., since the interaction of spinless particles can depend only on their separation r, so that the scattering amplitude and hence the cross-section can depend only on θ, as discussed previously.

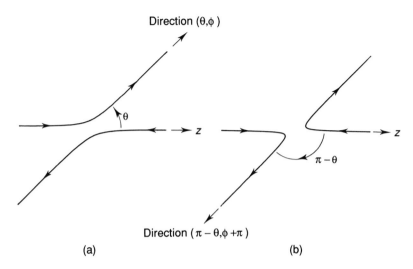

Fig. 11.1   The contributions $f(\theta, \phi)$ and $f(\pi - \theta, \phi + \pi)$ to the scattering amplitudes $[f(\theta, \phi) \pm f(\pi - \theta, \phi + \pi)]$, Eq. (11.41).

Eq. (11.43) is the differential cross-section for particles to be scattered into the solid angle $d\Omega$ in the direction (θ, ϕ). That this equation contains three terms reflects the fact that the outgoing-wave terms in Eq. (11.41) represent *amplitudes* for particles to be scattered into the direction (θ, ϕ). The first two terms in Eq. (11.43) are what one would expect classically; they represent the contributions to the cross-section for scattering into the direction (θ, ϕ) from particles incident from $z = -\infty$ and $z = +\infty$ respectively. However, for identical particles one must add amplitudes, not intensities, and this results in the third term in Eq. (11.43), which is a typical quantum-mechanical interference term.

Next we consider the scattering of two identical spin $\frac{1}{2}$ particles. We shall assume that the particles interact through a spin-independent central potential. Since the total spin is a good quantum number, we consider separately the scattering when the particles are in singlet and triplet spin states. For spin $\frac{1}{2}$ particles, the states must be antisymmetric under interchange of particle labels. Since the singlet spin function is antisymmetric in the spin coordinates, the space wave function for scattering in the singlet spin state must be symmetric. Hence we must use the upper (positive) signs in Eq. (11.41), just as for spinless particles, and obtain for the differential cross-section for scattering in the singlet state

$$\sigma_s(\theta) \, d\Omega = |f(\theta) + f(\pi - \theta)|^2 \, d\Omega. \qquad (11.44a)$$

(With our assumptions about the interaction potential, the scattering amplitude f is independent of ϕ and is the same for singlet and triplet spin states.) Since the triplet spin states are symmetric in the spin coordinates, it follows that the scattering cross-section, when the particles are in a triplet state, is obtained from the antisymmetric scattering state $\psi_A(\mathbf{r})$, Eq. (11.41), and is given by*

$$\sigma_t(\theta) \, d\Omega = |f(\theta) - f(\pi - \theta)|^2 \, d\Omega. \tag{11.44b}$$

Eqs. (11.44) assume that the colliding particles are in a definite spin state. In most scattering experiments, the spins are not correlated but are randomly oriented. (In analogy with light, one talks of unpolarized beams of particles.) We must then average over these random spin orientations. This corresponds to the assumption that in the colliding beams the four spin states occur with equal frequency and averaging incoherently over them. Since there are three independent triplet states and one singlet state, one obtains in this way the unpolarized differential scattering cross-section

$$\sigma_{\text{unpol}}(\theta) \, d\Omega = \tfrac{1}{4}\sigma_s(\theta) \, d\Omega + \tfrac{3}{4}\sigma_t(\theta) \, d\Omega. \tag{11.45}$$

11.5 PARTIAL WAVES AND PHASE SHIFTS

The angular momentum of a particle in a central potential $V(r)$ is conserved. A particle incident on this potential with a given value of the angular momentum will retain this value after the collision. Expressed in terms of waves, this means that we expand the incident plane wave e^{ikz} into a superposition of angular momentum eigenstates, each with a definite angular momentum quantum number l, and we can then treat separately the distortion of each of these partial waves by the potential $V(r)$, independently of the others.

In principle the partial wave analysis provides a complete solution of the scattering problem, but one easily shows that in practice it is an approximation appropriate for low energies and so complements the Born approximation which is valid at high energies. For a potential of range a (we shall assume $V(r) = 0$, for $r > a$, but with more careful wording the same argument applies to any short-range potential), a particle will only be scattered if it enters the region $r < a$. Classically this means that only particles with impact parameters $b < a$ are scattered (Fig. 11.2). A particle of energy E and momentum $p = \hbar k = (2mE)^{1/2}$ has angular momentum $pb = \hbar l$ and so will only be scattered if

$$\hbar l < ap = a(2mE)^{1/2}. \tag{11.46}$$

A particle of low energy (and therefore low momentum p) can only have a large angular momentum if its impact parameter b is large: it cannot penetrate

---

* In the more general case of spin-dependent forces, i.e. a potential of the form $V_1(r) + V_2(r)\boldsymbol{\sigma}_1 \cdot \boldsymbol{\sigma}_2$, Eqs. (11.44a) and (11.44b) would contain different scattering amplitudes $f_s(\theta)$ and $f_t(\theta)$ respectively.

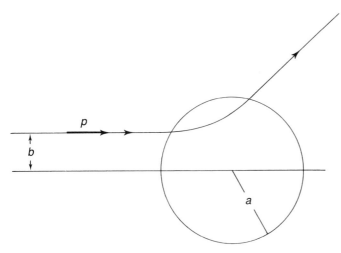

Fig. 11.2   Scattering by a potential of range a. Only particles with impact parameter $b < a$ will be scattered.

to sufficiently small distances r where the potential is effective.* Thus for a given energy E, only a finite number of partial waves, with $l = 0, 1, \ldots$ up to a certain value, will be scattered. At sufficiently low energies, only s-wave scattering will occur.

The expansion of the plane wave e^{ikz} into angular momentum eigenstates takes the form

$$e^{ikz} = e^{ikr\cos\theta} = \sum_{l=0}^{\infty} u_l(kr) Y_l^0(\theta, \phi). \qquad (11.47)$$

No spherical harmonics Y_l^m with $m \neq 0$ occur in (11.47) since these depend on ϕ. The expansion coefficients $u_l(kr)$ are given by

$$u_l(kr) = \int d\Omega\, Y_l^0(\theta, \phi) e^{ikr\cos\theta}. \qquad (11.48)$$

For $l > 0$, the evaluation of (11.48) leads to spherical Bessel functions, and we shall only consider $l = 0$, s-wave scattering. For $l = 0$, Eq. (11.48) gives

$$u_0(kr) = (4\pi)^{1/2}\, \frac{\sin kr}{kr} \qquad (11.49)$$

---

* The same conclusion follows from quantum mechanics. The solution of the radial Schrödinger equation (2.78) behaves like r^l at the origin [Eq. (2.84)], i.e. the repulsive centrifugal potential forces the wave function to vanish more rapidly at the origin for larger l, and the particle has less chance of penetrating into the region $r < a$.

and the plane wave expansion (11.47) becomes

$$e^{ikz} = \frac{\sin kr}{kr} + \text{partial waves with } l > 0. \tag{11.50}$$

The s-wave part of the plane wave can be written

$$\frac{1}{2ik}\left(-\frac{e^{-ikr}}{r} + \frac{e^{ikr}}{r}\right) \tag{11.51}$$

i.e. it consists of a superposition of a spherical ingoing and a spherical outgoing wave; the amplitudes of these waves have the same absolute value, giving rise to the same fluxes of ingoing and outgoing particles.

We now consider the scattering of the $l = 0$ partial wave (11.51) by the potential $V(r)$. Earlier we found that $V(r)$ distorts the plane wave e^{ikz} into the scattering state $\psi(\mathbf{r})$, characterized by the asymptotic form

$$\psi(\mathbf{r}) \simeq e^{ikz} + \frac{1}{r}e^{ikr}f(\theta, \phi) \tag{11.52}$$

which comprises the incident plane wave plus the spherical outgoing scattered wave. In the same way, the s-wave part (11.51) of the plane wave e^{ikz} is augmented by a spherical outgoing scattered s-wave, and the s-wave part $\psi_0(r)$ of the complete wave function $\psi(\mathbf{r})$ has the asymptotic form

$$\psi_0(r) \simeq \frac{\sin kr}{kr} + \frac{1}{r}e^{ikr}f_0 \tag{11.53}$$

which defines the s-wave scattering amplitude f_0. Since we are dealing with s-waves, f_0 does not depend on angles; it does of course depend on the scattering energy.

We rewrite Eq. (11.53) in terms of ingoing and outgoing waves as

$$\psi_0(r) \simeq \frac{1}{2ik}\left[-\frac{e^{-ikr}}{r} + \frac{e^{ikr}}{r}(1 + 2ikf_0)\right]. \tag{11.54}$$

Since the fluxes of ingoing and outgoing $l = 0$ particles must be the same, we must have

$$|1 + 2ikf_0| = 1, \tag{11.55}$$

i.e. $(1 + 2ik f_0)$ is a phase factor which we shall write

$$1 + 2ikf_0 = e^{2i\delta_0}. \tag{11.56}$$

The real parameter δ_0 is called the s-wave phase shift. (The factor 2 in the exponent in (11.56) ensures the conventional definition of δ_0.) δ_0, like f_0, depends on the scattering energy. Substituting Eq. (11.56) in Eq. (11.53), we obtain

$$\psi_0(r) \simeq \frac{\sin kr}{kr} + \frac{e^{ikr}}{r}\left(\frac{e^{2i\delta_0} - 1}{2ik}\right) \tag{11.57a}$$

$$= e^{i\delta_0} \frac{\sin(kr + \delta_0)}{kr}. \tag{11.57b}$$

At sufficiently low energies, so that only s-wave scattering occurs, the general expression (11.14) for the differential cross-section in terms of the scattering amplitude reduces to

$$\sigma(\theta)\, d\Omega = |f_0|^2\, d\Omega = \frac{d\Omega}{k^2}\sin^2\delta_0. \tag{11.58}$$

As expected, this cross-section is isotropic. For the total cross-section for pure s-wave scattering we therefore obtain

$$\sigma_{\text{tot}} = \frac{4\pi}{k^2}\sin^2\delta_0. \tag{11.59}$$

It is important to remember that Eqs. (11.58) and (11.59) only apply if there is s-wave scattering only. In this case, Eq. (11.59) provides an upper bound to the cross-section: since $|\sin\delta_0| \ll 1$, we must have $\sigma_{\text{tot}} \leqslant 4\pi/k^2$.

To solve the s-wave scattering problem, one must determine the phase shift δ_0. Before showing how this is done, I want to indicate how our results generalize to higher partial waves.* In the expansion (11.47) of the plane wave e^{ikz}, the partial wave corresponding to each angular momentum l consists of a superposition of a spherical ingoing and a spherical outgoing wave, whose amplitudes have equal magnitudes. The effect of the potential $V(r)$ is to change the phase but not the magnitude of the amplitude of the outgoing-wave part of each partial wave. (This follows from angular momentum conservation: the outgoing and ingoing fluxes of particles with angular momentum l are equal.) For the $l = 0$ partial wave, this led to the wave function $\psi_0(r)$ with the asymptotic form

$$\psi_0(r) \simeq \frac{\sin kr}{kr} + \frac{1}{r}e^{ikr}f_0, \quad f_0 = \frac{1}{k}e^{i\delta_0}\sin\delta_0, \tag{11.60}$$

---

* I shall omit the derivation which involves Bessel functions and only state the results. For the derivation see, for example, Merzbacher, section 11.5.

[see Eqs. (11.53) and (11.56)]. When taking all partial waves into account, one obtains for the complete wave function the asymptotic form (11.52), i.e.

$$\psi(\mathbf{r}) \simeq e^{ikz} + \frac{1}{r} e^{ikr} f(\theta), \tag{11.61}$$

with the scattering amplitude $f(\theta)$ given by*

$$f(\theta) = \frac{\sqrt{(4\pi)}}{k} \sum_{l=0}^{\infty} (2l + 1)^{1/2} e^{i\delta_l} \sin \delta_l Y_l^0(\theta, \phi). \tag{11.62}$$

The real parameter δ_l is called the l-wave phase shift. The $l = 0$ term in (11.62) reduces to our above result for the s-wave scattering amplitude f_0.

From Eqs. (11.13) and (11.62) follows the differential cross-section per unit solid angle

$$\sigma(\theta) = \frac{4\pi}{k^2} \sum_{l=0}^{\infty} \sum_{l'=0}^{\infty} [(2l + 1)(2l' + 1)]^{1/2} \exp[i(\delta_l - \delta_{l'})]$$

$$\times \sin \delta_l \sin \delta_{l'} Y_l^0(\theta, \phi) Y_{l'}^0(\theta, \phi) \tag{11.63}$$

and hence, using the orthonormality (2.57) of the spherical harmonics, the total cross-section

$$\sigma_{tot} = \frac{4\pi}{k^2} \sum_{l=0}^{\infty} (2l + 1) \sin^2 \delta_l. \tag{11.64}$$

The phase shifts, which are functions of the scattering energy, represent an excellent way of parametrizing the experimental and theoretical analyses of a scattering process. The differential cross-section (11.63) is a power series in $\cos \theta$ [since $Y_l^0(\theta, \phi)$ is a polynomial of degree l in $\cos \theta$, see Eqs. (2.54), (2.55) and (2.50)]. For a short-range potential, only the partial waves up to a certain angular momentum $l = L$, which depends on the scattering energy, will contribute to the scattering and the angular distribution (11.63) will reduce to a polynomial of degree $2L$ in $\cos \theta$. The object of a scattering experiment, at a given energy, is to measure the differential cross-section sufficiently accurately to allow the determination of the phase shifts from Eq. (11.63). The task of the theory is to calculate the phase shifts for the potential $V(r)$. To this end one must solve the radial Schrödinger equation (2.78) for each value of l, and the asymptotic forms of the solutions then give the phase shifts. (In section 11.5.1 this procedure will be illustrated for s-wave scattering.) We have considered potential scattering. However, the phase shift analysis formalism can be extended. It can then be applied to collision phenomena generally.

---

* This result holds provided $V(r)$ tends to zero faster than $1/r$ as $r \rightarrow \infty$. In particular, it is modified for the Coulomb potential. That the scattering amplitude is independent of the angle ϕ is due to the fact that we are considering a central potential, as discussed previously.

11.5.1   s-wave scattering by an attractive square-well potential

We illustrate how to find the s-wave phase shift for the attractive square-well potential

$$V(r) = \begin{cases} -V_0, & (V_0 > 0), & r < a \\ 0, & & r > a. \end{cases} \tag{11.65}$$

For s-wave scattering at the energy $E = \hbar^2 k^2 / 2m$, we require the isotropic solution $\psi_0(r)$ of the Schrödinger equation

$$\left[-\frac{\hbar^2}{2m} \nabla^2 + V(r) \right] \psi_0(r) = E\psi_0(r) \tag{11.66}$$

which at the origin satisfies the boundary condition

$$\psi_0(0) = \text{constant} \neq 0 \tag{11.67}$$

[compare Eq. (2.84)] and which has the asymptotic form (11.57). For $r < a$, Eq. (11.66) possesses the s-wave solutions $e^{\pm iKr}/r$, where

$$K = \left[\frac{2m}{\hbar^2}(E + V_0) \right]^{1/2}. \tag{11.68}$$

Hence the solution which satisfies the boundary condition (11.67) is given by

$$\psi_0(r) = A \frac{\sin Kr}{Kr}, \qquad r \leqslant a, \tag{11.69a}$$

where A is a constant. Since $V(r) = 0$ for $r > a$, Eq. (11.57b) gives $\psi_0(r)$ not only asymptotically but for all values $r \geqslant a$:

$$\psi_0(r) = e^{i\delta_0} \frac{\sin(kr + \delta_0)}{kr}, \qquad r \geqslant a. \tag{11.69b}$$

The solutions (11.69a) and (11.69b) must be matched smoothly at $r = a$, i.e. so that $\psi_0(r)$ and $d\psi_0(r)/dr$ are continuous at $r = a$ [compare Eqs. (2.11)]. Equivalently one can match $[r\psi_0(r)]$ smoothly at $r = a$. This simplifies the algebra a little and gives

$$\tan(ka + \delta_0) = \frac{k}{K} \tan(Ka) \tag{11.70}$$

whence

$$\tan \delta_0 = \frac{\dfrac{k}{K}\tan(Ka) - \tan(ka)}{1 + \dfrac{k}{K}\tan(Ka)\tan(ka)}. \tag{11.71}$$

From this equation one can find $\sin \delta_0$, and hence the cross-section from Eq. (11.59) if only s-wave scattering occurs.

At energies sufficiently low so that $ka \ll 1$, it follows from Eq. (11.46) that only s-waves are scattered. Furthermore, we can approximately take $\tan (ka) \approx ka$, so that Eq. (11.71) becomes

$$\tan \delta_0 = \frac{ka\left[\dfrac{\tan (Ka)}{Ka} - 1\right]}{1 + (ka)^2 \dfrac{\tan (Ka)}{Ka}}. \tag{11.72}$$

It follows that $\tan \delta_0 = 0$, there is no s-wave scattering and the cross-section (11.59) vanishes, if the scattering energy E is such that

$$\frac{\tan (Ka)}{Ka} = 1. \tag{11.73}$$

The phenomenon here described, known as the Ramsauer–Townsend effect, occurs in the scattering of slow electrons by rare gas atoms. The observed cross-sections possess a sharp minimum (and nearly vanish) at about 1 eV. At these energies the higher partial waves, with $l > 0$, make only very small contributions to the cross-section, and calculations of the s-wave phases using approximate potentials to represent the target atoms confirm the above explanation of the minima in the cross-sections.

PROBLEMS 11

11.1 Verify that Eq. (11.23) possesses the solutions (11.24).

11.2 Eq. (11.37) states a condition for the validity of the Born approximation for scattering by the square-well potential (11.36). Derive, from Eq. (11.37), a criterion for the validity of the Born approximation at low energies ($ka \ll 1$).

Discuss qualitatively the existence of bound states for a three-dimensional attractive square-well potential [$V_0 < 0$ in Eq. (11.36)], and hence interpret the criterion for the validity of the Born approximation at low energies.

What conclusions can be drawn for the case of a three-dimensional repulsive square-well potential ($V_0 > 0$)?

11.3 Determine the total cross-section for scattering of particles of mass m by the potential

$$V(r) = \begin{cases} \infty, & r < a, \\ 0, & r > a, \end{cases}$$

at energies sufficiently low for only s-wave scattering to occur. (This potential is called a hard-sphere potential: for $r \leqslant a$, the wave function must vanish; the incident particle cannot penetrate into the region $r < a$.)

11.4 In section 11.4, Eqs. (11.44–45), we expressed the unpolarized differential cross-section for the scattering of two identical spin $\frac{1}{2}$ particles, interacting through a spin-independent central potential, in terms of the unsymmetrized scattering amplitude $f(\theta)$. Derive the corresponding results for particles of spin $s (= 0, \frac{1}{2}, 1, \frac{3}{2}, \ldots)$.

11.5 A beam of spin $\frac{1}{2}$ particles is scattered by a target of spin 0 particles, the interaction being through a spin-dependent central force. It can be shown that if the incident beam is specified by the wave function $\chi \exp(i\mathbf{k} \cdot \mathbf{r})$, with χ a general spin state, then the wave function of the elastically scattered beam has the asymptotic form

$$\psi(\mathbf{r}) \simeq \frac{e^{ikr}}{r} [f_1(\theta) + f_2(\theta)\mathbf{n} \cdot \boldsymbol{\sigma}]\chi, \tag{11.74}$$

where $\mathbf{n}$ denotes the unit vector normal to the plane of scattering (i.e. $\mathbf{n} = \mathbf{k} \wedge \mathbf{k}'/|\mathbf{k}|^2$, $\mathbf{k}$ and $\mathbf{k}'$ being the wave vectors of the incident and scattered particles), f_1 and f_2 are complex functions of the scattering angle θ and $\boldsymbol{\sigma} = (\sigma_x, \sigma_y, \sigma_z)$ are the Pauli spin matrices. The polarization $P(\theta)$ of the particles scattered through θ is defined as

$$P(\theta) = a(\mathbf{n}) - a(-\mathbf{n}),$$

where $a(\pm\mathbf{n})$ are the fractions of particles scattered through θ with spins parallel to $\pm\mathbf{n}$ respectively. If the incident beam is unpolarized, show that

$$P(\theta) = \frac{2 \, \mathcal{R}e[f_1(\theta)f_2^*(\theta)]}{|f_1|^2 + |f_2|^2}. \tag{11.75}$$

CHAPTER

The Dirac formalism*

The development of quantum mechanics in terms of wave functions—i.e. wave mechanics, the way we started in this book—is one approach. There are others. Dirac invented a very general formulation of quantum mechanics which unifies these different approaches, facilitates switching between them and gives new insights into their meaning. In section 5.1, I introduced the Dirac notation as a versatile and transparent way of writing elaborate mathematical expressions. In this chapter, I want to go beyond mere notation and develop the mathematics of the Dirac formalism and its interpretation. I shall not develop quantum mechanics from first principles in this way, as Dirac did, but shall utilize results and ideas obtained previously in terms of wave functions, when this helps to explain the Dirac formulation and its relation to wave mechanics.[†]

12.1 KETS

The principle of the linear superposition of wave functions is basic to quantum mechanics. The corresponding mathematical theory is that of linear vector spaces. In vector analysis in ordinary three-dimensional space, if $\mathbf{u}_1$ and

---

* This chapter presupposes only material from Chapters 1 to 3, including the starred section 2.6.2.

[†] Comprehensive treatments of the Dirac formalism will be found in most more advanced books on quantum mechanics. Some of these are listed in the bibliography. An excellent, mathematically simple, conceptually penetrating account of the basic ideas and how to apply them in simple situations is given in Feynman III.

$\mathbf{u}_2$ are two vectors, so is

$$\mathbf{u} = c_1\mathbf{u}_1 + c_2\mathbf{u}_2 \tag{12.1a}$$

where c_1 and c_2 are arbitrary real numbers. In quantum mechanics wave functions specify the states of a system. Dirac recognized that it is not just the wave functions which obey the linear superposition principle, but the states themselves which do so. Dirac denotes states by the symbol $|\ \rangle$ which he calls a state ket or, more briefly, just a *ket*. Instead of specifying states by wave functions $\psi_a, \psi_b, \ldots$ we could specify them by kets $|\psi_a\rangle, |\psi_b\rangle, \ldots$, meaning the states whose wave functions are $\psi_a, \psi_b, \ldots$. Alternatively, we could write $|a\rangle$, $|b\rangle, \ldots$ where $a, b, \ldots$ label the states with the wave functions $\psi_a, \psi_b, \ldots$ in any convenient way. For example, for a particle in a central potential with the wave function $\psi_{nlm}(\mathbf{r})$, Eq. (2.92), we could write $|nlm\rangle$ instead of $|\psi_{nlm}\rangle$.

Kets satisfy the linear superposition principle. If $|u_1\rangle$ and $|u_2\rangle$ are two kets, so is

$$|u\rangle = c_1|u_1\rangle + c_2|u_2\rangle \tag{12.1b}$$

where c_1 and c_2 are two arbitrary *complex* numbers. Just as $\mathbf{u}_1, \mathbf{u}_2$, with the property (12.1a), are ordinary vectors (I shall use the word 'ordinary' to denote vectors in three-dimensional space), so we can think of $|u_1\rangle, |u_2\rangle$ as vectors in a sort of space of their own: ket space.* Accordingly, one also refers to kets as ket vectors or state vectors. There are many similarities between ordinary vectors and ket vectors. One can think of the latter as a generalization of ordinary vectors. The ideas of ket space enable one to visualize much of the mathematics of kets in a geometrical way, by analogy with the geometry of ordinary vectors.

Ket space is in some respects different from ordinary space. One difference we have just met: the coefficients c_1, c_2 in Eq. (12.1a) in ordinary space are real numbers; the corresponding coefficients in Eq. (12.1b) in ket space are complex numbers.

Another difference is that in ordinary space, given any three linearly independent vectors $\mathbf{u}_1, \mathbf{u}_2$ and $\mathbf{u}_3$, *any* other vector $\mathbf{u}$ can be written as a linear superposition

$$\mathbf{u} = \sum_{n=1}^{3} c_n\mathbf{u}_n. \tag{12.2a}$$

Ordinary space is three-dimensional. In contrast, ket space is infinitely dimensional: it requires an infinite set of linearly independent kets $|u_1\rangle, |u_2\rangle, \ldots$ to be

---

* Mathematicians call this space Hilbert space, after the great mathematician David Hilbert, who was largely responsible for the rigorous mathematical theory of this vector space, long before quantum mechanics.

able to express *any* ket $|u\rangle$ in the form

$$|u\rangle = \sum_{n=1}^{\infty} c_n |u_n\rangle. \quad * \tag{12.2b}$$

Clearly, not every infinite set of linearly independent kets has this property. For example, if we omit $|u_1\rangle$ from the above set, then $|u_1\rangle$ itself cannot be written in the form (12.2b). The set of kets $|u_1\rangle$, $|u_2\rangle$, ... is called *complete* if any arbitrary ket $|u\rangle$ can be expanded in the form (12.2b). It may help the reader to appreciate these and some of the following considerations for kets, which at this stage may appear somewhat abstract, to think of the corresponding situations for ordinary vectors on the one hand and for wave functions on the other; for example, to compare the expansion theorem (12.2b) for a ket with the expansion (12.2a) of a vector or with the expansion of an arbitrary wave function in terms of a complete set of eigenfunctions [compare Eq. (1.31)].

As for ordinary vectors, we require the concept of a scalar product for kets. With any two ordinary vectors **a** and **b**, one associates a scalar product $\mathbf{a} \cdot \mathbf{b}$ which is a real number. Similarly, with any two kets $|a\rangle$ and $|b\rangle$, we associate a *complex* number, called the scalar product of $|a\rangle$ with $|b\rangle$ and denoted by $\langle a|b\rangle$. The scalar product has the following three properties.

(a) $$\langle b|a\rangle^* = \langle a|b\rangle: \tag{12.3}$$

the scalar products of $|a\rangle$ with $|b\rangle$ and of $|b\rangle$ with $|a\rangle$ are complex conjugates of each other.[†] This is in contrast to ordinary vectors for which the scalar product is real and $\mathbf{b} \cdot \mathbf{a} = \mathbf{a} \cdot \mathbf{b}$.

(b) The scalar product of any ket $|v\rangle$ with

$$|u\rangle = c_1 |u_1\rangle + c_2 |u_2\rangle \tag{12.4}$$

is given by

$$\langle v|u\rangle = c_1 \langle v|u_1\rangle + c_2 \langle v|u_2\rangle. \tag{12.5a}$$

One obtains $\langle u|v\rangle$ from this equation by taking its complex conjugate and using Eq. (12.3):

$$\langle u|v\rangle = c_1^* \langle u_1|v\rangle + c_2^* \langle u_2|v\rangle. \tag{12.5b}$$

---

* The definition of linear independence is the usual one. The kets $|u_1\rangle$, $|u_2\rangle$, ... $|u_p\rangle$ are linearly independent if the equation

$$c_1 |u_1\rangle + c_2 |u_2\rangle + \cdots + c_p |u_p\rangle = 0$$

is only satisfied for $c_1 = c_2 = \cdots = c_p = 0$.

[†] Nevertheless, one often refers rather imprecisely to $\langle a|b\rangle$ or $\langle b|a\rangle$ as the scalar product of $|a\rangle$ with $|b\rangle$, or of $|a\rangle$ and $|b\rangle$, it being clear from the context which is meant.

(c) It follows from Eq. (12.3) that the scalar product of a ket with itself, $\langle a|a \rangle$, is real. We shall postulate that

$$\langle a|a \rangle > 0 \qquad (12.6)$$

for any ket $|a\rangle$, other than the null ket $|\text{null}\rangle$. The latter is defined by

$$|a\rangle + |\text{null}\rangle = |a\rangle \qquad (12.7a)$$

for any ket $|a\rangle$, whence

$$|\text{null}\rangle = 0 \quad \text{and} \quad \langle \text{null}|\text{null}\rangle = 0. \qquad (12.7b)$$

The symbol $\langle \ \ |$ which occurs as the left-hand side of scalar products is called a *bra*. The names bra and ket originate, of course, from $\langle \ \rangle$ = bra(c)ket. It is possible, but not necessary, to think of bras as vectors in their own right, forming a linear vector space of their own, called bra space. If $\langle a|$ and $\langle b|$ are two vectors of bra space, so is any linear combination

$$c_1 \langle a| + c_2 \langle b|.$$

This bra space is called the dual space of the ket space and is constructed from the latter by associating with every ket $|a\rangle$ a bra $\langle a|$ in such a way that associated with the ket

$$|u\rangle = c_1|u_1\rangle + c_2|u_2\rangle \qquad (12.8a)$$

is the bra

$$\langle u| = c_1^* \langle u_1| + c_2^* \langle u_2|. \qquad (12.8b)$$

A scalar product is now always formed between a bra vector and a ket vector. In particular, from Eq. (12.8b), we have

$$\langle u|v \rangle = c_1^* \langle u_1|v \rangle + c_2^* \langle u_2|v \rangle \qquad (12.9)$$

in agreement with Eq. (12.5b).

The definitions of normalization and orthogonality now follow at once. A ket $|a\rangle$ is normalized (to unity) if

$$\langle a|a \rangle = 1. \qquad (12.10)$$

Two kets $|a\rangle$ and $|b\rangle$ are orthogonal if

$$\langle a|b \rangle = \langle b|a \rangle = 0. \qquad (12.11)$$

These results enable us to construct complete sets of orthonormal kets from a complete set of kets. (This is the Schmidt orthogonalization process, discussed in problem 1.4.)

If $|u_1\rangle$, $|u_2\rangle$, ... is a complete set of orthonormal kets, then

$$\langle u_m|u_n \rangle = \delta_{mn} \qquad (12.12a)$$

and any ket $|u\rangle$ can be written

$$|u\rangle = \sum_{n=1}^{\infty} c_n |u_n\rangle. \tag{12.13}$$

Taking the scalar product of $|u\rangle$ with $|u_m\rangle$ gives the expansion coefficients

$$c_m = \langle u_m | u \rangle, \tag{12.14}$$

so that the expansion (12.13) can also be written in the rather neat form

$$|u\rangle = \sum_{n=1}^{\infty} |u_n\rangle \langle u_n | u \rangle. \tag{12.15a}$$

This equation holds for any complete orthonormal set $|u_1\rangle$, $|u_2\rangle$, ... and any ket $|u\rangle$. It is the general statement of the expansion theorem of a state (i.e. a ket) in terms of a complete set of orthonormal states. This set is called the *basis* for the expansion. We can think of the basis as defining an orthogonal coordinate system in ket space, analogous to three mutually orthogonal unit vectors $\mathbf{e}_1, \mathbf{e}_2$ and $\mathbf{e}_3$ in ordinary space, for which

$$\mathbf{e}_m \cdot \mathbf{e}_n = \delta_{mn}, \qquad m, n = 1, 2, 3 \tag{12.12b}$$

and

$$\mathbf{u} = \sum_{n=1}^{3} \mathbf{e}_n (\mathbf{e}_n \cdot \mathbf{u}) \tag{12.15b}$$

for any vector $\mathbf{u}$.

When developing quantum mechanics in terms of wave functions in Chapters 1 to 3, we saw how to generate complete orthonormal sets of wave functions as the eigenfunctions of Hermitian operators which represent observables. The expansion in terms of these eigenfunctions was seen to be closely related to the probability interpretation in quantum mechanics. In order to translate these results into the language of kets, we must next consider operators in ket space.

12.2 OPERATORS

In section 1.2 an operator was defined as a prescription for transforming wave functions. We now define an operator as a prescription for transforming kets. An operator A operating on the ket $|u\rangle$ transforms it into $|v\rangle = A|u\rangle$. The ket into which A transforms $|u\rangle$ has been labelled v. Alternatively we can label it Au and write

$$|v\rangle = |Au\rangle = A|u\rangle. \tag{12.16}$$

An operator A is linear if operating on any ket of the form

$$|v\rangle = c_1 |u_1\rangle + c_2 |u_2\rangle \tag{12.17a}$$

it transforms it into

$$A|v\rangle = c_1 A|u_1\rangle + c_2 A|u_2\rangle. \tag{12.17b}$$

With one exception (the operation of time reversal which we shall not consider in this book), all operators of quantum mechanics are linear.

Taking the scalar product of (12.16) with $|w\rangle$ gives

$$\langle w|v\rangle = \langle w|Au\rangle = \langle w|A|u\rangle. \tag{12.18}$$

It follows from the last equation and Eq. (12.3) that

$$\langle w|A|u\rangle^* = \langle w|Au\rangle^* = \langle Au|w\rangle. \tag{12.19}$$

Note, we do not have $\langle A|u|w\rangle$ on the right-hand side of this equation, which would be meaningless. $\langle w|A|u\rangle$ is called the *matrix element* of the operator A between the states $|w\rangle$ and $|u\rangle$.

An example of an operator which is most useful in practice is afforded by Eq. (12.15a). This equation holds for *any* ket $|u\rangle$. We can therefore interpret

$$I = \sum_{n=1}^{\infty} |u_n\rangle\langle u_n| \tag{12.20}$$

on the right-hand side of Eq. (12.15a) as the identity operator with the property: I operating on *any* ket $|u\rangle$ leaves it unaltered:

$$I|u\rangle = \sum_{n=1}^{\infty} |u_n\rangle\langle u_n|u\rangle = |u\rangle. \tag{12.21}$$

This equation has a simple geometrical interpretation. It represents the state vector $|u\rangle$ as the sum of its projections on to the orthogonal unit vectors $|u_n\rangle$. (Eq. (12.15b) admits a similar interpretation in ordinary space.)

Eq. (12.20) is called a *completeness relation*. It holds only if the set $|u_1\rangle, |u_2\rangle, \ldots$ form a *complete* set of *orthonormal* kets; only in this case is operating with

$$\sum_{n=1}^{\infty} |u_n\rangle\langle u_n|$$

on *any* ket $|u\rangle$ equivalent to the identity transformation $I|u\rangle = |u\rangle$. Later in this chapter, we shall have examples of the usefulness of the completeness relation in manipulating equations involving kets, bras and operators.

An operator A is defined as *Hermitian* if for any two kets $|u\rangle$ and $|v\rangle$

$$\langle v|Au\rangle = \langle Av|u\rangle. \tag{12.22}$$

On account of Eqs. (12.18) and (12.19), this can be written

$$\langle v|A|u\rangle = \langle u|A|v\rangle^*. \tag{12.23}$$

A useful generalization can be introduced at this point. The *adjoint* operator $A^\dagger$ of the operator A is defined as follows: for any two states $|u\rangle$ and $|v\rangle$

$$\langle v|A^\dagger|u\rangle = \langle u|A|v\rangle^*. \tag{12.24}$$

Eq. (12.24) can equivalently be written

$$\langle u|Av\rangle = \langle A^\dagger u|v\rangle. \tag{12.24a}$$

It follows from Eq. (12.23) that for a Hermitian operator

$$A^\dagger = A. \tag{12.25}$$

Hence a Hermitian operator is also called *self-adjoint*. The adjoint operator can also be defined directly in terms of kets and bras. If

$$|w\rangle = A|v\rangle \tag{12.26a}$$

then Eq. (12.24) can be written

$$\langle v|A^\dagger|u\rangle = \langle u|w\rangle^* = \langle w|u\rangle.$$

Since this equation holds for any state $|u\rangle$, it follows that

$$\langle v|A^\dagger = \langle w|, \tag{12.26b}$$

i.e. if A, operating on $|v\rangle$ to the right, transforms $|v\rangle$ into $A|v\rangle$, then $A^\dagger$, operating on $\langle v|$ to the left, transforms $\langle v|$ into $\langle v|A^\dagger$.

It follows from Eq. (12.23) that for a Hermitian operator A

$$\langle u|A|u\rangle^* = \langle u|A|u\rangle: \tag{12.27}$$

the expectation value $\langle u|A|u\rangle$ of a Hermitian operator A is real for any state $|u\rangle$. The development of the theory of Hermitian operators in terms of kets closely mimics that in terms of wave functions, given in Chapter 1. I shall only briefly state the main results, leaving it to the reader to fill in the details as practice in the Dirac formalism.

We shall assume that a Hermitian operator A possesses a complete set of eigenkets (also called eigenstates) $|1\rangle, |2\rangle, \ldots$ and corresponding eigenvalues $a_1, a_2, \ldots$:

$$A|n\rangle = a_n|n\rangle, \qquad n = 1, 2, \ldots. \tag{12.28}$$

The eigenkets can be taken as normed, so that

$$\langle n|A|n\rangle = a_n. \tag{12.29}$$

It follows from Eq. (12.27) that the eigenvalues a_n are real. Furthermore, one easily proves that the eigenkets belonging to different eigenvalues are orthogonal. (See problem 12.2.)

From eigenkets belonging to the same eigenvalue, one can always form orthonormal linear combinations. Hence the complete set of eigenkets of A can always be chosen to be orthonormal:

$$\langle m|n \rangle = \delta_{mn}, \tag{12.30}$$

If the eigenvalues $a_1, a_2, \ldots$ of A are non-degenerate, we can specify the eigenstates by the eigenvalues and write them $|a_1\rangle, |a_2\rangle, \ldots$. In the case of degeneracy, we need further labels to specify the states. We know from section 3.1, where this question was discussed for wave functions, that this is most conveniently done by introducing a complete set of commuting Hermitian operators $A, B, \ldots, Q$. Their simultaneous eigenstates, labelled by their eigenvalues, are then uniquely determined and constitute a complete set of orthogonal states which can be taken as normalized.*

12.3   PROBABILITY AMPLITUDES

To obtain the physical interpretation of the state vector formalism, we employ the same line of reasoning which we used for wave functions in section 1.2. The states of a system are represented by normalized kets, observables by Hermitian operators. We shall assume that corresponding to each observable A there exists a Hermitian operator A which possesses a complete set of orthonormal eigenkets $|1\rangle, |2\rangle, \ldots$, belonging to eigenvalues $a_1, a_2, \ldots$.

As in section 1.2, we take as our basic interpretative postulate: for a system in the state $|u\rangle$ (which is assumed normed), the expectation value of the observable A is given by

$$\langle A \rangle = \langle u|A|u \rangle. \tag{12.31}$$

Substituting the expansion of $|u\rangle$ in terms of the eigenstates of A,

$$|u\rangle = \sum_{n=1}^{\infty} |n\rangle\langle n|u\rangle, \tag{12.32}$$

in Eq. (12.31) gives

$$\langle A \rangle = \sum_{n=1}^{\infty} a_n |\langle n|u\rangle|^2. \tag{12.33}$$

If the eigenvalues a_n are non-degenerate, this last expression leads to the interpretation of $|\langle n|u\rangle|^2$ as the probability that measuring A on the system in the state $|u\rangle$ gives the result a_n. If a is a degenerate eigenvalue of A, the probability of obtaining the result a is given by

$$\sum_{a_n = a} |\langle n|u\rangle|^2$$

where the summation is over all eigenstates $|n\rangle$ with eigenvalues $a_n = a$.

---

* Strictly speaking, the eigenkets are only determined up to a phase factor $e^{i\alpha}$, where α is a real number. However, just as for wave functions, such a phase factor is not significant: $|u\rangle$ and $|u\rangle e^{i\alpha}$ represent the same physical state, with the same expectation values, etc.

If the system is in the eigenstate $|n\rangle$, measuring A is certain to give the result a_n. Hence we interpret $|\langle n|u\rangle|^2$ as the probability that the system when in the state $|u\rangle$ is also in the state $|n\rangle$; for short, we shall call $|\langle n|u\rangle|^2$ the probability that the state $|n\rangle$ is present in the state $|u\rangle$, with $\langle n|u\rangle$ the corresponding *probability amplitude*. This interpretation, which we here gave for a state $|u\rangle$ and an eigenstate $|n\rangle$, holds generally: if $|u\rangle$ and $|v\rangle$ are normed states of a system, we interpret $|\langle v|u\rangle|^2$ as the probability and $\langle v|u\rangle$ as the probability amplitude that the system when in the state $|u\rangle$ is also in the state $|v\rangle$, i.e. that the state $|v\rangle$ is present in the state $|u\rangle$.

In section 1.2, probabilities and probability amplitudes were expressed in terms of overlap integrals of two wave functions, i.e. of functions of the position coordinates. In the more general formulation now, probability amplitudes are given by scalar products of two state vectors. In the next section, we shall see how to derive the formulation in terms of wave functions—i.e. wave mechanics—from that in terms of kets.

With the interpretation of scalar products as probability amplitudes, all our earlier results, in Chapters 1 and 3, about standard deviations, compatible observables, etc., can be formulated in terms of state vectors instead of wave functions. Their derivations are left as exercises for the reader.

12.4 WAVE MECHANICS REGAINED

So far in this chapter we have considered observables with discrete eigenvalue spectra. There are, of course, observables with continuous spectra; for example, the position coordinate $\mathbf{r}$ of a particle may assume all values. The corresponding Hermitian operator $\hat{\mathbf{r}}$ possesses a completely continuous spectrum of eigenvalues; $|\mathbf{r}\rangle$ is an eigenstate of $\hat{\mathbf{r}}$, belonging to the eigenvalue $\mathbf{r}$, for all values of $\mathbf{r}$, i.e.

$$\hat{\mathbf{r}}|\mathbf{r}\rangle = \mathbf{r}|\mathbf{r}\rangle \tag{12.34}$$

for all $\mathbf{r}$.* We choose the normalization of these states such that

$$\langle \mathbf{r}'|\mathbf{r}\rangle = \delta^{(3)}(\mathbf{r} - \mathbf{r}'). \tag{12.35}$$

We shall see below that this normalization is appropriate. Eq. (12.35) is the analogue of Eq. (12.30) for the discrete case.

The expansion of a normed state $|a\rangle$ of the particle in terms of the position eigenstates $|\mathbf{r}\rangle$ takes the form

$$|a\rangle = \int d^3\mathbf{r}|\mathbf{r}\rangle\langle \mathbf{r}|a\rangle, \tag{12.36}$$

---

* We shall use the circumflex accent to denote the position and momentum operators $\hat{\mathbf{r}}$ and $\hat{\mathbf{p}}$, to avoid possible confusion with the ordinary variables $\mathbf{r}$ and $\mathbf{p}$.

where we have an integral instead of a sum over eigenstates because we are dealing with a continuous spectrum.

We know the interpretation of $\langle \mathbf{r}|a \rangle$ from the last section. It is the probability amplitude that a particle in the state $|a\rangle$ is located at $\mathbf{r}$; more accurately,

$$|\langle \mathbf{r}|a \rangle|^2 \, d^3\mathbf{r} \tag{12.37a}$$

is the probability that a particle in the state $|a\rangle$ is located in the volume $d^3\mathbf{r}$ at $\mathbf{r}$. Hence $\langle \mathbf{r}|a \rangle$ is precisely what we have previously called the wave function of the particle in the state, labelled a, and denoted by, say, $\psi_a(\mathbf{r})$:

$$\langle \mathbf{r}|a \rangle = \psi_a(\mathbf{r}). \tag{12.38}$$

The probability expression (12.37a) now assumes the more familiar form

$$|\psi_a(\mathbf{r})|^2 \, d^3\mathbf{r}. \tag{12.37b}$$

This is an important result. The interpretation of (12.38) as a probability amplitude (from which the probability interpretation of Eqs. (12.37) follows) is merely a particular case of the general interpretation of a scalar product as a probability amplitude. In our original development, in Chapter 1, of quantum mechanics in terms of wave functions—that is, of wave mechanics—the probability interpretation of (12.37b) had to be introduced as a *separate* postulate. The description of states by wave functions is called the $\mathbf{r}$-*representation* (or the *coordinate representation*), and wave mechanics is the form quantum mechanics assumes in the $\mathbf{r}$-representation, i.e. when all state vectors are expanded using the complete set of orthonormal position eigenstates as basis for the expansions.

We next justify our choice (12.35) for the normalization of the position eigenstates. Taking the scalar product of Eq. (12.36) with $|\mathbf{r}'\rangle$ gives

$$\psi_a(\mathbf{r}') = \int d^3\mathbf{r} \langle \mathbf{r}'|\mathbf{r} \rangle \psi_a(\mathbf{r}). \tag{12.39}$$

Since $|a\rangle$ is an arbitrary state vector and therefore $\psi_a(\mathbf{r})$ an arbitrary function of $\mathbf{r}$, Eq. (12.39) can only hold if $\langle \mathbf{r}'|\mathbf{r} \rangle$ is the Dirac delta function $\delta^{(3)}(\mathbf{r} - \mathbf{r}')$ [see Eq. (2.130)], in agreement with Eq. (12.35). It also follows that, for Eq. (12.36) to hold for arbitrary states $|a\rangle$, the operator

$$I = \int d^3\mathbf{r} |\mathbf{r}\rangle\langle \mathbf{r}| \tag{12.40}$$

must be the identity operator, and (12.40) the completeness relation for the orthonormal position eigenstates $|\mathbf{r}\rangle$.

With the identification (12.38), the scalar product of two states $|a\rangle$ and $|b\rangle$ is easily expressed in terms of wave functions. From Eq. (12.36)

$$\langle b|a \rangle = \int d^3\mathbf{r}\langle b|\mathbf{r} \rangle\langle \mathbf{r}|a \rangle = \int d^3\mathbf{r}\psi_b^*(\mathbf{r})\psi_a(\mathbf{r}), \tag{12.41}$$

which expresses $\langle b|a \rangle$ in the familiar form of an overlap integral of wave functions.

To complete the identification between the Dirac formalism and wave mechanics, we must express operators in the r-representation. We see from Eqs. (12.34) and (12.35) that

$$\langle \mathbf{r}'|\hat{\mathbf{r}}|\mathbf{r} \rangle = \mathbf{r}\langle \mathbf{r}'|\mathbf{r} \rangle = \mathbf{r}\delta^{(3)}(\mathbf{r} - \mathbf{r}')$$

and similarly for any function $f(\mathbf{r})$ which can be expanded in a power series in $\mathbf{r}$:

$$\langle \mathbf{r}'|f(\hat{\mathbf{r}})|\mathbf{r} \rangle = f(\mathbf{r})\delta^{(3)}(\mathbf{r} - \mathbf{r}'). \tag{12.42}$$

It follows that

$$\langle \mathbf{r}|f(\hat{\mathbf{r}})|a \rangle = \int d^3 r' \langle \mathbf{r}|f(\hat{\mathbf{r}})|\mathbf{r}' \rangle \langle \mathbf{r}'|a \rangle$$

$$= f(\mathbf{r})\langle \mathbf{r}|a \rangle = f(\mathbf{r})\psi_a(\mathbf{r}). \tag{12.43}$$

Thus in the r-representation, the operator $f(\hat{\mathbf{r}})$ operating on a state $|a \rangle$ corresponds to multiplying the wave function $\psi_a(\mathbf{r})$ by $f(\mathbf{r})$.

Lastly, we need the form of the momentum operator $\hat{\mathbf{p}}$ and of functions of $\hat{\mathbf{p}}$ in the r-representation. $\hat{\mathbf{p}}$ has a completely continuous spectrum with all values $\mathbf{p}$ as eigenvalues:

$$\hat{\mathbf{p}}|\mathbf{p} \rangle = \mathbf{p}|\mathbf{p} \rangle. \tag{12.44a}$$

The momentum eigenstates $|\mathbf{p} \rangle$ satisfy the orthonormality relation

$$\langle \mathbf{p}'|\mathbf{p} \rangle = \delta^{(3)}(\mathbf{p}' - \mathbf{p}) \tag{12.44b}$$

and the completeness relation

$$\int d^3 p |\mathbf{p} \rangle \langle \mathbf{p}| = I. \tag{12.44c}$$

In the r-representation, the momentum eigenfunctions are given by the plane waves (2.131):

$$\langle \mathbf{r}|\mathbf{p} \rangle = u_\mathbf{p}(\mathbf{r}) = \frac{1}{(2\pi\hbar)^{3/2}} e^{i\mathbf{p}\cdot\mathbf{r}/\hbar}. \tag{12.45}$$

Expanding an arbitrary state $|a \rangle$ in momentum eigenstates $|\mathbf{p} \rangle$ gives

$$|a \rangle = I|a \rangle = \int d^3 p |\mathbf{p} \rangle \langle \mathbf{p}|a \rangle \tag{12.46}$$

whence

$$\langle \mathbf{r}|a \rangle = \int d^3 p \langle \mathbf{r}|\mathbf{p} \rangle \langle \mathbf{p}|a \rangle$$

i.e.

$$\psi_a(\mathbf{r}) = \frac{1}{(2\pi\hbar)^{3/2}} \int d^3p \, e^{i\mathbf{p}\cdot\mathbf{r}/\hbar} \phi_a(\mathbf{p}) \tag{12.47a}$$

where $\psi_a(\mathbf{r}) = \langle \mathbf{r}|a\rangle$ and $\phi_a(\mathbf{p}) = \langle \mathbf{p}|a\rangle$. Conversely

$$\langle \mathbf{p}|a\rangle = \int d^3r \langle \mathbf{p}|\mathbf{r}\rangle \langle \mathbf{r}|a\rangle$$

i.e.

$$\phi_a(\mathbf{p}) = \frac{1}{(2\pi\hbar)^{3/2}} \int d^3r \, e^{-i\mathbf{p}\cdot\mathbf{r}/\hbar} \psi_a(\mathbf{r}). \tag{12.47b}$$

Eqs. (12.47a–b) are the usual pair of Fourier transforms connecting position and momentum probability amplitudes $\psi_a(\mathbf{r})$ and $\phi_a(\mathbf{p})$. [Compare the corresponding one-dimensional equations (2.115–116).]

From Eqs. (12.46) and (12.45) we obtain

$$\langle \mathbf{r}|\hat{\mathbf{p}}|a\rangle = \int d^3p \langle \mathbf{r}|\hat{\mathbf{p}}|\mathbf{p}\rangle \langle \mathbf{p}|a\rangle$$

$$= \int d^3p \, \mathbf{p} \langle \mathbf{r}|\mathbf{p}\rangle \langle \mathbf{p}|a\rangle$$

$$= \frac{1}{(2\pi\hbar)^{3/2}} \int d^3p \, \mathbf{p} \, e^{i\mathbf{p}\cdot\mathbf{r}/\hbar} \phi_a(\mathbf{p}) = -i\hbar\nabla\psi_a(\mathbf{r}) \tag{12.48}$$

where the last step follows from Eq. (12.47a). We conclude that in the r-representation the operator $\hat{\mathbf{p}}$ is given by $-i\hbar\nabla$. * Analogously to the derivation of Eq. (12.48), one shows that for any function $f(\mathbf{p})$, which can be expanded in a power series in $\mathbf{p}$,

$$\langle \mathbf{r}|f(\hat{\mathbf{p}})|a\rangle = f(-i\hbar\nabla)\psi_a(\mathbf{r}). \tag{12.49}$$

Combining the results (12.43) and (12.49), it follows that an operator $f(\hat{\mathbf{r}}, \hat{\mathbf{p}})$ of a one-particle system is given in the r-representation by $f(\mathbf{r}, -i\hbar\nabla)$. More precisely, if the system is in the state $|a\rangle$, with wave function $\psi_a(\mathbf{r}) = \langle \mathbf{r}|a\rangle$, then

$$\langle \mathbf{r}|f(\hat{\mathbf{r}}, \hat{\mathbf{p}})|a\rangle = f(\mathbf{r}, -i\hbar\nabla)\psi_a(\mathbf{r}). \tag{12.50}$$

These results enable us to rewrite equations involving state vectors in the r-representation. For example, the eigenvalue equation

$$H(\hat{\mathbf{r}}, \hat{\mathbf{p}})|n\rangle = E_n|n\rangle \tag{12.51a}$$

---

* This is, of course, not a derivation of the momentum operator, since the momentum eigenstates (12.45) were obtained in Chapter 1 by postulating that $\hat{\mathbf{p}} = -i\hbar\nabla$. It is merely a way of establishing a connection between wave mechanics and the ket formalism.

for the Hamiltonian $H(\hat{\mathbf{r}}, \hat{\mathbf{p}})$ becomes, on taking the scalar product with $|\mathbf{r}\rangle$,

$$\langle \mathbf{r}|H(\hat{\mathbf{r}}, \hat{\mathbf{p}})|n\rangle = E_n\langle \mathbf{r}|n\rangle.$$

From Eq. (12.50), this takes on the more familiar form

$$H(\mathbf{r}, -i\hbar\nabla)\psi_n(\mathbf{r}) = E_n\psi_n(\mathbf{r}) \tag{12.51b}$$

where $\psi_n(\mathbf{r}) = \langle \mathbf{r}|n\rangle$ is the energy eigenfunction corresponding to the energy eigenket $|n\rangle$.

12.5 HARMONIC OSCILLATOR

An interesting and important application of the operator methods of this chapter is to the harmonic oscillator, which we treated in terms of wave functions in section 2.7.

The oscillator Hamiltonian is

$$H = \frac{1}{2m}p^2 + \tfrac{1}{2}m\omega^2 x^2, \tag{12.52}$$

with the position and momentum operators x and p satisfying the commutation relation

$$[x, p] = i\hbar. \tag{12.53}$$

We introduce the operators

$$\left.\begin{array}{c} a \\ a^\dagger \end{array}\right\} = \frac{1}{(2\hbar m\omega)^{1/2}}(m\omega x \pm ip) \tag{12.54}$$

and the Hermitian operator

$$N = a^\dagger a. \tag{12.55}$$

It follows from Eqs. (12.53) and (12.54) that

$$[a, a^\dagger] = 1, \tag{12.56}$$

and that the Hamiltonian (12.52) can be written

$$H = \hbar\omega(N + \tfrac{1}{2}). \tag{12.57}$$

To obtain the energy eigenvalues of H, we require the eigenvalues of the operator N. If $|\lambda\rangle$ is a normed eigenstate of N, with the eigenvalue λ, i.e.

$$N|\lambda\rangle = \lambda|\lambda\rangle, \quad \langle \lambda|\lambda\rangle = 1, \tag{12.58}$$

then it follows from (12.55) that

$$\lambda = \langle \lambda|N|\lambda\rangle = \langle \lambda|a^\dagger a\lambda\rangle = \langle a\lambda|a\lambda\rangle \geqslant 0,$$

i.e. the eigenvalues of N are non-negative and there must exist a smallest eigenvalue λ_0. From Eqs. (12.58) and (12.55–56):

$$Na^\dagger|\lambda\rangle = a^\dagger aa^\dagger|\lambda\rangle = a^\dagger(a^\dagger a + 1)|\lambda\rangle = (\lambda + 1)a^\dagger|\lambda\rangle \qquad (12.59a)$$

and similarly one shows that

$$Na|\lambda\rangle = (\lambda - 1)a|\lambda\rangle. \qquad (12.59b)$$

Thus, if $|\lambda\rangle$ is an eigenstate of N belonging to the eigenvalue λ, then $a^\dagger|\lambda\rangle$ and $a|\lambda\rangle$ are eigenstates of N belonging to the eigenvalues $(\lambda + 1)$ and $(\lambda - 1)$ respectively. Operating repeatedly with a on $|\lambda\rangle$ produces a sequence of eigenstates with eigenvalues $(\lambda - 1), (\lambda - 2), \ldots$. Since there exists a lowest eigenvalue λ_0, we must have

$$a|\lambda_0\rangle = 0 \qquad (12.60)$$

so that the procedure of generating successive eigenstates with lower eigenvalues breaks off at $|\lambda_0\rangle$. From Eq. (12.60)

$$N|\lambda_0\rangle = a^\dagger a|\lambda_0\rangle = 0, \qquad (12.61)$$

so that $|\lambda_0\rangle$ belongs to the eigenvalue $\lambda_0 = 0$, and the eigenvalues of N are $n = 0, 1, 2, \ldots$:

$$N|n\rangle = n|n\rangle, \qquad n = 0, 1, 2, \ldots. \qquad (12.62)$$

From Eq. (12.57) the energy eigenvalues E_n of the oscillator Hamiltonian follow:

$$E_n = \hbar\omega(n + \tfrac{1}{2}), \qquad n = 0, 1, \ldots, \qquad (12.63)$$

in agreement with Eq. (2.152). It is also easy to recover the spatial eigenfunctions from the eigenstates $|n\rangle$ by going over to the coordinate representation (see problem 12.4).

It follows from Eqs. (12.62–63) that we can interpret $|n\rangle$ as a state with n quanta present, each with energy $\hbar\omega$. We see from Eqs. (12.59) that $a^\dagger|n\rangle$ is a state with $(n + 1)$ quanta, i.e. $a^\dagger$ creates a quantum, and that $a|n\rangle$ is a state with $(n - 1)$ quanta, i.e. a absorbs a quantum. a and $a^\dagger$ are accordingly known as absorption (or annihilation) and creation operators respectively. The operator N is called the occupation number operator; its eigenvalues $n (= 0, 1, \ldots)$ specify the number of quanta present in the state $|n\rangle$.

The importance of the above analysis is due to the fact that one can represent very complicated systems as a superposition of normal modes. For example, for the electromagnetic field in an enclosure the different normal modes correspond to electromagnetic waves with different wave vectors $\mathbf{k}$ and definite polarization. Each normal mode is described, independently of the others, by a harmonic oscillator equation. In the quantized theory, each mode corresponds to photons of a specific kind (i.e. with a definite wave vector $\mathbf{k}$ and a definite state of polarization). For each kind of photon, we have creation and annihilation operators, and occupation number operators whose eigenvalues $(= 0, 1, \ldots)$ specify the number of photons of that type present in the corresponding eigenstate. I have here attempted to indicate in barest outline the ideas underlying the quantization of the electromagnetic field or, for that matter, of other fields.*

12.6   REPRESENTATIONS

Eqs. (12.51a) and (12.51b) state the same physics. The relation between them is analogous to that between an equation expressed in terms of vectors or in terms of components referred to a particular coordinate system. For example, Newton's law

$$\mathbf{F} = m\frac{\mathrm{d}^2}{\mathrm{d}t^2}\mathbf{r} \tag{12.64a}$$

expressed in terms of components with respect to a coordinate system defined by three orthogonal unit vectors $\mathbf{e}_1, \mathbf{e}_2$ and $\mathbf{e}_3$, becomes

$$\mathbf{e}_s \cdot \mathbf{F} = m\frac{\mathrm{d}^2}{\mathrm{d}t^2}\mathbf{e}_s \cdot \mathbf{r}, \qquad s = 1, 2, 3.$$

This can also be written

$$F_s = m\frac{\mathrm{d}^2}{\mathrm{d}t^2}r_s \tag{12.64b}$$

where $F_s = \mathbf{e}_s \cdot \mathbf{F}$ and $r_s = \mathbf{e}_s \cdot \mathbf{r}$. Eqs. (12.64a–b) for ordinary vectors correspond to Eqs. (12.51a–b) for kets.

Just as there are many ways of expressing a vector equation in terms of components by choosing different coordinate systems, so a ket equation can be written in many different representations by choosing different complete sets of kets as basis states for expansions. The r-representation resulted from using the complete set of orthonormal position eigenvectors $|\mathbf{r}\rangle$ as basis for expansions. More generally, if $|\xi_1\rangle, |\xi_2\rangle, \ldots$ is a complete set of orthonormal kets—they will usually be the eigenstates of a complete set of commuting observables—and we expand all state vectors with respect to this set of kets, we obtain the representation with $\{|\xi_1\rangle, |\xi_2\rangle, \ldots\}$ as basis. For example, if $|u\rangle$, $|v\rangle, \ldots$ are arbitrary kets and $A, B, \ldots$ are operators, then

$$|u\rangle = \sum_n |\xi_n\rangle\langle\xi_n|u\rangle = \sum_n |\xi_n\rangle u_n \tag{12.65}$$

---

* For a simple discussion of the quantization of the electromagnetic field, along the above lines, see Mandl and Shaw, Chapter 1.

and

$$\langle v|u\rangle = \sum_n \langle v|\xi_n\rangle\langle\xi_n|u\rangle = \sum_n v_n^* u_n \qquad (12.66)$$

where $u_n = \langle\xi_n|u\rangle$ and $v_n = \langle\xi_n|v\rangle$. Similarly

$$|v\rangle = A|u\rangle \qquad (12.67)$$

becomes

$$\langle\xi_m|v\rangle = \sum_n \langle\xi_m|A|\xi_n\rangle\langle\xi_n|u\rangle \qquad (12.68a)$$

i.e.

$$v_m = \sum_n A_{mn} u_n \qquad (12.68b)$$

where $A_{mn} = \langle\xi_m|A|\xi_n\rangle$. The right-hand side of Eq. (12.66) is the scalar product $\langle v|u\rangle$ expressed in the $\{|\xi_1\rangle, |\xi_2\rangle, \ldots\}$ representation, and Eqs. (12.68) express (12.67) in this representation. One also calls it a matrix representation, since Eqs. (12.66) and (12.68) are standard matrix equations. In them, the ket $|u\rangle$ is represented by a column vector with elements $u_1, u_2, \ldots$, the bra $\langle v|$ is represented by a row vector with elements $v_1^*, v_2^*, \ldots$, and the operator A by the matrix A_{mn}.

I have been careful to make a sharp distinction between a state vector $|a\rangle$ and its wave function $\psi_a(\mathbf{r}) = \langle\mathbf{r}|a\rangle$. As we have seen, there exists an exact correspondence between the two descriptions; they are completely equivalent. Once this point is appreciated, it should not be misleading to refer to both $|a\rangle$ and $\psi_a(\mathbf{r})$ as the state of a system and even write $|a\rangle = \psi_a(\mathbf{r})$. This or something like it is often done, although strictly speaking it would be better to use a different symbol, say $\doteq$, to mean 'is equivalent to' and write $|a\rangle \doteq \psi_a(\mathbf{r})$. [This situation is analogous to that with ordinary vectors. One might, for example, speak of the position vector $\mathbf{r}$ as the vector (x, y, z), meaning the vector with these components referred to a particular coordinate system.] Similar comments apply to representations other than the $\mathbf{r}$-representation.

Different representations are extensively employed both in the general theory and in specific applications of quantum mechanics. For this reason, the theory of representations and of the transformations between different representations plays an important role in more advanced quantum mechanics. Some of the underlying ideas are illustrated in the problems at the end of this chapter.

PROBLEMS 12

12.1 $|\langle v|u\rangle|^2$ is interpreted as the probability that the normed state $|v\rangle$ is present in the normed state $|u\rangle$. Show that

$$|\langle v|u\rangle|^2 \leqslant 1, \qquad (12.69)$$

as required by this probability interpretation.

Hint: Evaluate $\langle w|w \rangle$ where

$$|w\rangle = |v\rangle\langle v|u\rangle - |u\rangle. \tag{12.70}$$

12.2 The Hermitian operator A possesses the complete set of eigenstates $|1\rangle, |2\rangle, \ldots$, belonging to the eigenvalues $a_1, a_2, \ldots$. Prove that the states $|m\rangle$ and $|n\rangle$ are orthogonal if $a_m \neq a_n$.

12.3 The representation in which the momentum eigenstates $|\mathbf{p}\rangle$, with the properties (12.44a–c), form the basis is known as the $\mathbf{p}$-representation or the momentum representation. In this representation, the state $|a\rangle$ is represented by $\langle \mathbf{p}|a\rangle = \phi_a(\mathbf{p})$ which is called the momentum space wave function, in analogy to the coordinate space wave function, $\langle \mathbf{r}|a\rangle = \psi_a(\mathbf{r})$.

(a) Show that in the $\mathbf{p}$-representation the operator $\hat{\mathbf{r}}$ for the position coordinate is given by

$$\hat{\mathbf{r}} = i\hbar \nabla_{\mathbf{p}} \tag{12.71}$$

where $\nabla_{\mathbf{p}} = (\partial/\partial p_x, \partial/\partial p_y, \partial/\partial p_z)$ is the gradient operator in momentum space, and obtain $\langle b|\hat{\mathbf{r}}|a\rangle$ in terms of the momentum space wave functions of the states $|a\rangle$ and $|b\rangle$.

(b) Write down the Hamiltonian of the one-dimensional harmonic oscillator in the $\mathbf{p}$-representation and *hence* obtain the momentum probability distribution of the oscillator ground state.

12.4 (a) By writing Eq. (12.60) in the coordinate representation, obtain the wave function of the harmonic oscillator ground state.

(b) It follows from Eqs. (12.59) and (12.62) that the oscillator eigenstates $|n \pm 1\rangle$ are proportional to $a^\dagger|n\rangle$ and $a|n\rangle$ respectively. Find the constants of proportionality so that $|n \pm 1\rangle$ are normed to unity if $|n\rangle$ is normed to unity. Hence express the normed state $|n\rangle$ in terms of the normed state $|0\rangle$.

12.5 An operator U is called unitary, if its inverse U^{-1} equals its adjoint, $U^{-1} = U^\dagger$, i.e. if

$$U^\dagger U = UU^\dagger = I. \tag{12.72}$$

(a) Show that a unitary transformation U leaves the scalar product invariant, i.e. if U transforms two arbitrary states $|u\rangle$ and $|v\rangle$ into

$$|u'\rangle = U|u\rangle, \quad |v'\rangle = U|v\rangle,$$

then

$$\langle v'|u'\rangle = \langle Uv|Uu\rangle = \langle v|u\rangle. \tag{12.73}$$

(b) Let $|\xi_1\rangle, |\xi_2\rangle, \ldots$ be a complete set of orthonormal states. Show that the states $|\eta_1\rangle, |\eta_2\rangle, \ldots$, where

$$|\eta_n\rangle = U|\xi_n\rangle, \quad n = 1, 2, \ldots \tag{12.74}$$

also form a complete set of orthonormal states.

(c) From the matrix representation of an operator A, referred to the states $|\xi_1\rangle, |\xi_2\rangle, \ldots$ as basis, obtain the matrix representation of A, referred to the states (12.74) as basis.

12.6 Obtain the eigenstates $|1\rangle, |2\rangle, \ldots$ and the corresponding eigenvalues $a_1, a_2, \ldots$ of a Hermitian operator A, given its representation $A_{rs} = \langle \xi_r|A|\xi_s \rangle$ referred to the complete set of orthonormal states $|\xi_1\rangle, |\xi_2\rangle, \ldots$ as basis.

Hints and solutions to problems

PROBLEMS 1

1.1 (a) $(\pi a_0^3)^{-1/2}$ (b) 0.32 (c) The probability that the electron lies within a spherical shell with radii r and $r + dr$ is

$$P(r)\, dr = 4\pi r^2\, dr\, |\psi(r)|^2.$$

The maximum of $P(r)$ gives the most probable value a_0. (d) $1.5a_0$ (e) $\langle -e^2/4\pi\varepsilon_0 r\rangle = -e^2/4\pi\varepsilon_0 a_0$ (f) $\Delta r = 0.87a_0$. (g) For a bound state, we must have $\langle p_x\rangle = 0$. $\langle p_x\rangle \neq 0$ implies the electron wanders off to infinity. This result also follows directly from $\langle -i\hbar\, \partial/\partial x\rangle$.

$\phi(r)$ must be orthogonal to $\psi(r)$, whence $\lambda = -1/2a_0$.

1.2 (i) 0.5 (from symmetry) (ii) 0.82.

1.3 (i) From symmetry

$$\int |xf(r)|^2\, d^3\mathbf{r} = \int |yf(r)|^2\, d^3\mathbf{r} = \int |zf(r)|^2\, d^3\mathbf{r},$$

hence $C = 1/\sqrt{2}$.

(ii) $\hat{l}_z\psi_m(\mathbf{r}) = -i\hbar(x\, \partial/\partial y - y\, \partial/\partial x)\psi_m(\mathbf{r}) = \hbar m\psi_m(\mathbf{r})$, $m = 1, 0, -1$; hence $\lambda_m = \hbar m$, $m = 1, 0, -1$.

(iii)
$$\phi(\mathbf{r}) = \frac{1}{\sqrt{2}}\psi_0(\mathbf{r}) - \frac{i}{2}[\psi_1(\mathbf{r}) + \psi_{-1}(\mathbf{r})].$$

From Eq. (1.38): $P(0) = \frac{1}{2}$, $P(\hbar) = P(-\hbar) = \frac{1}{4}$.

The significance of these results—introduced here as an exercise about eigenstates—will become apparent in section 2.4 on the Stern–Gerlach experiment.

1.4
$$\psi_1(\mathbf{r}) = \phi_1(\mathbf{r})/\sqrt{I_{11}}, \qquad \text{where } I_{mn} = \int \phi_m^*(\mathbf{r})\phi_n(\mathbf{r})\, d^3\mathbf{r}.$$

Choose λ so that $\chi_2(\mathbf{r}) = \phi_2(\mathbf{r}) - \lambda\phi_1(\mathbf{r})$ is orthogonal to $\phi_1(\mathbf{r})$:

$$\int \phi_1^*(\mathbf{r})\chi_2(\mathbf{r})\, d^3\mathbf{r} = I_{12} - \lambda I_{11} = 0$$

whence $\lambda = I_{12}/I_{11}$. Normalizing $\chi(\mathbf{r})$ gives

$$\psi_2(\mathbf{r}) = \frac{\phi_2(\mathbf{r}) - \lambda\phi_1(\mathbf{r})}{\sqrt{(I_{22} - |\lambda|^2 I_{11})}}.$$

There are infinitely many ways of choosing two orthonormal linear combinations of ϕ_1 and ϕ_2; ψ_1 and ψ_2 is one particular choice. The analogous problem in ordinary vector analysis is the following: given two non-collinear vectors $\mathbf{v}_1$ and $\mathbf{v}_2$, to construct linear combinations of $\mathbf{v}_1$ and $\mathbf{v}_2$ which are orthogonal unit vectors $\mathbf{e}_1$ and $\mathbf{e}_2$. $\mathbf{v}_1$ and $\mathbf{v}_2$ define a plane; any two orthogonal unit vectors in this plane can be chosen for $\mathbf{e}_1$ and $\mathbf{e}_2$.

The above process—known as the Schmidt orthogonalization process—can be extended to construct orthonormal sets of functions from more than two linearly independent functions.

1.5 The analysis closely parallels the classical case. The momentum operators of the centre-of-mass and relative motions are defined by

$$\mathbf{P} = \mathbf{p}_1 + \mathbf{p}_2, \qquad \mathbf{p} = m\left(\frac{1}{m_1}\mathbf{p}_1 - \frac{1}{m_2}\mathbf{p}_2\right) \tag{1}$$

where $m = m_1 m_2/(m_1 + m_2)$ is the reduced mass of the particles and $\mathbf{p}_j = -i\hbar\nabla_j = -i\hbar(\partial/\partial x_j, \partial/\partial y_j, \partial/\partial z_j), j = 1, 2$. Changing from $\mathbf{r}_1, \mathbf{r}_2$ to $\mathbf{R}, \mathbf{r}$ as independent variables in (1), one obtains

$$\mathbf{P} = -i\hbar\nabla_R, \qquad \mathbf{p} = -i\hbar\nabla, \tag{2}$$

where ∇_R and ∇ are the gradient operators with respect to the vectors $\mathbf{R}$ and $\mathbf{r}$ respectively. Eqs. (1) give $\mathbf{P}$ and $\mathbf{p}$ in terms of $\mathbf{p}_1$ and $\mathbf{p}_2$, and hence

$$\frac{1}{2m_1}\mathbf{p}_1^2 + \frac{1}{2m_2}\mathbf{p}_2^2 = \frac{1}{2M}\mathbf{P}^2 + \frac{1}{2m}\mathbf{p}^2. \tag{3}$$

With (3) substituted in the Schrödinger equation (1.77), it separates into

$$i\hbar\frac{\partial\Phi(\mathbf{R}, t)}{\partial t} = -\frac{\hbar^2}{2M}\nabla_R^2\Phi(\mathbf{R}, t), \qquad i\hbar\frac{\partial\phi(\mathbf{r}, t)}{\partial t} = \left[-\frac{\hbar^2}{2m}\nabla^2 + V(r)\right]\phi(\mathbf{r}, t).$$

The first of these equations is the Schrödinger equation of a free particle of mass M; it represents the motion of the centre-of-mass of the two-particle system. The second equation is the Schrödinger equation for the motion of a particle of mass m in the potential $V(r)$; it represents the relative motion of the two particles. This separation into centre-of-mass and relative motions is the same as occurs in classical mechanics.

1.6 $$-\frac{d}{dt}\int_V \rho(\mathbf{r}, t)\, d^3r = -\int_V \frac{\partial}{\partial t}|\psi(\mathbf{r}, t)|^2\, d^3r$$

$$= -\int_V \frac{i\hbar}{2m}(\psi^*\nabla^2\psi - \psi\nabla^2\psi^*)\, d^3r = \int_V (\nabla\cdot\mathbf{j})\, d^3r \tag{1}$$

where the second step follows using the Schrödinger equation, and the third from the vector identity

$$\nabla\cdot[f(\mathbf{r})\nabla g(\mathbf{r})] = \nabla f(\mathbf{r})\cdot\nabla g(\mathbf{r}) + f(\mathbf{r})\nabla^2 g(\mathbf{r}).$$

Using Gauss's theorem, one transforms the right-hand side of Eq. (1) into

$$\int_S \mathbf{j} \cdot d\mathbf{S}. \tag{2}$$

$\rho(\mathbf{r}, t)$ is the probability density that the particle is at $\mathbf{r}$ at time t, and $\mathbf{j}(\mathbf{r}, t)$ is the probability current density, i.e. the probability per unit time and per unit area that the particle crosses a surface element perpendicular to $\mathbf{j}$ in the direction of $\mathbf{j}$. The left-hand side of Eq. (1) is the rate at which the probability decreases that the particle is within the volume V; the integral (2) is the probability per unit time that the particle crosses outwards through the surface S.

Eq. (1) can be written

$$\int_V \left\{ \frac{\partial}{\partial t} \rho(\mathbf{r}, t) + \mathbf{V} \cdot \mathbf{j}(\mathbf{r}, t) \right\} d^3\mathbf{r} = 0$$

and since this holds for *any* region V, it follows that

$$\frac{\partial}{\partial t} \rho(\mathbf{r}, t) + \mathbf{V} \cdot \mathbf{j}(\mathbf{r}, t) = 0. \tag{3}$$

Eq. (3) is the analogue of the classical equations of continuity; for example, in hydrodynamics or electromagnetism.* Like these, it states a conservation theorem; in this case that of probability.

PROBLEMS 2

2.1
$$\psi(x, t) = \frac{1}{\sqrt{2}} [\phi_1(x) \exp(-i\mathscr{E}_1 t/\hbar) + \phi_2(x) \exp(-i\mathscr{E}_2 t/\hbar)]$$

with $\mathscr{E}_n$ given by Eq. (2.26). Hence

$$P_+(t) = \int_0^a |\psi(x, t)|^2 \, dx = \frac{1}{2} + \frac{4}{3\pi} \cos \omega t, \qquad P_-(t) = 1 - P_+(t),$$

where $\omega = (\mathscr{E}_2 - \mathscr{E}_1)/\hbar = 3\hbar\pi^2/(8ma^2)$.

$P_+(t)$ and $P_-(t)$ oscillate with the period $T = 2\pi/\omega$ between the values 0.92 and 0.08, i.e. the particle is alternately much more likely to be in the right-hand half and the left-hand half of the well. The system oscillates with period T between the states $[\phi_1(x) \pm \phi_2(x)]/\sqrt{2}$. The reader should sketch the wave functions $\phi_1(x)$, $\phi_2(x)$, $\psi(x, 0)$ and $\psi(x, \pi/\omega)$. Compare also the discussion at the end of section 1.3.

2.2 We continue to denote the eigenfunctions of the two energy levels E_1 and E_2 in the potential (2.163) by $\phi_1(x)$ and $\phi_2(x)$; they are of course no longer given by Eqs. (2.25). Since the potential (2.163) is symmetric about the origin [i.e. $V(-x) = V(x)$], $\phi_1(x)$ and $\phi_2(x)$ will be even and odd functions of x with no and one node respectively. Inside the barrier region $|x| < b$, these wave functions will be of the form $\cosh(\kappa x)$ and $\sinh(\kappa x)$ respectively, with

$$\kappa = [(2m/\hbar^2)(V_0 - E)]^{1/2}.$$

---

* See, for example, Grant and Phillips, section 10.1.

The higher the barrier, the less the wave functions will penetrate inside the barrier region $|x| < b$; the particle will preferentially be found in the outer regions $-a < x < -b$ and $b < x < a$. For a high barrier, the barrier transmission coefficient is small and the particle has difficulty in tunnelling through the barrier: the period of oscillation $T = 2\pi\hbar/(E_2 - E_1)$ will be long and the energy difference $E_2 - E_1$ small.

For $V_0 \to \infty$, the particle cannot penetrate into the barrier region. A particle initially in the right-hand region $b < x < a$ stays there for all times: $T = \infty$ and we must have $E_2 = E_1$. This is easily verified since the eigenstates $\phi_1(x)$ and $\phi_2(x)$ now are

$$\phi_1(x) = \phi_2(x) = \frac{1}{\sqrt{(a-b)}} \sin \frac{\pi(a-x)}{(a-b)}, \quad b < x < a$$

$$\phi_1(x) = \phi_2(x) = 0, \quad |x| < b \text{ or } x > a$$

$$\phi_1(-x) = \phi_1(x), \quad \phi_2(-x) = -\phi_2(x).$$

Both states have the same energy

$$E_1 = E_2 = \frac{\hbar^2}{2m} \left(\frac{\pi}{a-b} \right)^2.$$

2.3   The even bound states are obtained from Eq. (2.18b), i.e. from $\xi \tan \xi = \eta$, where ξ and η are defined in Eqs. (2.19). The solutions of this equation are found from the intersection of the curves

$$y = \tan \xi, \quad y = \frac{1}{\xi} \left(\frac{2mV_0 a^2}{\hbar^2} - \xi^2 \right)^{1/2}.$$

The reader should sketch these curves and hence derive conditions analogous to Eqs. (2.22).

2.4   With $\psi(r) = u(r)/r$, $u(r)$ satisfies

$$\frac{d^2 u}{dr^2} + \frac{2m}{\hbar^2} [E - V(r)] u = 0$$

and the boundary conditions $u(0) = u(\infty) = 0$ [see Eqs. (2.88) and (2.9)]. The solutions are

$$u(r) = B \sin qr, \quad r < a, \quad u(r) = C e^{-\kappa r}, \quad r > a,$$

with q and κ defined by Eqs. (2.4), and B/C chosen so that u and du/dr are continuous at $r = a$. This is the problem for the odd bound states of the one-dimensional square-well potential which was solved in section 2.1.

Note that the three-dimensional square-well potential does not necessarily possess any bound states; it does so only if it is sufficiently attractive. The same is true for the odd bound states of the one-dimensional square-well. In contrast there always exists at least one even bound state for the latter potential (see the previous problem).

2.5   Using the results of the last problem and of section 2.1, we have from Eq. (2.18a)

$$\cot qa = -\kappa/q.$$

κ and q are given by Eq. (2.4), where m is now the reduced mass of the neutron–proton system (see problem 1.5): $m = M/2$. With the assumption $\mathscr{E} \ll V_0$, Eqs. (2.4) become

$$\kappa = (M\mathscr{E})^{1/2}/\hbar, \qquad q = [M(V_0 - \mathscr{E})]^{1/2}/\hbar \approx (MV_0)^{1/2}/\hbar,$$

and $\kappa/q \approx (\mathscr{E}/V_0)^{1/2} \approx 0$. Hence $\cot qa \approx 0$ and $qa \approx \pi/2$. (It follows from the node theorem that for the ground state, with no internal node, $qa \approx \pi/2$, rather than $qa \approx 3\pi/2$, etc.) Hence

$$qa \approx \frac{(MV_0)^{1/2}}{\hbar} \frac{\hbar}{\mu c} = \frac{M}{\mu} \left(\frac{V_0}{Mc^2} \right)^{1/2} \approx \frac{\pi}{2}$$

and $V_0 \approx 50 \text{ MeV}$, consistent with $\mathscr{E} \ll V_0$.

2.6   We match the solutions (2.30), (2.31) and (2.32b) so that ϕ and $d\phi/dx$ are continuous at $x = a$ and $x = -a$. Matching at $x = a$ gives two equations from which B and B' can be expressed in terms of C:

$$2\kappa B \, e^{-\kappa a} = (\kappa - ik)C \, e^{ika}, \qquad 2\kappa B' \, e^{\kappa a} = (\kappa + ik)C \, e^{ika}.$$

Matching at $x = -a$ gives two equations from which A can be expressed in terms of B and B':

$$2ikA = B(ik - \kappa) \, e^{(\kappa + ik)a} + B'(ik + \kappa) \, e^{(-\kappa + ik)a}.$$

Substituting for B and B' in this equation, one obtains Eq. (2.37) for C/A.

2.7   (a) $E > V_0$:

$$\phi(x) = A \, e^{ikx} + B \, e^{-ikx}, \quad x < 0; \qquad \phi(x) = C \, e^{iqx}, \quad x > 0; \tag{1}$$

with $k = (2mE/\hbar^2)^{1/2}$, $q = [2m(E - V_0)/\hbar^2]^{1/2}$. Matching ϕ and $d\phi/dx$ at $x = a$ gives B/A and C/A and hence, from Eqs. (2.36),

$$T = \frac{q}{k} \left| \frac{C}{A} \right|^2 = \frac{4kq}{(k+q)^2}, \qquad R = \left| \frac{B}{A} \right|^2 = \left| \frac{k-q}{k+q} \right|^2. \tag{2}$$

$T + R = 1$, as required.

   (b) $E < V_0$: For $x > 0$, we must now take

$$\phi(x) = C \, e^{-\kappa x}, \qquad x > 0, \tag{3}$$

with $\kappa = -iq$. Hence the second of Eqs. (2) gives

$$R = \left| \frac{k - i\kappa}{k + i\kappa} \right|^2 = 1, \qquad \text{whence } T = 1 - R = 0,$$

i.e. $j_T = 0$: for the wave function (3) there is no transmitted current. We cannot simply put $q = i\kappa$ in Eq. (2) for T. Eqs. (2.34a–b) for the currents are correct for plane waves only. In general, the current density $\mathbf{j}$ is defined by Eq. (1.79). For a wave function such as (3), which is real except possibly for an overall phase factor, Eq. (1.79) gives $\mathbf{j} = 0$.

2.8 $$\psi(\mathbf{r}) = \frac{1}{\sqrt{2}} \psi_x^0(\mathbf{r}) + \frac{i}{2} [\psi_x^{-1}(\mathbf{r}) - \psi_x^1(\mathbf{r})]$$

where $\psi_x^m(\mathbf{r})$ are the normed angular momentum eigenstates of l_x, Eqs. (2.69). Hence from Eq. (1.38): $P(0) = \frac{1}{2}$, $P(\hbar) = P(-\hbar) = \frac{1}{4}$.

2.9 Since the energy levels of the one-dimensional oscillator are

$$\mathcal{E}(n) = \hbar\omega(n + \tfrac{1}{2}), \qquad n = 0, 1, 2, \ldots.$$

[see Eq. (2.152)], those of the three-dimensional oscillator are

$$E_N = \mathcal{E}(n_x) + \mathcal{E}(n_y) + \mathcal{E}(n_z) = \hbar\omega(N + \tfrac{3}{2})$$

where

$$N = n_x + n_y + n_z = 0, 1, 2, \ldots, \qquad n_x, n_y, n_z = 0, 1, \ldots. \tag{1}$$

The degeneracy g_N of E_N equals the number of ways of satisfying Eq. (1). Hence $g_0 = 1,\ g_1 = 3,\ g_2 = 6$.

The energy eigenfunctions can from Eqs. (2.153) and (2.140) be written

$$\Phi_{n_x n_y n_z}(\mathbf{r}) = \text{const. } e^{-\alpha^2 r^2/2} H_{n_x}(\alpha x) H_{n_y}(\alpha y) H_{n_z}(\alpha z);$$

from Eq. (2.92) they can alternatively be written

$$\psi_{nlm}(\mathbf{r}) = R_{nl}(r) Y_l^m(\theta, \phi).$$

The $(2l + 1)$ states with $m = l, \ldots, -l$ and given values of l and n are degenerate. E_0 corresponds to $l = 0$, with $\psi_{100}(r) = \text{const. } \exp(-\alpha^2 r^2/2)$. E_1 corresponds to $l = 1$: from Eqs. (2.154) and (2.56) Φ_{001} and $(\Phi_{100} \pm i\Phi_{010})$ reduce to

$$\psi_{11m}(\mathbf{r}) = \text{const. } r\, e^{-\alpha^2 r^2/2} Y_1^m(\theta, \phi), \qquad m = 1, 0, -1.$$

E_2, with $g_2 = 6$, can in various ways be seen to comprise an s- and a d-state:

$$\Phi_{200}(\mathbf{r}) + \Phi_{020}(\mathbf{r}) + \Phi_{002}(\mathbf{r}) = \text{const. } e^{-\alpha^2 r^2/2}(\alpha^2 r^2 - \tfrac{3}{2})$$

is a function of r only and contains one node, i.e. it is the first excited s-state $\psi_{200}(r)$, and

$$2\Phi_{002}(\mathbf{r}) - \Phi_{200}(\mathbf{r}) - \Phi_{020}(\mathbf{r})$$
$$= \text{const. } r^2 e^{-\alpha^2 r^2/2}(3\cos^2\theta - 1) = \text{const. } r^2 e^{-\alpha^2 r^2/2} Y_2^0(\theta, \phi)$$

is the lowest-lying d-state $\psi_{120}(\mathbf{r})$ (no node). Hence the remaining four linearly independent states must also be d-states and

$$\psi_{12m}(\mathbf{r}) = \text{const. } r^2 e^{-\alpha^2 r^2/2} Y_2^m(\theta, \phi), \qquad m = 2, \ldots, -2.$$

2.10 With $f = \exp(-\xi^2/2)$, Eq. (2.165) gives

$$D(e^{-\xi^2}) = e^{-\xi^2/2}(D - \xi)e^{-\xi^2/2}.$$

Calculating $D^n(e^{-\xi^2})$ by repeating this procedure n times leads to the required result.

2.11 From the three-dimensional analogue of Eq. (2.116)

$$\phi(\mathbf{p}) = \frac{1}{(2\pi\hbar)^{3/2}} \int d^3r\, e^{-i\mathbf{p}\cdot\mathbf{r}/\hbar} \frac{e^{-r/a_0}}{\sqrt{(\pi a_0^3)}} = \frac{1}{\pi}\left(\frac{8a_0^3}{\hbar^3}\right)^{1/2} \frac{1}{(1 + p^2 a_0^2/\hbar^2)^2}. \tag{1}$$

Analogous to Eq. (2.126), the momentum distribution, i.e. the probability that the electron has a momentum in the range d^3p at $\mathbf{p}$, is given by $|\phi(\mathbf{p})|^2 d^3p$. (1) is a function of $p = |\mathbf{p}|$ only: $\phi(\mathbf{p}) = \phi(p)$; this is a consequence of the fact that ψ is a function of r only. It follows that the momentum distribution is isotropic in momentum space, and the probability that the magnitude of the momentum lies in the range p to $p + dp$ is

$$|\phi(p)|^2 4\pi p^2 \, dp = \frac{32a_0^3}{\pi\hbar^3} p^2 \, dp \frac{1}{(1 + p^2 a_0^2/\hbar^2)^4}.$$

PROBLEMS 3

3.1 The commutation relations (3.44) follow from those for **r** and **p**, Eqs. (3.6), using the commutator identities (3.5). Some typical proofs are:

$$[l_x, x] = [yp_z - zp_y, x] = 0,$$

$$[l_x, y] = [yp_z - zp_y, y] = -z[p_y, y] = i\hbar z,$$

$$[l_x, l_y] = [yp_z - zp_y, l_y] = y[p_z, l_y] - [z, l_y]p_y$$
$$= y(-i\hbar p_x) - (-i\hbar x)p_y = i\hbar l_z,$$

$$[l_x, \mathbf{l}^2] = [l_x, l_y^2] + [l_x, l_z^2]$$
$$= [l_x, l_y]l_y + l_y[l_x, l_y] + [l_x, l_z]l_z + l_z[l_x, l_z]$$
$$= (i\hbar l_z)l_y + l_y(i\hbar l_z) + (-i\hbar l_y)l_z + l_z(-i\hbar l_y) = 0.$$

3.2 Since $\partial V_1(\mathbf{r})/\partial z = \partial V_2(\mathbf{r})/\partial x = \partial V_2(\mathbf{r})/\partial y = 0$, Eq. (3.30) gives

$$[p_z, V_1(\mathbf{r})] = 0, \qquad [p_x, V_2(\mathbf{r})] = [p_y, V_2(\mathbf{r})] = 0.$$

Similar to the derivation of Eq. (3.30) for $\mathbf{p} = -i\hbar\nabla$, one derives

$$[l_z, V(\mathbf{r})] = -i\hbar \, \partial V(\mathbf{r})/\partial\phi.$$

In spherical polar coordinates $\rho = r \sin\theta$, $z = r \cos\theta$; hence

$$[l_z, V_1(\mathbf{r})] = 0, \quad [l_z, V_2(\mathbf{r})] = 0.$$

Also $V_3(\mathbf{r})$ commutes with $\mathbf{l}^2$ and l_x; and the kinetic energy operator $\mathbf{p}^2/2m$ commutes with $\mathbf{p}$, $\mathbf{l}^2$ and $\mathbf{l}$. Hence: (i) p_z and l_z are constants of the motion for the potential $V_1(\mathbf{r})$; (ii) p_x, p_y and l_z for $V_2(\mathbf{r})$; and (iii) $\mathbf{l}^2$ and l_x for $V_3(\mathbf{r})$. In addition, energy is conserved in all three cases since the Hamiltonians are time-independent.

3.3

$$[x_n, p_{nx}] = [y_n, p_{ny}] = [z_n, p_{nz}] = i\hbar, \qquad n = 1, 2, \Big\}$$
$$\text{all other commutators of } \mathbf{r}_m, \mathbf{p}_n \ (m, n = 1, 2) \text{ vanish.} \Big\} \tag{1}$$

With $r^2 = (x_1 - x_2)^2 + (y_1 - y_2)^2 + (z_1 - z_2)^2$, we have (cf. Eq. (3.30))

$$[\mathbf{p}_n, V(r)] = -i\hbar\nabla_n V(r) = -i\hbar\frac{dV(r)}{dr}\nabla_n r.$$

Since $\nabla_1 r = -\nabla_2 r$, $[\mathbf{P}, V(r)] = 0$. From (1), **P** also commutes with the kinetic operator in (3.45). Hence $[\mathbf{P}, H] = 0$: the total linear momentum (of an isolated system) is a constant of the motion, as in classical physics.

In terms of **R**, **r**, **P** and **p**, defined in problem 1.5 and its solution, and writing $R_x = X$, $r_x = x$, etc., Eqs. (1) become

$$[R_j, P_k] = [r_j, p_k] = i\hbar\delta_{jk}, \qquad j, k = x, y, z, \Big\}$$
$$\text{all other commutators of } \mathbf{R}, \mathbf{r}, \mathbf{P} \text{ and } \mathbf{p} \text{ vanish.} \Big\} \tag{2}$$

(Note that all commutators connecting centre-of-mass and relative coordinates vanish.) In problem 1.5, we showed that

$$H = \frac{1}{2M}\mathbf{P}^2 + \frac{1}{2m}\mathbf{p}^2 + V(r). \tag{3}$$

Hence from (2), $[\mathbf{P}, H] = 0$, as before.

3.4 Applying the Hermiticity condition (1.12) successively to A and B gives

$$\int \psi^*(iAB)\psi \, d^3\mathbf{r} = i\int (A\psi)^* B\psi \, d^3\mathbf{r} = i\int (BA\psi)^*\psi \, d^3\mathbf{r} = \int (-iBA\psi)^*\psi \, d^3\mathbf{r}. \tag{1}$$

Interchanging A and B in Eq. (1) and subtracting the resulting equation from (1), it follows that $C = i[A, B]$ is Hermitian. If C is a number, this implies that $C^* = C$.

3.5 From $\psi(-x) = \psi(x)$, it follows that $\langle x \rangle = \langle p_x \rangle = 0$. Hence

$$E = \langle H \rangle = \frac{1}{2m}\langle p_x^2 \rangle + \frac{1}{2}m\omega^2\langle x^2 \rangle = \frac{1}{2m}(\Delta p_x)^2 + \frac{1}{2}m\omega^2(\Delta x)^2. \tag{1}$$

In the ground state we expect the product $(\Delta x)(\Delta p_x)$ to assume the minimum value consistent with the uncertainty principle (3.34), i.e. $\Delta p_x = \frac{1}{2}\hbar/(\Delta x)$. Substituting this expression in (1) and minimizing E with respect to $(\Delta x)^2$ gives the estimate $E = \frac{1}{2}\hbar\omega$. This happens to be the exact ground state energy, a consequence of the particular form of the harmonic oscillator ground state wave function (2.137).

3.6 We have $\langle x \rangle = 0$, $\langle x^2 \rangle = 1/2a$; hence $\Delta x = 1/\sqrt{(2a)}$.

$$p_x\psi = -i\hbar \, d\psi/dx = i\hbar ax\psi; \quad \text{hence} \quad \langle p_x \rangle = 0.$$

$$p_x^2\psi = -i\hbar \, d(i\hbar ax\psi)/dx = \hbar^2 a\psi - \hbar^2 a^2 x^2\psi;$$

hence $\langle p_x^2 \rangle = \hbar^2 a - \hbar^2 a^2\langle x^2 \rangle = \frac{1}{2}\hbar^2 a$ and $\Delta p_x = \frac{1}{2}\hbar\sqrt{(2a)}$.

Conclusions: (1) $\Delta x \, \Delta p_x = \frac{1}{2}\hbar$: for a Gaussian wave packet the product $\Delta x \, \Delta p_x$ attains the minimum value allowed by the uncertainty principle. For other forms of wave packet, $\Delta x \, \Delta p_x$ is larger. (2) As a increases, $\Delta x = 1/\sqrt{(2a)}$ decreases, i.e. the wave packet becomes more localized. This occurs at the price of a bigger momentum spread: $\Delta p_x = \frac{1}{2}\hbar\sqrt{(2a)}$.

The ground state wave function (2.137) of the one-dimensional harmonic oscillator has the form (3.46), and this explains the result of the previous problem 3.5.

3.7 The expansion of $\psi(x)$ in momentum eigenstates is given by Eqs. (2.115)–(2.116), resulting in

$$|\phi(p)|^2 = \frac{1}{\hbar\sqrt{(\pi a)}} e^{-p^2/a\hbar^2}.$$

Like $|\psi(x)|^2$, this is a Gaussian distribution. It is centred on $p = 0$ and has the standard deviation $\Delta p_x = \frac{1}{2}\hbar\sqrt{(2a)}$, as found in problem 3.6.

PROBLEMS 4

4.1 Proceeding as in section 4.3, Eqs. (4.49)–(4.51), one obtains Eq. (4.78) with

$$L_z = -i\hbar \sum_{n=1}^{N} \partial/\partial\phi_n = \sum_{n=1}^{N} l_{nz}.$$

Clearly L_z is the z-component of the resultant angular momentum **L** of the system of particles. Similar results hold for the other components. Invariance of $f(\mathbf{r}_1, \ldots, \mathbf{r}_N)$ under all rotations means $L_j f = 0$, $j = x, y, z$, etc., analogously to the results for one particle obtained in section 4.3.

4.2
$$[L_x, L_y] = [l_{1x} + l_{2x}, l_{1y} + l_{2y}]$$
$$= [l_{1x}, l_{1y}] + [l_{2x}, l_{2y}] = i\hbar(l_{1z} + l_{2z}) = i\hbar L_z. \tag{4.80a}$$

$$[\mathbf{L}^2, l_{1x}] = [\mathbf{l}_1^2 + \mathbf{l}_2^2 + 2\mathbf{l}_1 \cdot \mathbf{l}_2, l_{1x}] = 2\sum_j [l_{1j}, l_{1x}]l_{2j}$$
$$= 2(-i\hbar l_{1z}l_{2y} + i\hbar l_{1y}l_{2z}) \neq 0, \tag{1}$$

which proves (4.80c). Interchanging the particle labels 1 and 2 in Eq. (1) and adding the resulting equation to Eq. (1) gives the second of Eqs. (4.80a). Lastly

$$[\mathbf{l}_1^2, L_j] = [\mathbf{l}_1^2, l_{1j}] = 0, \qquad [\mathbf{l}_1^2, \mathbf{L}^2] = 2[\mathbf{l}_1^2, \mathbf{l}_1 \cdot \mathbf{l}_2] = 0. \tag{4.80b}$$

4.3 The criteria for invariance under translations or rotations reproduce the results given for problem 3.2, except that conservation of $\mathbf{l}^2$ for $V_3(\mathbf{r})$ cannot be deduced in this way: V_3 is invariant under rotations about the x-axis only; nevertheless $\mathbf{l}^2$ commutes with V_3 and so is conserved.

In addition it follows by considering inversion that parity is conserved for V_1 and V_3 but not for V_2.

4.4 The wave function of a $(1s)^2(nl)$ state is of the form

$$\psi_{1s}(r_1)\psi_{1s}(r_2)R_{nl}(r_3)Y_l^m(\theta_3, \phi_3). \tag{1}$$

Under inversion $(r, \theta, \phi) \to (r, \pi - \theta, \phi + \pi)$, and the wave function (1) is multiplied by $(-1)^l$, i.e. it has parity $(-1)^l$. The ground state $(1s)^2(2s)$ has positive parity. Hence Laporte's rule (4.30) allows transitions $(1s)^2(nl) \to (1s)^2(2s)$ for $l =$ odd, e.g. from p-states, and forbids transitions for $l =$ even, e.g. from s- or d-states. Actually the angular momentum selection rule (4.36) restricts this further for electric dipole transitions to $(1s)^2(np) \to (1s)^2(2s)$ only.

4.5 Under the transformation $\mathbf{r}_j \to \mathbf{r}_j' = -\mathbf{r}_j$, $j = 1, \ldots, Z$, the operator $\sum x_j z_j$ remains invariant. Hence if ψ_i and ψ_f have the parities Π_i and Π_f $(= \pm$ each) then $Q_{fi} = \Pi_f\Pi_iQ_{fi}$ and $\Pi_f = \Pi_i$ for $Q_{fi} \neq 0$: parity does not change in an electric quadrupole transition.

PROBLEMS 5

5.2 σ_z is already diagonal. For a general Hermitian matrix A the eigenvalues a follow from

$$\det(A_{ij} - a\delta_{ij}) = 0. \tag{1}$$

For each of the Pauli matrices, Eq. (1) gives $a = \pm 1$ (as it must!). The eigenstates $\begin{bmatrix} \lambda \\ \mu \end{bmatrix}$ are solutions of

$$\sigma_x \begin{bmatrix} \lambda \\ \mu \end{bmatrix} = a \begin{bmatrix} \lambda \\ \mu \end{bmatrix}, \quad \text{etc.}$$

E.g. the normed eigenstate belonging to the eigenvalue $+1$ of σ_x is $\begin{bmatrix} 1/\sqrt{2} \\ 1/\sqrt{2} \end{bmatrix}$.

5.3 Proceeding from Eq. (5.56), one obtains $\frac{1}{2}\hbar \cos \theta$.

5.4 In section 5.5.2, Eqs. (5.83a–c), we obtained the states $|1, 1, L, L\rangle$, $L = 2, 1, 0$. $|1, 1, 2, -2\rangle$ is obvious. Since the $L = 2$ and $L = 1$ states are respectively symmetric and antisymmetric in the particle labels 1 and 2, the states $|1, 1, 2, \pm 1\rangle$, $|1, 1, 1, 0\rangle$ and $|1, 1, 1, -1\rangle$ follow at once; e.g.

$$|1, 1, 2, 1\rangle = \frac{1}{\sqrt{2}}\{Y_1^0(\Omega_1)Y_1^1(\Omega_2) + Y_1^1(\Omega_1)Y_1^0(\Omega_2)\}.$$

The state $|1, 1, 2, 0\rangle$ must have the form

$$|1, 1, 2, 0\rangle = aY_1^{-1}(\Omega_1)Y_1^1(\Omega_2) + bY_1^0(\Omega_1)Y_1^0(\Omega_2) + aY_1^1(\Omega_1)Y_1^{-1}(\Omega_2).$$

The conditions $\langle 1, 1, 0, 0|1, 1, 2, 0\rangle = 0$ and $\langle 1, 1, 2, 0|1, 1, 2, 0\rangle = 0$ give $b = 2a$ and $a = 1/\sqrt{6}$.

5.5 Denoting the eigenstates by $|j, m\rangle$, where $j = \frac{3}{2}$ or $\frac{1}{2}$, we have:

(a) $j = \frac{3}{2}$: $|\frac{3}{2}, \frac{3}{2}\rangle = \psi_1\alpha.$ (1)

Employing Eq. (5.28b), we obtain from (1)

$$|\tfrac{3}{2}, \tfrac{1}{2}\rangle = \frac{1}{\hbar\sqrt{3}}j_-|\tfrac{3}{2}, \tfrac{3}{2}\rangle = \frac{1}{\hbar\sqrt{3}}(l_- + s_-)\psi_1\alpha = \frac{1}{\sqrt{3}}(\sqrt{2}\psi_0\alpha + \psi_1\beta).   \quad (2)$$

The states $|\frac{3}{2}, -M\rangle$ follow most simply from $|\frac{3}{2}, M\rangle$, Eqs. (1) and (2), by replacing ψ_m by ψ_{-m} and interchanging α and β. (They can of course also be derived by use of the lowering operator j_-. The states obtained by the two procedures may differ by an overall phase factor -1.)

(b) $j = \frac{1}{2}$: The states $|\frac{1}{2}, \pm\frac{1}{2}\rangle$ must be orthogonal to $|\frac{3}{2}, \pm\frac{1}{2}\rangle$. Hence from (2)

$$|\tfrac{1}{2}, \tfrac{1}{2}\rangle = \frac{1}{\sqrt{3}}(\psi_0\alpha - \sqrt{2}\psi_1\beta), \qquad |\tfrac{1}{2}, -\tfrac{1}{2}\rangle = \frac{1}{\sqrt{3}}(\psi_0\beta - \sqrt{2}\psi_{-1}\alpha).   \quad (3)$$

5.6 Combining the proton spins gives three triplet states ($S_{pp} = 1$, $M_{pp} = 1, 0, -1$) and a singlet state ($S_{pp} = 0$). Combining the neutron spin with the triplet pp states gives $S = \frac{3}{2}$, with $M = \pm\frac{1}{2}, \pm\frac{3}{2}$, or $S = \frac{1}{2}$, with $M = \pm\frac{1}{2}$. Combining the neutron spin with the pp singlet state gives $S = \frac{1}{2}$, $M = \pm\frac{1}{2}$.

Let 1 and 2 label the protons, 3 the neutron.

(a) $S = \frac{1}{2}$ states arising from $S_{pp} = 0$:

$$|\tfrac{1}{2}, \tfrac{1}{2}\rangle = \frac{1}{\sqrt{2}}(\alpha_1\beta_2 - \beta_1\alpha_2)\alpha_3, \qquad |\tfrac{1}{2}, -\tfrac{1}{2}\rangle = \frac{1}{\sqrt{2}}(\alpha_1\beta_2 - \beta_1\alpha_2)\beta_3.   \quad (1)$$

(b) $S = \frac{3}{2}$ states: $|\tfrac{3}{2}, \tfrac{3}{2}\rangle = \alpha_1\alpha_2\alpha_3$, etc.   (2)

(With $|\frac{3}{2}, \frac{3}{2}\rangle$ symmetric in the particle labels, the other $S = \frac{3}{2}$ states must also be symmetric.)

(c) $S = \frac{1}{2}$ states arising from $S_{pp} = 1$: We shall denote these states by $|\frac{1}{2}, M\rangle'$, in order to distinguish them from the states (1) which arose from $S_{pp} = 0$. The state $|\frac{1}{2}, \frac{1}{2}\rangle'$ must be of the form

$$|\tfrac{1}{2}, \tfrac{1}{2}\rangle' = A_1\beta_1\alpha_2\alpha_3 + A_2\alpha_1\beta_2\alpha_3 + A_3\alpha_1\alpha_2\beta_3.   \quad (3)$$

The orthogonality and normalization conditions

$$\langle\tfrac{3}{2}, \tfrac{1}{2}|\tfrac{1}{2}, \tfrac{1}{2}\rangle' = \langle\tfrac{1}{2}, \tfrac{1}{2}|\tfrac{1}{2}, \tfrac{1}{2}\rangle' = 0, \qquad '\langle\tfrac{1}{2}, \tfrac{1}{2}|\tfrac{1}{2}, \tfrac{1}{2}\rangle' = 1   \quad (4)$$

determine A_1, A_2 and A_3, giving

$$|\tfrac{1}{2}, \tfrac{1}{2}\rangle' = \frac{1}{\sqrt{3}} \left\{ \sqrt{2}\alpha_1\alpha_2\beta_3 - \frac{1}{\sqrt{2}} (\alpha_1\beta_2 + \beta_1\alpha_2)\alpha_3 \right\}. \tag{5}$$

Interchanging the spin up and spin down states α and β in Eq. (5) gives $|\tfrac{1}{2}, -\tfrac{1}{2}\rangle'$. (Compare Eqs. (3) in the solution of the last problem with Eq. (5) and $|\tfrac{1}{2}, -\tfrac{1}{2}\rangle'$. Except possibly for a phase factor (-1) they agree: we are in both cases combining angular momenta 1 and $\tfrac{1}{2}$ to get the resultant $\tfrac{1}{2}$. A similar correspondence exists between the states with angular momentum $\tfrac{3}{2}$ of these two problems.)

5.7 Since $\mathbf{s}_1 \cdot \mathbf{s}_2 = \tfrac{1}{2}(\mathbf{S}^2 - \tfrac{3}{2}\hbar^2)$ and $\mathbf{S}^2$ has the eigenvalues 0 and $2\hbar^2$ in singlet and triplet states respectively, one has

$$H|0, 0\rangle = -\tfrac{3}{4}A\hbar^2|0, 0\rangle, \qquad H|1, M\rangle = \tfrac{1}{4}A\hbar^2|1, M\rangle, \qquad M = 1, 0, -1.$$

Hence the singlet and triplet interaction energies are $E_s = -\tfrac{3}{4}A\hbar^2$ and $E_t = \tfrac{1}{4}A\hbar^2$.
From: $(s_{1z} - s_{2z})\alpha_1\alpha_2 = 0$, $(s_{1z} - s_{2z})\beta_1\beta_2 = 0$,

$$(s_{1z} - s_{2z})\alpha_1\beta_2 = \hbar\alpha_1\beta_2, \qquad (s_{1z} - s_{2z})\beta_1\alpha_2 = -\hbar\beta_1\alpha_2,$$

it follows that

$$H'|1, \pm 1\rangle = H|1, \pm 1\rangle = E_t|1, \pm 1\rangle,$$

$$H'|1, 0\rangle = E_t|1, 0\rangle + \lambda B\hbar|0, 0\rangle, \qquad H'|0, 0\rangle = E_s|0, 0\rangle + \lambda B\hbar|1, 0\rangle. \tag{1}$$

Hence $|1, \pm 1\rangle$ are eigenstates of H' with the same energy E_t as before, but $|1, 0\rangle$ and $|0, 0\rangle$ are no longer eigenstates of H'. The two remaining eigenstates of H' are of the form

$$|E\rangle = a|1, 0\rangle + b|0, 0\rangle \tag{2}$$

with the constants a and b determined from

$$H'|E\rangle = E|E\rangle. \tag{3}$$

Substituting (2) and (1) in (3) and taking the scalar products of the resulting equation with $|0, 0\rangle$ and $|1, 0\rangle$ leads to

$$a\lambda B\hbar + b(E_s - E) = 0, \qquad a(E_t - E) + b\lambda B\hbar = 0.$$

Equating the values of a/b from these equations gives a quadratic equation in E whose solutions are the two energy eigenvalues.

5.8 Let $\mathbf{S} = \mathbf{s}_1 + \mathbf{s}_2$, $\mathbf{L} = \mathbf{r} \wedge \mathbf{p}$ (where $\mathbf{r}$ and $\mathbf{p}$ are the relative position and momentum of the two nucleons), $\mathbf{J} = \mathbf{L} + \mathbf{S}$. Since $\mathbf{s}_1 \cdot \mathbf{s}_2 = \tfrac{1}{2}(\mathbf{S}^2 - \tfrac{3}{2}\hbar^2)$, V and therefore the deuteron Hamiltonian $H (= -\hbar^2\nabla^2/2\mu + V$, μ = reduced mass of deuteron) commute with $\mathbf{S}^2, S_z, \mathbf{L}^2$ and L_z: S, L, M_S and M_L are good quantum numbers, as is the parity Π since H is invariant under inversion. (In this case $\Pi = (-1)^L$ i.e. parity is determined by L.)

The augmented Hamiltonian $H' = H + V'$ is a scalar and therefore commutes with $\mathbf{J}^2$ and J_z. Because of its dependence on $\mathbf{r}$, H' does not commute with $\mathbf{L}^2$ or L_z. V', and therefore H', are symmetric in the spin variables $\mathbf{s}_1$ and $\mathbf{s}_2$. Hence the spin states must be symmetric or antisymmetric in the particle labels 1 and 2, i.e. they are triplet or singlet states: $S (= 1$ or 0) continues to be a good quantum number (but M_S is not). Parity continues to be conserved, since H' is invariant under inversion. Hence S, J, M_J and Π are good quantum numbers.

280 Hints and solutions to problems

(A more algebraic derivation of the last result is possible. From

$$2(\mathbf{s}_1 \cdot \mathbf{r})(\mathbf{s}_2 \cdot \mathbf{r}) = (\mathbf{S} \cdot \mathbf{r})^2 - (\mathbf{s}_1 \cdot \mathbf{r})^2 - (\mathbf{s}_2 \cdot \mathbf{r})^2 = (\mathbf{S} \cdot \mathbf{r})^2 - \tfrac{1}{2}\hbar^2 r^2$$

and expressing $\mathbf{s}_1 \cdot \mathbf{s}_2$ in terms of $\mathbf{S}^2$, V' can be written

$$V' = V_3(r)\frac{1}{2}\left\{\frac{3(\mathbf{S} \cdot \mathbf{r})^2}{r^2} - \mathbf{S}^2\right\}$$

whence $[V', \mathbf{S}^2] = 0$, $[V', S_z] \neq 0$.)

5.9 From Eqs. (5.107) and (5.88) with $l_1 = l_2 = l$

$$\Psi_{LM_l}(\mathbf{r}_2, \mathbf{r}_1) = (-1)^L \Psi_{LM_l}(\mathbf{r}_1, \mathbf{r}_2),$$

whence the result follows.

PROBLEMS 6

6.1 Hund's rule, applied to the configuration $(2p)^3$, gives $S = \tfrac{3}{2}$, with completely symmetric spin states $\alpha_1\alpha_2\alpha_3$, etc. Hence the space wave function must be completely antisymmetric in $\mathbf{r}_1$, $\mathbf{r}_2$ and $\mathbf{r}_3$. From three wave functions $\psi_{2,1,m}(\mathbf{r})$ only one antisymmetric wave function can be formed; hence it must be an S-state $(L = 0)$ and the term is $^4S_{3/2}$. The only completely antisymmetric space wave function is the determinant

$$\det(\psi_{2,1,1}(\mathbf{r}_1), \psi_{2,1,0}(\mathbf{r}_2), \psi_{2,1,-1}(\mathbf{r}_3)),$$

and this must be multiplied by the symmetric quartet spin states $\alpha_1\alpha_2\alpha_3$ etc.

6.2 Six (np) orbitals give rise to 20 antisymmetric $(np)^3$ states. In the last problem we saw that ^4S is the only $S = \tfrac{3}{2}$ multiplet; all others must have $S = \tfrac{1}{2}$.

Combining $l_1 = l_2 = 1$ gives resultants $L_{12} = 0, 1, 2$. Combining these with $l_3 = 1$ gives only one $L = 0$ state (from $L_{12} = 1$), responsible for the above ^4S multiplet; i.e. there is no ^2S multiplet. Ignoring the Pauli principle, this leaves the possibilities of ^2P, ^2D and ^2F multiplets, comprising 6, 10 and 14 states respectively. Four of the 20 states comprise the ^4S multiplet, leaving 16 to be accounted for. Hence the other multiplets must be a ^2P and a ^2D multiplet.

6.3 The transitions allowed by the electric dipole selection rules (6.24a) to (6.24d) are (a), (b), (d), (f) and (h). Of these, (b) and (f) have $\Delta S \neq 0$, i.e. they violate the selection rule (6.24e) and only occur through breakdown in LS coupling due to the presence of spin–orbit interaction.

6.4 $\Pi_f = \Pi_i$, $\Delta J = 0, \pm 1$, $\Delta M = 0, \pm 1$; $J_i = 0 \to J_f = 0$ strictly forbidden.

With $\mathbf{J} = \mathbf{L} + \mathbf{S}$, we have $[\mathbf{J}, H] = 0$, where H is the atomic Hamiltonian. Hence

$$0 = \langle f|JH - HJ|i\rangle = (E_i - E_f)\langle f|J|i\rangle$$

and $\langle f|\mathbf{J}|i\rangle = 0$ for $E_f \neq E_i$. Hence the magnetic dipole matrix element (6.32) reduces to $\langle f|\mathbf{S}|i\rangle$. If LS coupling holds exactly, $[\mathbf{S}, H] = 0$ and $\langle f|\mathbf{S}|i\rangle = 0$ for $E_f \neq E_i$.

6.5 Without loss of generality, we can choose axes such that $\hat{\mathbf{a}} = \hat{\mathbf{z}} = (0, 0, 1)$ and $\hat{\mathbf{b}} = \hat{\mathbf{n}} = (\sin\theta, 0, \cos\theta)$. Denoting the eigenstates of $\boldsymbol{\sigma}_2 \cdot \hat{\mathbf{z}}$ by α_2, β_2, and of $\boldsymbol{\sigma}_2 \cdot \hat{\mathbf{n}}$ by α_2', β_2', we have

$$\alpha_2 = \alpha_2' c - \beta_2' s, \quad \beta_2 = \alpha_2' s + \beta_2' c,$$

where $c \equiv \cos(\frac{1}{2}\theta)$ and $s \equiv \sin(\frac{1}{2}\theta)$. These equations follow from Eqs. (5.55), where we identify α_2, β_2 with $|\alpha\rangle, |\beta\rangle$, and α'_2, β'_2 with $|\hat{n}+\rangle$ and $|\hat{n}-\rangle$. Direct calculation gives

$$\chi_0 = \frac{1}{\sqrt{2}}(\alpha_1\alpha'_2 s + \alpha_1\beta'_2 c - \beta_1\alpha'_2 c + \beta_1\beta'_2 s). \tag{1}$$

The probability $\langle \hat{a}+, \hat{b}+\rangle$ that particles 1 and 2 have spins parallel to $\hat{a}$ and $\hat{b}$ respectively is given by the square of the coefficient of $\alpha_1\alpha'_2$ in (1), i.e.

$$\langle \hat{a}+, \hat{b}+\rangle = \tfrac{1}{2}\sin^2(\tfrac{1}{2}\theta),$$

in agreement with Eq. (6.30).

6.6 With axes chosen as in the last problem

$$\sigma_1 \cdot \hat{a} = \sigma_{1z}, \quad \sigma_2 \cdot \hat{b} = \sigma_{2x}\sin\theta + \sigma_{2z}\cos\theta.$$

$$\begin{aligned}
\langle\chi_0|(\sigma_1 \cdot \hat{a})(\sigma_2 \cdot \hat{b})|\chi_0\rangle \\
&= \tfrac{1}{2}\langle\alpha_1\beta_2 - \beta_1\alpha_2|\sigma_{1z}(\sigma_{2x}\sin\theta + \sigma_{2z}\cos\theta)|\alpha_1\beta_2 - \beta_1\alpha_2\rangle \\
&= \tfrac{1}{2}\{\langle\alpha_1|\sigma_{1z}|\alpha_1\rangle\langle\beta_2|\sigma_{2z}\cos\theta|\beta_2\rangle + \langle\beta_1|\sigma_{1z}|\beta_1\rangle\langle\alpha_2|\sigma_{2z}\cos\theta|\alpha_2\rangle\} \\
&= -\cos\theta.
\end{aligned}$$

The same result follows from the solution of the last problem:

$$\begin{aligned}
\langle\chi_0|(\sigma_1 \cdot \hat{a})(\sigma_2 \cdot \hat{b})|\chi_0\rangle \\
&= \langle\hat{a}+, \hat{b}+\rangle + \langle\hat{a}-, \hat{b}-\rangle - \langle\hat{a}+, \hat{b}-\rangle - \langle\hat{a}-, \hat{b}+\rangle = -\cos\theta.
\end{aligned}$$

This correlation leads to another formulation of Bell's inequality (see D. Harrison, *Am. J. of Phys.*, **50** (1982), 811).

PROBLEMS 7

7.1 We treat the difference between the electrostatic energy of the electron in the field of the finite-size proton and of the point proton as a perturbation

$$V(r) = \begin{cases} \dfrac{-3e^2}{8\pi\varepsilon_0 R^3}(R^2 - \tfrac{1}{3}r^2) + \dfrac{e^2}{4\pi\varepsilon_0 r}, & r < R, \\[2ex] 0, & r > R. \end{cases}$$

For the ground state $\psi_{1s}(r) = e^{-r/a_0}(\pi a_0^3)^{-1/2}$, this gives the energy shift

$$\Delta E^{(1)} = \langle\psi_{1s}|V(r)|\psi_{1s}\rangle = \int_0^R 4\pi r^2\,dr\,V(r)\,e^{-2r/a_0}(\pi a_0^3)^{-1}.$$

Since $R \ll a_0$, we can put $e^{-2r/a_0} = 1$, whence

$$\Delta E^{(1)} = \frac{4}{5}\left(\frac{R}{a_0}\right)^2 \text{Ry}; \quad \text{for } R \approx 10^{-15}\,\text{m}: \quad \Delta E^{(1)} \simeq 4 \times 10^{-9}\,\text{eV}.$$

7.2 The perturbation is $V(x) = -eEx$. The first-order shift

$$\Delta E_n^{(1)} = -eE\langle n|x|n\rangle = 0, \tag{1}$$

since $V(x) = V(-x)$ and the eigenvalues $E_n^{(0)}$ are non-degenerate. [(1) also follows from Eq. (7.76).]

The second-order level shift is, from Eq. (7.20),

$$E_n^{(2)} = \sum_{p \neq n} (-eE)^2 \frac{|\langle p|x|n \rangle|^2}{\hbar\omega(n - p)}.$$

From (7.76), only the terms $p = n \pm 1$ contribute, giving

$$E_n^{(2)} = \frac{e^2 E^2}{\hbar\omega} \frac{\hbar}{2m\omega} \left(\frac{n+1}{-1} + \frac{n}{1} \right) = \frac{-e^2 E^2}{2m\omega^2}.$$

The Hamiltonian, with the electric field present, is

$$H = -\frac{\hbar^2}{2m} \frac{d^2}{dx^2} + \frac{1}{2} m\omega^2 x^2 - eEx.$$

Let $\xi = x - eE/(m\omega^2)$, then

$$H = -\frac{\hbar^2}{2m} \frac{d^2}{d\xi^2} + \frac{1}{2} m\omega^2 \xi^2 - \frac{e^2 E^2}{2m\omega^2},$$

i.e. the levels $E_n^{(0)}$ are shifted by *exactly* $-e^2 E^2/(2m\omega^2)$.

7.3 From Eqs. (7.41a–b)

$$E_{nl}^{(1)} = \langle \phi_{nl}|V(r)|\phi_{nl} \rangle = \sum_{m, m'} c_m^* c_{m'} \langle u_{nlm}|V(r)|u_{nlm'} \rangle.$$

The wave functions $u_{nlm}(\mathbf{r})$ are of the form $R_{nl}(r)Y_l^m(\theta, \phi)$, so that

$$\langle u_{nlm}|V(r)|u_{nlm'} \rangle = \int_0^\infty r^2\, dr\, |R_{nl}(r)|^2 V(r) \delta_{mm'}.$$

Since $\sum_m |c_m|^2 = 1$, it follows that

$$E_{nl}^{(1)} = \int_0^\infty r^2\, dr\, |R_{nl}(r)|^2 V(r)$$

which is independent of the coefficients c_m.

7.4 Since V, like H_0, is invariant under rotations, the level shift does not depend on the magnetic quantum number m and is given by

$$E_{mass}^{(1)}(nl) = \langle nlm|V|nlm \rangle$$

$$= \frac{-1}{2\mu c^2} \langle nlm|H_0^2 - H_0 V_c - V_c H_0 + V_c^2|nlm \rangle$$

$$= \frac{-1}{2\mu c^2} \langle nlm|E_n^2 - 2E_n V_c + V_c^2|nlm \rangle$$

$$= \frac{-1}{2\mu c^2} \left[E_n^2 + \frac{2E_n e^2}{4\pi\varepsilon_0} \langle nlm|\frac{1}{r}|nlm \rangle + \left(\frac{e^2}{4\pi\varepsilon_0} \right)^2 \langle nlm|\frac{1}{r^2}|nlm \rangle \right].$$

Substituting $e^2/(4\pi\varepsilon_0) = 2n^2 a_0 |E_n|$ and Eqs. (7.78) gives

$$E^{(1)}_{\text{mass}}(nl) = -\frac{|E_n|^2}{2\mu c^2}\left(\frac{8n}{2l+1} - 3\right) \tag{1}$$

and writing $|E_n| = (e^2/4\pi\varepsilon_0\hbar c)^2 \mu c^2/(2n^2)$ for one of the factors $|E_n|$ in (1), we obtain the required result (7.77).

7.5 The spectrum is shown in Fig. 7.2. The Landé g factors of the three terms are $g(^2S_{1/2}) = 2$, $g(^2P_{1/2}) = \frac{2}{3}$ and $g(^2P_{3/2}) = \frac{4}{3}$. Denoting the individual atomic levels by $(^{2S+1}L_J, M)$, the frequencies for the transitions $(^2P_{1/2}, M') \to (^2S_{1/2}, M)$ are, from Eq. (7.70a), given by

$$\nu[(^2P_{1/2}, M') \to (^2S_{1/2}, M)] = \nu_1 + \frac{\mu_B B}{h}[g(^2P_{1/2})M' - g(^2S_{1/2})M]$$

$$= \nu_1 + \frac{\mu_B B}{h}(\tfrac{2}{3}M' - 2M).$$

Hence the D_1 line splits into four spectral lines with frequencies

$$\nu_1 \pm \frac{2}{3}\frac{\mu_B B}{h}, \quad \nu_1 \pm \frac{4}{3}\frac{\mu_B B}{h}.$$

Similarly one finds that the D_2 line splits into six lines with frequencies

$$\nu_2 \pm \frac{1}{3}\frac{\mu_B B}{h}, \quad \nu_2 \pm \frac{\mu_B B}{h}, \quad \nu_2 \pm \frac{5}{3}\frac{\mu_B B}{h}.$$

7.6 The solution to this problem, i.e. an energy level diagram analogous to Fig. 7.2, will be found in Kuhn, p. 290, Fig. V.8.

7.7 The level separation $\delta = h\Delta\nu = hc\Delta\lambda/\lambda^2$. Hence from Eq. (7.71)

$$g = \frac{\delta}{\mu_B B} = \frac{hc\Delta\lambda}{\mu_B B\lambda^2} = 1.0.$$

Since the ground state is $^1S_0$, three lines implies $J = 1$. From the definition of g, Eq. (7.70b), one deduces $L = 1, S = 0$, i.e. the upper term is $^1P_1$. (The transition is the electric dipole transition $(6s)(6p)^1P_1 \to (6s)^2\,^1S_0$.)

7.8 Spin–orbit interaction produces three levels, with $J = \frac{1}{2}, \frac{3}{2}$ and $\frac{5}{2}$, $(2J + 1)$-fold degenerate and shifted by $-\frac{5}{2}A\hbar^2$, $-A\hbar^2$ and $\frac{3}{2}A\hbar^2$ respectively.

In a weak field: Each of these levels is split into $(2J + 1)$ equidistant levels, with level separations $\frac{8}{3}\mu_B B$, $\frac{26}{15}\mu_B B$ and $\frac{8}{5}\mu_B B$ for $J = \frac{1}{2}, \frac{3}{2}$ and $\frac{5}{2}$ respectively.

In a strong field: Neglecting spin–orbit interaction, the ^4P level splits into nine equidistant levels, with level spacing $\mu_B B$; of these, three (which three?) are doubly degenerate. The spin–orbit interaction leaves four of the nine levels unaltered, and shifts and/or splits the others, leaving one of them twofold degenerate.

PROBLEMS 8

8.1 Take $\psi_T = C \exp(-ax^2)$. This trial function has the required properties: no nodes, even parity, and is localized. C is determined by normalization, the parameter a by minimizing $\langle \psi_T | H | \psi_T \rangle$. The relevant integrals follow from

$$\int_{-\infty}^{\infty} dx\, e^{-2ax^2} = \left(\frac{\pi}{2a} \right)^{1/2}$$

by repeated differentiation with respect to (2a). The calculation is simplified by using a result analogous to Eq. (8.14) for the kinetic energy integral. One obtains the upper bound

$$\left(\frac{3Am}{4\hbar^2} \right)^{1/3} \left(\frac{3\hbar^2}{4m} \right).$$

8.2 The energy eigenstates must be parity eigenstates, and the nth excited state must have n internal nodes. Hence take trial functions

$$\psi_1(x) = C_1 x\, e^{-\alpha x^2/2}, \quad \psi_2(x) = C_2(1 + \gamma x^2)\, e^{-\beta x^2/2}$$

for the first and second excited states respectively, with C_1, C_2, α, β and γ to be determined. We require

$$\langle \psi_0 | \psi_1 \rangle = 0, \quad \langle \psi_2 | \psi_1 \rangle = 0, \quad \langle \psi_2 | \psi_0 \rangle = 0.$$

The first two of these equations follow from parity; the last is used to determine γ. C_1 and C_2 follow from normalization, α and β from the variational method. Since our trial functions have the form of the actual energy eigenstates, one obtains the exact energy eigenvalues $\frac{3}{2}\hbar\omega$ and $\frac{5}{2}\hbar\omega$ for the upper bounds. (Compare section 2.7.)

8.3 $$\psi_T(r) = A \exp(-\beta r/2a)$$

is an s-state with the correct behaviour at $r = 0$ and $r = \infty$. A is determined from normalization. Minimizing

$$\langle \psi_T | H | \psi_T \rangle = \frac{\hbar^2}{2m} \left(\frac{\beta}{2a} \right)^2 - \frac{4\hbar^2}{3ma^2} \left(\frac{\beta}{\beta + 1} \right)^3$$

with respect to β gives $\beta = 1$ and hence the upper bound $-\hbar^2/(24ma^2)$ for the ground state energy.

8.4 If $|\psi_1\rangle$ and $|\psi_2\rangle$ are the normed ground states in the potentials V_1 and V_2, i.e. $H_i|\psi_i\rangle = E_i|\psi_i\rangle$, $i = 1, 2$, then

$$E_1 = \langle \psi_1 | H_1 | \psi_1 \rangle \leqslant \langle \psi_2 | H_1 | \psi_2 \rangle \leqslant \langle \psi_2 | H_2 | \psi_2 \rangle = E_2,$$

where the first inequality follows from the variational principle of section 8.1, and the second from $V_1(\mathbf{r}) \leqslant V_2(\mathbf{r})$.

From this result it follows, for example, that for a particle moving in a central potential $V(r)$ the energy of the lowest-lying p-state lies above that of the lowest-lying s-state, etc.

PROBLEMS 9

9.1 The equation of motion for the spin is

$$i\hbar \frac{\partial}{\partial t} |\psi(t)\rangle = -\tfrac{1}{2}\gamma\hbar B\sigma_z |\psi(t)\rangle. \tag{1}$$

We write the general spin state $|\psi(t)\rangle$ as a superposition of the spin states

$$|\lambda\rangle = \frac{1}{\sqrt{2}}\begin{bmatrix} 1 \\ 1 \end{bmatrix}, \quad |\mu\rangle = \frac{1}{\sqrt{2}}\begin{bmatrix} 1 \\ -1 \end{bmatrix}$$

which are the spin-up and spin-down eigenstates of σ_x:

$$|\psi(t)\rangle = a(t)|\lambda\rangle + b(t)|\mu\rangle. \tag{2}$$

Substituting (2) into (1) gives

$$i\hbar\dot{a}|\lambda\rangle + i\hbar\dot{b}|\mu\rangle = -\tfrac{1}{2}\gamma\hbar B(a|\mu\rangle + b|\lambda\rangle)$$

and projecting out coefficients of $|\lambda\rangle$ and $|\mu\rangle$:

$$i\dot{a} = -\tfrac{1}{2}\gamma Bb, \quad i\dot{b} = -\tfrac{1}{2}\gamma Ba.$$

Solving these equations for the initial conditions $a(0) = 1$, $b(0) = 0$, one obtains

$$a(t) = \cos(\tfrac{1}{2}\gamma Bt), \quad b(t) = i\sin(\tfrac{1}{2}\gamma Bt).$$

Hence
(a) $\Pr(s_x = \tfrac{1}{2}\hbar, t) = |a(t)|^2 = \cos^2(\tfrac{1}{2}\gamma Bt)$;
(b) $\Pr(s_x = -\tfrac{1}{2}\hbar, t) = |b(t)|^2 = \sin^2(\tfrac{1}{2}\gamma Bt)$;
(c) In terms of the spin-up and spin-down eigenstates $|\alpha\rangle$ and $|\beta\rangle$ of σ_z:

$$|\psi(t)\rangle = \frac{1}{\sqrt{2}}(a + b)|\alpha\rangle + \frac{1}{\sqrt{2}}(a - b)|\beta\rangle$$

whence

$$\Pr(s_z = \tfrac{1}{2}\hbar, t) = \tfrac{1}{2}|a + b|^2 = \tfrac{1}{2}|\exp(\tfrac{1}{2}i\gamma Bt)|^2 = \tfrac{1}{2}.$$

9.2 The system specified by Eqs. (9.74a–b) was studied in section 9.2, but this problem does not presuppose a knowledge of section 9.2.

Let $|u_1\rangle$ and $|u_2\rangle$ be the spin-up and spin-down eigenstates of σ_z. In these states, H_0 has the eigenvalues $E_1 = -\tfrac{1}{2}\gamma\hbar B_0$ and $E_2 = \tfrac{1}{2}\gamma\hbar B_0$. At $t = 0$, the system is in the state $|u_2\rangle$. In first-order perturbation theory, the probability amplitude $c_1^{(1)}(t)$ that at time t the system is in the spin-up state $|u_1\rangle$ is, from Eq. (9.34b), given by

$$c_1^{(1)}(t) = \frac{1}{i\hbar}\int_0^t V_{12}(t')\exp(i\omega_{12}t')\,dt', \quad \omega_{12} = (E_1 - E_2)/\hbar = -\gamma B_0, \tag{1}$$

where

$$V_{12}(t') = \langle u_1 | V(t') | u_2 \rangle = -\tfrac{1}{2}\gamma\hbar B_1 \exp(-i\omega t'). \tag{2}$$

Hence

$$c_1^{(1)}(t) = \frac{i\gamma B_1}{(\omega + \gamma B_0)}e^{-i(\omega + \gamma B_0)/2}\sin[\tfrac{1}{2}(\omega + \gamma B_0)t]. \tag{3}$$

The resulting expression for $|c_1^{(1)}(t)|^2$ agrees with Eq. (9.31).

For $\omega = \omega_{12}$, one obtains from Eq. (3)—or directly from (1) and (2)—

$$c_1^{(1)}(t) = \tfrac{1}{2}i\gamma B_1 t, \quad |c_1^{(1)}(t)|^2 = \tfrac{1}{4}(\gamma B_1)^2 t^2.$$

For $t > 2/|\gamma B_1|$, the probability $|c_1^{(1)}(t)|^2$ becomes larger than one, which is obviously nonsense. For perturbation theory to hold we must have $|c_1^{(1)}(t)|^2 \ll 1$ i.e. $t \ll 2/|\gamma B_1|$.

9.3 Eliminating $c_2(t)$ from Eqs. (9.24) gives

$$\ddot{c}_1 + i\Omega\dot{c}_1 = -(\tfrac{1}{2}\gamma B_1)^2 c_1 \tag{1}$$

where $\Omega = \omega - \omega_{12}$. $c_1 = \exp{(ipt)}$ is a solution of (1) for

$$p = -\tfrac{1}{2}\Omega \pm \tfrac{1}{2}\Delta$$

where

$$\Delta = + [(\gamma B_1)^2 + \Omega^2]^{1/2} = [(\gamma B_1)^2 + (\omega + \gamma B_0)^2]^{1/2}.$$

The solution of (1) for which $c_1(0) = 0$ is

$$c_1(t) = a\, e^{-i\Omega t/2} \sin{(\tfrac{1}{2}\Delta t)} \tag{2}$$

and the constant a is determined by the other initial condition $c_2(0) = 1$. From the first of Eqs. (9.24) and Eq. (2)

$$c_2(t) = \frac{a}{i\gamma B_1}\, e^{i\Omega t/2}[\Delta \cos{(\tfrac{1}{2}\Delta t)} - i\Omega \sin{(\tfrac{1}{2}\Delta t)}], \tag{3}$$

and $c_2(0) = 1$ gives $a = i\gamma B_1/\Delta$. From Eqs. (2) and (3), with this value of a, follow Eqs. (9.27) and (9.28).

9.4 Take the direction of the electric field as z-axis. The probability amplitude for the transition $(1s) \rightarrow (2p, m = 0)$ is, from Eq. (9.34b),

$$c(t) = \frac{1}{i\hbar} \int_0^t \langle 2p, m = 0| V(t')|1s \rangle\, e^{i\omega t'}\, dt' \tag{1}$$

where $\omega = (E_{2p} - E_{1s})/\hbar$ and

$$V(t) = -e(-\mathscr{E}_0 z\, e^{-t/\tau}) = e\mathscr{E}_0 r \cos\theta\, e^{-t/\tau}.$$

With the hydrogenic states (see Chapter 2)

$$|1s\rangle = \frac{1}{\sqrt{(\pi a_0^3)}}\, e^{-r/a_0}, \qquad |2p, m = 0\rangle = \frac{r}{\sqrt{(32\pi a_0^5)}}\, e^{-r/2a_0} \cos\theta,$$

$$\langle 2p, m = 0|V(t)|1s\rangle = e\mathscr{E}_0\, e^{-t/\tau}\langle 2p, m = 0|r \cos\theta|1s\rangle$$

$$= \frac{e\mathscr{E}_0\, e^{-t/\tau}}{\pi a_0^4 4\sqrt{2}} \int_0^\infty dr\, r^4\, e^{-3r/2a_0} \int d\Omega \cos^2\theta$$

$$= \frac{e\mathscr{E}_0\, e^{-t/\tau}}{\pi a_0^4 4\sqrt{2}}\left[\left(\frac{2a_0}{3}\right)^5 4!\right]\left[\frac{4\pi}{3}\right] \equiv A\, e^{-t/\tau}. \tag{2}$$

From (1) and (2)

$$c(t) = \frac{A}{i\hbar} \int_0^t \exp{(-t'/\tau + i\omega t')}\, dt' = \frac{A}{i\hbar}\frac{e^{i\omega t - t/\tau} - 1}{i\omega - 1/\tau}$$

and

$$|c(\infty)|^2 = \left|\frac{A}{i\hbar(i\omega - 1/\tau)}\right|^2 = \frac{2^{15}}{3^{10}}\frac{(e\mathscr{E}_0 a_0)^2}{(E_{2p} - E_{1s})^2 + (\hbar/\tau)^2}.$$

9.5 We treat

$$V(t) = -Fx \text{ for } 0 < t < \tau, \quad V(t) = 0 \text{ otherwise,}$$

as a perturbation. The probability amplitude for the transition from the ground state $|0\rangle$ of the oscillator to its first excited state $|1\rangle$ is, from Eq. (9.34b), given by

$$c(\tau) = \frac{-F}{i\hbar} \langle 1|x|0\rangle \int_0^\tau dt\, e^{i\omega t} = \frac{F}{\hbar\omega}\left(\frac{\hbar}{2m\omega}\right)^{1/2}(e^{i\omega\tau} - 1),$$

where we carried out the integration and used Eq. (7.76) for the matrix element $\langle 1|x|0\rangle$. $|c(\tau)|^2$ attains its maximum value $2F^2/(\hbar m\omega^3)$ for $\tau\omega = \pi, 3\pi, \ldots$.

9.6 The general state of the K meson can be written

$$|t\rangle = a_1(t)|K_1\rangle + a_2(t)|K_2\rangle. \tag{1}$$

It follows from $|t = 0\rangle = |K^0\rangle$ and Eq. (9.75) that

$$a_1(0) = a_2(0) = 1/\sqrt{2}. \tag{2}$$

Since K_1 and K_2 are unstable particles with lifetimes τ_1 and τ_2, it follows that in the rest frame of the meson (where its energy equals its rest mass multiplied by c^2) the amplitudes $a_j(t)$ are of the form [compare Eq. (9.69)]

$$a_j(t) = a_j(0)\exp\left(-im_jc^2t/\hbar - t/2\tau_j\right), \qquad j = 1, 2. \tag{3}$$

The probability amplitude $\langle \overline{K}^0|t\rangle$ that at time t the meson is a $\overline{K}^0$ meson is from Eqs. (9.75) and (1) to (3) given by

$$\langle \overline{K}^0|t\rangle = \left[\frac{1}{\sqrt{2}}(\langle K_1| - \langle K_2|)\right]\left[\sum_{j=1}^{2}\frac{1}{\sqrt{2}}\exp\left(-im_jc^2t/\hbar - t/2\tau_j\right)|K_j\rangle\right]$$

$$= \tfrac{1}{2}\exp\left(-im_1c^2t/\hbar - t/2\tau_1\right) - \tfrac{1}{2}\exp\left(-im_2c^2t/\hbar - t/2\tau_2\right)$$

(since $\langle K_i|K_j\rangle = \delta_{ij}$) and

$$|\langle \overline{K}^0|t\rangle|^2 = \tfrac{1}{4}\left\{e^{-t/\tau_1} + e^{-t/\tau_2} - 2\exp\left[-\left(\frac{1}{\tau_1} + \frac{1}{\tau_2}\right)\frac{t}{2}\right]\cos\frac{(m_1 - m_2)c^2t}{\hbar}\right\}.$$

Since the K^0 and $\overline{K}^0$ mesons have different decay modes, it is possible to study these oscillations between K^0 and $\overline{K}^0$ and so find the mass difference $|m_1 - m_2|$ by observing the decay products. These oscillations are known as strangeness oscillations since K^0 and $\overline{K}^0$ possess different strangeness quantum numbers. (For a discussion of strangeness oscillations and of the somewhat similar phenomenon of neutrino oscillations see, for example, Martin and Shaw (sections 9.2.5 and 10.2.1) or Perkins (sections 7.14 and 7.15). The K^0–$\overline{K}^0$ problem is also treated in Feynman III, section 11-5.)

PROBLEMS 10

10.1 The total cross-section (10.2) is calculated most easily by introducing $K = |\mathbf{K}|$ from Eqs. (10.18) as variable of integration instead of θ. From the differential cross-section (10.21) and writing $k = p/\hbar$, we obtain

$$\sigma_{\text{tot}} = \frac{2\pi}{k^2}\int_0^{2k} K\, dK\sigma(\theta) = \left(\frac{mZe^2}{\varepsilon_0\hbar^2}\right)^2 \frac{1}{\pi\mu^2(\mu^2 + 4k^2)}.$$

10.2 From Eq. (10.17)

$$\tilde{V}(K) = \frac{4\pi V_0}{K} \int_0^\infty e^{-ar} r \sin Kr \, dr = \frac{8\pi V_0 a}{(a^2 + K^2)^2}. \tag{1}$$

(The last integral is most easily evaluated by writing $\sin Kr$ in terms of exponentials.) From Eqs. (10.13) and (1)

$$\sigma(\theta) \, d\Omega = \left(\frac{4V_0 am}{\hbar^2}\right)^2 \frac{d\Omega}{(a^2 + K^2)^4}.$$

10.3 The electron density for the hydrogenic ground state is given by

$$\rho(r) = \frac{e^{-2r/a_0}}{\pi a_0^3} \tag{1}$$

and the form factor, from Eq. (10.50), by

$$F(K) = \frac{1}{(1 + K^2 a_0^2/4)^2}. \tag{2}$$

Hence, from Eq. (10.49),

$$\sigma(\theta) = \left(\frac{2}{a_0 K^2}\right)^2 \left[1 - \frac{1}{(1 + K^2 a_0^2/4)^2}\right]^2 = 4a_0^2 \frac{(8 + K^2 a_0^2)^2}{(4 + K^2 a_0^2)^4},$$

$$\sigma_{\text{tot}} = \frac{2\pi}{k^2} \int_0^{2k} K \, dK \sigma(\theta) = \frac{8\pi}{k^2 a_0^2} \int_0^{2k} \frac{K \, dK}{K^4} \left[1 - \frac{1}{(1 + K^2 a_0^2/4)^2}\right]^2$$

$$= \pi a_0^2 \frac{7(ka_0)^4 + 18(ka_0)^2 + 12}{3(1 + k^2 a_0^2)^3}$$

where $\hbar k$ is the momentum of the incident electron.

10.4 According to Eq. (10.40), the wave function (10.58) leads to the electron density

$$\rho(r) = \frac{2}{\pi b^3} e^{-2r/b}$$

and hence the form factor

$$F(K) = \frac{2}{(1 + K^2 b^2/4)^2}.$$

Comparing the last two equations with Eqs. (1) and (2) of the solution to problem 10.3, it follows that the Born-approximation cross-sections for electron–helium scattering are obtained from those for electron–hydrogen scattering by multiplying the latter cross-section expressions by a factor 4 and replacing a_0 by $b = (16/27)a_0$. Thus

$$\sigma(\theta) = 16b^2 \frac{(8 + K^2 b^2)^2}{(4 + K^2 b^2)^4}, \qquad \sigma_{\text{tot}} = 4\pi b^2 \frac{7(kb)^4 + 18(kb)^2 + 12}{3(1 + k^2 b^2)^3}.$$

10.5 The differential cross-section per unit solid angle for the electron to be scattered through an angle θ, accompanied by the (1s) → (2s) transition, is given by

$$\sigma(\theta) = \frac{p'}{p} \frac{2^7 a_0^2}{(9/4 + K^2 a_0^2)^6},$$

where a_0 is the Bohr radius, p and p' are the momenta of the incident and scattered electrons and $\hbar K$ is the momentum transfer; i.e. if E is the energy of the incident electron, then

$$p = (2mE)^{1/2}, \quad \frac{p'^2}{2m} = E + E_{1s} - E_{2s} = E - \tfrac{3}{4} \text{ Ry},$$

$$(\hbar K)^2 = p^2 + p'^2 - 2pp' \cos \theta.$$

PROBLEMS 11

11.1 For a function of r only

$$\nabla^2 f(r) = \frac{1}{r^2} \frac{d}{dr} \left(r^2 \frac{d}{dr} \right) f(r).$$

Hence

$$(\nabla^2 + k^2)(e^{\pm ikr}/r) = e^{\pm ikr} \nabla^2 \frac{1}{r} = e^{\pm ikr}[-4\pi\delta^{(3)}(\mathbf{r})]$$

where the last step follows from Eqs. (11.17) and (11.20). Since $\delta^{(3)}(\mathbf{r}) = 0$ for $\mathbf{r} \neq 0$, and $\exp(\pm ikr) = 1$ for $r = 0$, we can omit the exponential from the right-hand side of the last equation.

11.2 Expanding (11.37) in powers of $\xi(\ll 1)$ gives the criterion

$$m|V_0|a^2/\hbar^2 \ll 1. \tag{1}$$

If a bound state exists in the attractive square-well potential ($V_0 < 0$), the particle is more or less confined to the region $r < a$. From the uncertainty principle, its momentum must be at least of the order of $\hbar/a$ and its kinetic energy of the order of $\hbar^2/(2ma^2)$. For a bound state to exist, we must have $|V_0| > \hbar^2/(2ma^2)$. Thus the criterion (1) states that the potential is not sufficiently attractive to support bound states.

The criterion (1) depends on $|V_0|$. It follows that this criterion will hold for a three-dimensional repulsive square-well potential ($V_0 > 0$), provided it is sufficiently weak for the corresponding attractive potential (in which V_0 is replaced by $-V_0$) not to be capable of supporting bound states.

11.3 For $r \geqslant a$, the $l = 0$ partial wave $\psi_0(r)$ is given by Eq. (11.69b). For the hard-sphere potential, ψ_0 must satisfy the boundary condition $\psi_0(a) = 0$, so that $\delta_0 = -ka$. If only s-wave scattering occurs, the cross-section is given by

$$\sigma_{\text{tot}} = \frac{4\pi}{k^2} \sin^2 (ka).$$

In the extreme low-energy limit, $ka \ll 1$, this reduces to $\sigma_{\text{tot}} = 4\pi a^2$.

11.4 For two particles of spin s, there are $(2s + 1)^2$ independent spin states $\chi_1(m_1)\chi_2(m_2)$, $m_1, m_2 = s, \ldots, -s$. These comprise $(2s + 1)$ symmetric states $\chi_1(m)\chi_2(m)$, $m = s, \ldots, -s$, and $(2s + 1)2s$ linear combinations

$$\chi_1(m_1)\chi_2(m_2) \pm \chi_1(m_2)\chi_2(m_1), \qquad m_1 \neq m_2,$$

i.e. there are $(2s + 1)s$ antisymmetric spin states and $(2s + 1)(s + 1)$ symmetric spin states.

We know from section 4.4 that for particles with integer spin $(s = 0, 1, \ldots)$, the states must be symmetric under interchange of particle labels; for half-integer spin $(s = \frac{1}{2}, \frac{3}{2}, \ldots)$ they must be antisymmetric. Hence

$$\sigma_{\text{unpol}}(\theta) = |f(\theta)|^2 + |f(\pi - \theta)|^2$$

$$\pm \frac{2}{2s + 1} \mathcal{R}e[f(\theta)f^*(\pi - \theta)], \qquad \begin{cases} s = 0, 1, \ldots, \\ s = \frac{1}{2}, \frac{3}{2}, \ldots. \end{cases}$$

11.5 Taking the z-axis along $\mathbf{n}$, Eq. (11.74) becomes

$$\psi(\mathbf{r}) \simeq \frac{e^{ikr}}{r} [f_1(\theta) + f_2(\theta)\sigma_z]\chi.$$

If the incident beam is 100 per cent polarized with spins parallel to $\mathbf{n}$, i.e. in the spin state α, then the scattered wave is

$$\psi_\alpha(\mathbf{r}) \simeq \frac{e^{ikr}}{r} [f_1(\theta) + f_2(\theta)]\alpha;$$

all scattered particles have spin parallel to $\mathbf{n}$, with the cross-section

$$\sigma_\alpha(\theta) = |f_1(\theta) + f_2(\theta)|^2.$$

Similarly, if the incident beam is fully polarized with spins parallel to $-\mathbf{n}$, the scattered wave is

$$\psi_\beta(\mathbf{r}) \simeq \frac{e^{ikr}}{r} [f_1(\theta) - f_2(\theta)]\beta;$$

all scattered particles have spins parallel to $-\mathbf{n}$, with the cross-section

$$\sigma_\beta(\theta) = |f_1(\theta) - f_2(\theta)|^2.$$

The incident unpolarized beam can be thought of as an incoherent superposition of two beams with spins parallel to $\pm\mathbf{n}$ and of equal intensities. Hence

$$a(\pm\mathbf{n}) = \frac{|f_1(\theta) \pm f_2(\theta)|^2}{|f_1(\theta) + f_2(\theta)|^2 + |f_1(\theta) - f_2(\theta)|^2},$$

from which the result (11.75) follows. Note that for $f_2 = 0$, i.e. spin-independent forces, $P(\theta) = 0$: the scattered beam is unpolarized, as expected.

PROBLEMS 12

12.1 Substituting (12.70) in $\langle w|w \rangle \geq 0$ gives Eq. (12.69). More generally, if the kets $|a\rangle$ and $|b\rangle$ are not normed, Eq. (12.69) is replaced by

$$|\langle b|a \rangle|^2 \leq \langle a|a \rangle\langle b|b \rangle, \tag{1}$$

known as Schwarz's inequality. The equal sign in (1) holds only if $|b\rangle = \text{const.} |a\rangle$. The corresponding inequality for ordinary vectors is $(\mathbf{a} \cdot \mathbf{b})^2 \leqslant \mathbf{a}^2 \mathbf{b}^2$.

12.2 $\langle n|A|m\rangle = a_m \langle n|m\rangle.$

Taking the complex conjugate of this equation and interchanging the labels m and n gives

$$\langle n|A|m\rangle = a_n \langle n|m\rangle$$

since the eigenvalues of $A(= A^\dagger)$ are real. Hence

$$(a_m - a_n)\langle n|m\rangle = 0$$

and $\langle n|m\rangle = 0$ for $a_m \neq a_n$.

12.3   (a) The result (12.71) follows by evaluating $\langle \mathbf{p}|\hat{\mathbf{r}}|a\rangle$. Hence

$$\langle b|\hat{\mathbf{r}}|a\rangle = \int d^3\mathbf{p}\langle b|\mathbf{p}\rangle\langle \mathbf{p}|\hat{\mathbf{r}}|a\rangle = \int d^3\mathbf{p}\,\phi_b^*(\mathbf{p})i\hbar\nabla_\mathbf{p}\,\phi_a(\mathbf{p}).$$

(b) In the $\mathbf{p}$-representation, the oscillator Hamiltonian

$$H(\hat{x}, \hat{p}) = \frac{\hat{p}^2}{2m} + \tfrac{1}{2}m\omega^2\hat{x}^2$$

becomes

$$H\left(i\hbar\frac{d}{dp}, p\right) = -\tfrac{1}{2}m\omega^2\hbar^2\frac{d^2}{dp^2} + \frac{1}{2m}p^2.$$

This is of the same form as the oscillator Hamiltonian in the $\mathbf{r}$-representation, and the energy eigenstates in the two representations will be of the same form. In particular, the ground state solution of the Schrödinger equation in momentum space,

$$H\left(i\hbar\frac{d}{dp}, p\right)\phi_0(p) = \tfrac{1}{2}\hbar\omega\phi_0(p),$$

gives the probability distribution

$$|\phi_0(p)|^2 = \frac{e^{-p^2/m\hbar\omega}}{\sqrt{(\pi m\hbar\omega)}}.$$

12.4   (a) From Eq. (12.60), with $\lambda_0 = 0$, and Eq. (12.54) we have

$$\langle x|a|0\rangle = \langle x|\frac{1}{(2\hbar m\omega)^{1/2}}(m\omega\hat{x} + i\hat{p})|0\rangle = 0.$$

Hence from Eq. (12.50), writing $\langle x|0\rangle = \psi_0(x)$,

$$[m\omega x + i(-i\hbar\,\partial/\partial x)]\psi_0(x) = 0,$$

with the solution

$$\psi_0(x) = \text{const. } e^{-m\omega x^2/2\hbar},$$

in agreement with Eq. (2.137).

(b) From $\langle n|a^\dagger a|n\rangle = n\langle n|n\rangle = n$, it follows that the normed state $|n-1\rangle$ is given by

$$|n-1\rangle = \frac{1}{\sqrt{n}}\,a|n\rangle.$$

Similarly one obtains

$$|n+1\rangle = \frac{1}{\sqrt{(n+1)}}\,a^\dagger|n\rangle$$

whence

$$|n\rangle = \frac{(a^\dagger)^n}{\sqrt{n!}}\,|0\rangle.$$

12.5 (a)

$$\langle v|u\rangle = \langle v|U^\dagger U|u\rangle = \langle Uv|Uu\rangle = \langle v'|u'\rangle. \tag{12.73}$$

(b) It follows from Eq. (12.73) that $\langle v'|v'\rangle = \langle v|v\rangle$ and that $\langle v'|u'\rangle = 0$ if $\langle v|u\rangle = 0$, i.e. a unitary transformation preserves the orthonormality of states. (We can think of a unitary transformation as the analogue of a rotation in ordinary space.) Hence the states $|\eta_1\rangle$, $|\eta_2\rangle$,... are also orthonormal. If they are not complete, there must exist a state $|f\rangle$ such that $\langle \eta_n|f\rangle = 0$ for all n, i.e. $\langle \xi_n|U^\dagger f\rangle = 0$ for all n; this would contradict the assumption that $|\xi_1\rangle$, $|\xi_2\rangle$,... form a complete set. Hence $|\eta_1\rangle$, $|\eta_2\rangle$,... must also form a complete set of states.

(c) The matrix representation $\langle \eta_m|A|\eta_n\rangle$ is given by

$$\langle \eta_m|A|\eta_n\rangle = \langle \xi_m|U^\dagger A U|\xi_n\rangle$$
$$= \sum_{r,s} \langle \xi_m|U^\dagger|\xi_r\rangle\langle \xi_r|A|\xi_s\rangle\langle \xi_s|U|\xi_n\rangle.$$

12.6 To solve the eigenvalue problem

$$A|n\rangle = a|n\rangle, \tag{1}$$

given $A_{rs} = \langle \xi_r|A|\xi_s\rangle$, we expand

$$|n\rangle = \sum_s |\xi_s\rangle\langle \xi_s|n\rangle. \tag{2}$$

Substituting this expansion in Eq. (1), we obtain

$$\sum_s \langle \xi_r|A|\xi_s\rangle\langle \xi_s|n\rangle = a\langle \xi_r|n\rangle = \sum_s a\delta_{rs}\langle \xi_s|n\rangle$$

i.e.

$$\sum_s (A_{rs} - a\delta_{rs})\langle \xi_s|n\rangle = 0, \quad r = 1, 2, \dots. \tag{3}$$

This set of homogeneous equations only has non-trivial solutions if

$$\det(A_{rs} - a\delta_{rs}) = 0. \tag{4}$$

The roots of this equation in a are the eigenvalues $a_1, a_2, \dots$ of the operator A, and the solution $\langle \xi_r|n\rangle$, $r = 1, 2, \dots$, of Eqs. (3) for $a = a_n$ determines the corresponding eigenstate (2).

We can interpret the above solution of the eigenvalue problem as a unitary transformation U from the basis $|\xi_1\rangle, |\xi_2\rangle, \ldots$ to the basis of eigenkets $|1\rangle, |2\rangle, \ldots$:

$$|n\rangle = U|\xi_n\rangle = \sum_r |\xi_r\rangle U_{rn} \qquad (5)$$

where

$$U_{rn} = \langle \xi_r | U | \xi_n \rangle = \langle \xi_r | n \rangle \qquad (6)$$

(compare problem 12.5). From Eq. (1)

$$\langle m | A | n \rangle = a_m \delta_{mn}, \qquad (7a)$$

and from Eq. (5)

$$\langle m | A | n \rangle = \langle \xi_m | U^\dagger A U | \xi_n \rangle = \sum_{r,s} U^\dagger_{mr} A_{rs} U_{sn} \qquad (7b)$$

in agreement with the last problem. We see from Eqs. (7a–b) that U transforms the matrix $\langle \xi_r | A | \xi_s \rangle$ into the diagonal matrix $\langle \xi_m | U^\dagger A U | \xi_n \rangle$, whose elements along the principal diagonal are $a_1, a_2, \ldots$. We see from Eq. (6) that the components $\langle \xi_r | n \rangle$, $r = 1, 2, \ldots$, of the eigenstate $|n\rangle$ in the $\{|\xi_1\rangle, |\xi_2\rangle, \ldots\}$ representation form the nth column U_{rn}, $r = 1, 2, \ldots$, of the transformation matrix.

Bibliography

This bibliography lists the books to which I have referred in the text. Those on quantum mechanics form a selection suitable for complementary and further reading. They range in level from elementary to advanced, and a student of this book should be able to come to grips with and benefit from any of them.

Arfken, G., 1985, *Mathematical methods for physicists*, 3rd edn, Academic Press, San Diego, California.

Bell, J. S., 1987, *Speakable and unspeakable in quantum mechanics*, Cambridge University Press, Cambridge.

Bethe, H. A. and Jackiw, R., 1986, *Intermediate quantum mechanics*, 3rd edn, Benjamin/Cummings, Menlo Park, California.

Bransden, B. H. and Joachain, C. J., 1983, *Physics of atoms and molecules*, Longman, London.

Bransden, B. H., and Joachain, C. J., 1989, *Introduction to quantum mechanics*, Longman, London.

Cohen-Tannoudji, C., Diu, B. and Laloë, F., 1977, *Quantum mechanics*, vols. I and II, Wiley, New York.

Corney, A., 1977, *Atomic and laser spectroscopy*, Oxford University Press, Oxford.

Davydov, A. S., 1965, *Quantum mechanics*, Pergamon, Oxford.

Dirac, P. A. M., 1958, *The principles of quantum mechanics*, 4th edn, Oxford University Press, Oxford.

Fano, U. and Fano, L., 1959, *Basic physics of atoms and molecules*, Wiley, New York.

Fermi, E., 1950, *Nuclear physics*, University of Chicago Press, Chicago.

Feynman, R. P., Leighton, R. B. and Sands, M., 1963–65, *The Feynman lectures on physics*, vols. I to III, Addison-Wesley, Reading, Mass.

French, A. P. and Taylor, E. F., 1979, *An introduction to quantum physics*, Nelson, Sunbury-on-Thames, Middlesex, England.

Gasiorowicz, S., 1974, *Quantum physics*, Wiley, New York.

Gottfried, K., 1966, *Quantum mechanics*, Benjamin, New York.

Grant, I. S. and Phillips, W. R., 1990, *Electromagnetism*, 2nd edn (Manchester Physics Series), Wiley, Chichester, England.

Heisenberg, W., 1949, *The physical principles of the quantum theory*, Dover, New York.

Kuhn, H. G., 1969, *Atomic spectra*, 2nd edn, Longmans, London.

Landau, L. D. and Lifshitz, E. M., 1977 (reprinted with corrections, 1989), *Quantum mechanics, non-relativistic theory*, 3rd edn, Pergamon, Oxford.

Lipson, S. G. and Lipson, H., 1981, *Optical physics*, 2nd edn, Cambridge University Press, Cambridge.

Mandl, F., 1988, *Statistical physics*, 2nd edn (Manchester Physics Series), Wiley, Chichester, England.

Mandl, F. and Shaw, G., 1984, *Quantum field theory*, Wiley, Chichester, England.

Martin, B. R. and Shaw, G., 1992, *Particle physics* (Manchester Physics Series), Wiley, Chichester, England.

Merzbacher, E., 1970, *Quantum mechanics*, 2nd edn, Wiley, New York.

Messiah, A., 1965, *Quantum mechanics*, vols. I and II, North-Holland, Amsterdam.

Moiseiwitsch, B. L., 1966, *Variational principles*, Interscience (Wiley), London.

Perkins, D. H., 1987, *Introduction to high energy physics*, 3rd edn, Addison-Wesley, Menlo Park, California.

Polkinghorne, J. C., 1984, *The quantum world*, Longman, London.

Sakurai, J. J., 1985, *Modern quantum mechanics*, Benjamin/Cummings, Menlo Park, California.

Schiff, L. I., 1968, *Quantum mechanics*, 3rd edn, McGraw-Hill, New York.

Slater, J. C., 1960, *Quantum theory of atomic structure*, vols. I and II, McGraw-Hill, New York.

Index

References to sections are printed in *italics*.